D1053160

SEP 07 2006

IUFRO Research Series

The International Union of Forest Research Organizations (IUFRO), with its 14,000 scientists from 700 member institutions in over 100 countries, is organized into nearly 300 research units that annually hold approximately 80 conferences, workshops and other meetings. The individual papers, proceedings and other material arising from these units and meetings are often published but in a wide array of different journals and other publications. The object of the IUFRO Research Series is to offer a single, uniform outlet for high quality publications arising from IUFRO meetings and other products of IUFRO's research units.

The editing, publishing and dissemination experience of CABI Publishing and the huge spread of scientific endeavours of IUFRO combine here to make information widely available that is of value to policy makers, resource managers, peer scientists and educators. The Board of IUFRO forms the Editorial Advisory Board for the Series and provides the monitoring and uniformity that such a high quality series requires in addition to the editorial work of the conference organizers.

While adding a new body of information to the plethora currently dealing with forestry and related resources, this series seeks to provide a single, uniform forum and style that all forest scientists will turn to first as an outlet for the conference material and other products, and that the users of information will also see as a reliable and reputable source.

Although the official languages of IUFRO include English, French, German and Spanish, the majority of modern scientific papers are published in English. In this series, all books will be published in English as the main language, allowing papers occasionally to be in other languages. Guidelines for submitting and publishing material in this series are available from the Publisher, Books and Reference Works, CABI Publishing, CAB International, Wallingford, Oxfordshire OX10 8DE, UK, and the IUFRO Secretariat, Mariabrunn (BFW), Hauptstrasse 7, A-1140 Vienna – Hagersdorf, Austria.

IUFRO Board:

President and Senior Officers:
Risto Seppälä, Helsinki, Finland (President)
Don Koo Lee, Seoul, Republic of Korea (Vice President)
Eric Teissier du Cros, Avignon, France (Vice President)
Jeffery Burley, Oxford, UK (Immediate Past President)
Peter Mayer, Vienna, Austria (Executive Secretary)

Division Coordinators:
Dennis Dykstra, Portland, Oregon, USA
Alain Franc, Bordeaux, France
Niels Elers Koch, Hoersholm, Denmark
John Parrotta, Washington, DC, USA
Ladislav Paule, Zvolen, Slovakia
Kazuo Suzuki, Tokyo, Japan
Klaus von Gadow, Göttingen, Germany
Hsiu Hwa Wang, Pingtung, China-Taipei

General Board Members:
Susan Conard, Washington, DC, USA
Rueben Guevara, Lima, Peru
Vitor Alfonso Hoeflich, Curitiba, Brazil
John Innes, Vancouver, BC, Canada
Iba Kone, Nairobi, Kenya
Gordon Miller, Edmonton, Alberta, Canada
Abdul Razak, Kuala Lumpur, Malaysia
Victor Teplyakov, Moscow, Russia
Karel Vancura, Prague, Czech Republic
Yaoguo Xiong, Beijing, P.R. China

IUFRO Research Series

Titles available:

1. *Forest Dynamics in Heavily Polluted Regions*
   Edited by J.L. Innes and J. Oleksyn

2. *Forest History: International Studies on Socioeconomic and Forest Ecosystem Change*
   Edited by M. Agnoletti and S. Anderson

3. *Methods and Approaches in Forest History*
   Edited by M. Agnoletti and S. Anderson

4. *Air Pollution and the Forests of Developing and Rapidly Industrializing Countries*
   Edited by J.L. Innes and A.H. Haron

5. *Forests in Sustainable Mountain Development: a State of Knowledge Report for 2000*
   Edited by M. Price and N. Butt

6. *Forests and Landscapes: Linking Ecology, Sustainability and Aesthetics*
   Edited by S.R.J. Sheppard and H.W. Harshaw

7. *Criteria and Indicators for Sustainable Forest Management*
   Edited by J. Raison, A. Brown and D. Flinn

8. *The Impact of Carbon Dioxide and Other Greenhouse Gases on Forest Ecosystems*
   Edited by D.F. Karnosky, R. Ceulemans, G. Scarascia-Mugnozza and J.L. Innes

9. *Environmental Change and Geomorphic Hazards in Forests*
   Edited by R.C. Sidle

10. *Forest Biodiversity: Lessons from History for Conservation*
    Edited by O. Honnay, K. Verheyen, B. Bossuyt and M. Hermy

11. *Forestry and Environmental Change: Socioeconomic and Political Dimensions*
    Edited by J.L. Innes, G.M. Hickey and H.F. Hoen

Forestry and Environmental Change: Socioeconomic and Political Dimensions

Report No. 5 of the IUFRO Task Force on Environmental Change

Edited by

J.L. Innes and G.M. Hickey

Forest Resources Management, University of British Columbia, Vancouver, Canada

and

H.F. Hoen

Department of Forest Sciences, Agricultural University of Norway

CABI Publishing

In association with

The International Union of Forestry Research Organizations (IUFRO)

CABI *Publishing* is a division of CAB *International*

CABI Publishing
CAB International
Wallingford
Oxon OX10 8DE
UK

Tel: +44 (0)1491 832111
Fax: +44 (0)1491 833508
Email: cabi@cabi.org

CABI Publishing
10E 40th Street
Suite 3203
New York, NY 10016
USA

Tel: +1 212 481 7018
Fax: +1 212 686 7993
Email: cabi-nao@cabi.org

© CAB International 2005. All rights reserved. No part of this publication may be reproduced in any form or by any means, electronically, mechanically, by photocopying, recording or otherwise, without the prior permission of the copyright owners.

A catalogue record for this book is available from the British Library, London, UK.
A catalogue record for this book is available from the Library of Congress, Washington DC, USA

Published in Association with:

The International Union of Forestry Research Organizations (IUFRO)
c/o Federal Forest Research Centre
Seckendorff-Gudent-Weg 8
A-1131 Vienna
Austria

ISBN 085 1 990029

Printed and bound in the UK by Cromwell Press, Trowbridge, from copy supplied by the editors.

Contents

Contributors vii

Foreword viii

Acknowledgements viii

1 The changing context of forestry 1
J.L. Innes and H.F. Hoen

2 The emerging role of forest quality in setting perspectives in forest management 15
N. Dudley

3 Making value compromises in an efficient economy: efficiency as value dialogue 31
M.C. Farmer and A. Randall

4 An intergenerational perspective on effects of environmental changes: discounting the future's viewpoint 53
C. Price

5 Deforestation processes in temperate and tropical forests: an economic perspective 75
O. Hofstad

6 The application of forest zoning as an alternative to multiple-use forestry 97
C.R. Nitschke and J.L. Innes

7 Capital management – the forests in countries in transition – welfare impacts 125
P. Csóka

8 Certification of forest management and wood products 143
J.L. Innes and G.M. Hickey

9 How will new large-scale nature reserves in temperate and boreal forests affect the global structural wood products sector? 169
J.M. Perez-Garcia

10 The establishment of large-scale reserves in temperate and boreal forests: some experiences of the USA and prospects for Finland 183
R.A. Sedjo

11 The new planning approach and criteria used for sustainable forest management in Turkey 193
Ü. Asan

12 Application of criteria and indicators for sustainable forest management to a GIS-based multiple-criteria decision-making approach for forest conservation planning in the Kinabalu Region, Sabah, Malaysia 213
M.H. Phua and M. Minowa

13 Social revolutions in forest management 229
J. Romm

14 Recent developments in decision analysis for forest management 239
G.A. Mendoza

Index 257

Contributors

Ü. Asan, Faculty of Forestry, University of Istanbul, 80895 Bahçeköy, Istanbul, Turkey
P. Csóka, State Forest Service, Ministry of Agriculture and Regional Development, Széchenyi utca 14, H - 1054 Budapest, Hungary
N. Dudley, Equilibrium Consultants, 23 Bath Buildings, Montpelier, Bristol BS6 5PT, UK
M.C. Farmer, School of Public Policy, Georgia Institute of Technology, Atlanta, Georgia 30332 – 0345, USA
G.M. Hickey, Faculty of Forestry, University of British Columbia, 2424 Main Mall, Vancouver, British Columbia V6T 1Z4, Canada
H.F. Hoen, Department of Forest Sciences, Agricultural University of Norway, P.O. Box 5003, NO-1432 Ås, Norway
O. Hofstad, Department of Forest Sciences, Agricultural University of Norway, P.O. Box 5003, NO-1432 Ås, Norway
J.L. Innes, Faculty of Forestry, University of British Columbia, 2424 Main Mall, Vancouver, British Columbia V6T 1Z4, Canada
G.A. Mendoza, Department of Natural Resources and Environmental Sciences, University of Illinois, Urbana, Illinois, USA
M. Minowa, Department of Forest Science, Graduate School of Agricultural and Life Sciences, The University of Tokyo, 1-1-1 Yayoi Bunkyo-ku, Tokyo, Japan 113-8657
C.R. Nitschke, Faculty of Forestry, University of British Columbia, 2424 Main Mall, Vancouver, British Columbia V6T 1Z4, Canada
J.M. Perez-Garcia, College of Forest Resources, University of Washington, Box 352100, Seattle, WA 98195-2100, USA
M.H. Phua, Department of Forest Science, Graduate School of Agricultural and Life Sciences, The University of Tokyo, 1-1-1 Yayoi Bunkyo-ku, Tokyo, Japan 113-8657
C. Price, School of Agricultural and Forest Sciences, University of Wales, Bangor, Gwynedd LL57 2UW, UK
A. Randall, Department of Agricultural and Resource Economics, Ohio State University, 103 Agricultural Administration Building, 2120 Fyffe Road, Columbus, OH 43210, USA
J. Romm, Department of Environmental Science, Policy and Management, University of California, Berkeley, 130C Giannini Hall, Berkeley, CA 642-6499, USA
R.A. Sedjo, Resources for the Future, 1616 P Street, NW, Washington, DC, 20036, USA

Foreword

Environmental change continues to be an issue of enormous importance to the forest sector. Many forest scientists are working on a myriad of issues associated with environmental change, ranging from air pollution, through climate change to changes in the policy and economic environment. This book provides a welcome addition to this latter area.

In the past, there has been a tendency to focus on the biophysical aspects of environmental change. An understanding of how these changes impact forests was essential as a first step in understanding how they impact the forest sector. Today, with the growing maturity of environmental change studies, the critical importance of the policy environment has become apparent. We now need to have policies that enable both mitigation of the problems (e.g. pollution abatement) and adaptation strategies to occur.

The 2004 tragedy that affected the countries surrounding the Indian Ocean has drawn attention to the need for the rehabilitation of coastal forests, particularly mangroves. Never has the need been so great for a social and economic environment that places adequate value on protection forests and other forests whose primary function is not the production of timber. Any progress in this area of research is welcome, and this book represents a timely contribution to the issue.

Risto Seppälä
President IUFRO

Acknowledgements

The editors would like to take this opportunity to thank a number of anonymous referees who spent considerable time reviewing the papers presented in this book. We would also like to thank the Publications Committee of the International Union of Forest Research Organizations for their thoughtful and helpful comments on the structure and content of the book.

The Changing Context of Forestry 1

J.L. Innes[1] and H.F. Hoen[2]

[1] *Department of Forest Resources Management, University of British Columbia, Vancouver, Canada;* [2] *Department of Forest Sciences, Agricultural University of Norway*

Considerable emphasis has been placed on the interactions between environmental change and forests in recent years. Organizations such as the Intergovernmental Panel on Climate Change have produced reports detailing scenarios of forest development associated with particular changes in climate. Similarly, scenarios have been produced looking at likely trends in air quality and pollutant deposition. Despite these efforts, many studies have failed to recognize that some of the biggest changes affecting forests are related to the socioeconomic environment rather than the physical environment. The environmental, economic and social values that are associated with forests continue to evolve, and major changes in forestry practices have occurred as a result. Industrial logging, especially of old-growth forests, is seen as being increasingly unacceptable, and even intensive plantation forestry is becoming more sensitive as an issue.

In some countries, a major change that is occurring relates to forest tenures. In countries such as Australia, New Zealand and Mexico, indigenous populations are gaining more and more access to forest resources. Such trends are likely to continue in many developing countries, but will also be apparent in developed countries such as Canada, which has considerable indigenous populations. At the same time as greater control is being given to local groups, international policies are also becoming more significant. These range from bilateral issues such as the softwood timber dispute between the USA and Canada, to multinational issues associated with the globalization of the forest industry and the desire to see problems such as the loss of biological diversity dealt with on a global scale.

In view of these changes, theory and models attempting to demonstrate ways in which forests might respond to environmental change need to take into account the complex socioeconomic changes that are occurring at the same time.

© CAB International 2005.
Forestry and Environmental Change: Socioeconomic and Political Dimensions (eds J.L. Innes, G.M. Hickey and H.F. Hoen)

1.1 Introduction

Although physical and chemical environmental changes have had marked impacts on climate change, for many of those practising forestry, it is the changes in the economic and social environment that have forced the most rapid and extensive changes in the ways that we manage forests. This book sets out to examine some of the remarkable changes that have occurred in the social, cultural and economic environment in which forestry is practised. It complements a number of other volumes published by the Task Force on Environmental Change of the International Union of Forest Research Organizations (IUFRO) and attempts to draw attention to the importance of the human element in the debate surrounding the effects of environmental change on forests.

1.2 The changing physical environment

The last 50 years have seen increasing concerns being expressed about the effects of physical and chemical changes in the environment on forests. Throughout most of the 20th century, concerns focused on chemical changes in the form of atmospheric pollution and its deleterious effects. This concern peaked in the 1980s, when major fears about forest decline led to the development of large-scale research programmes on the impacts of atmospheric pollution on forests and other aspects of the environment. Examples of such programmes included the Norwegian research programme Impacts of Acid Rain on Forests and Fish (SNSF), which ran from 1972 to 1980 (Abrahamsen *et al.*, 1984) and was followed up by a national monitoring programme starting in 1985 (Horntvedt *et al.*, 1992), the Finnish Acidification Research Programme (HAPRO), which ran from 1985 to 1990 (Kauppi *et al.*, 1990), the research of the Bavarian Forest Toxicology Research Group in the Fichtelgebirge of Germany (Schulze *et al.*, 1989), the National Acidic Assessment Program in the USA (Anon., 1991) and the DEFORPA (Forest Decline and Air Pollution) Programme in France (Landmann, 1992).

During the 1990s, the lack of evidence for major, multi-species forest declines (Innes and Skelly, 2002) resulted in most work on forest decline being phased out. The European Commission continues to finance a few major projects, but the focus of attention in the environmental research community has shifted to climate change and its effects on the environment. In particular, the work of the Intergovernmental Panel on Climate Change has looked at the impacts that climate change could have on the forests of the world (Watson *et al.*, 1996, 1998, 2000; McCarthy *et al.*, 2001). While some still argue that climate change induced by anthropogenic factors has not been adequately demonstrated, the evidence from a variety of sources is that climate change is always occurring (e.g. Goudie, 1983), and today there is clear evidence of a general warming trend. This is particularly apparent at higher latitudes, and impacts on forestry are already being seen (such as reduced winter logging seasons because of earlier snowmelt).

The International Union of Forest Research Organizations has responded to these changes in a number of ways. Its first interdisciplinary Task Force was established (in 1985) to look at the impacts of air pollution on forests. This remit was extended in 1990 in a new Task Force to look at both air pollution and climate change. In 1995, the two areas were amalgamated into the Task Force on Environmental Change, which has published a series of reports dealing with air pollution and climate change (Innes and Haron, 2000; Innes and Oleksyn, 2000; Karnosky *et al.*, 2001; Sidle, 2003). This volume represents one of the last in this series. However, rather than looking at physicochemical changes, as in the other reports, this volume examines some of the social and economic changes that have occurred and the way these are influencing forest management.

In some cases, the different types of change cannot be neatly separated. For example, the destruction of tropical rainforest reflects economic conditions (see Chapter 5), and has major social implications (Stewart, 1994; Bowles and Prickett, 2001; Smouts, 2003), but there are also major implications for the physical environment, ranging from soil erosion and loss of fertility to impacts on local and regional climates. Berkes and Folke (1998) and Pierce Colfer and Byron (2001) provide detailed accounts of some of the linkages between social and ecological systems, and major programmes, such as the Millennium Ecosystem Assessment, are beginning to recognize these connections.

1.3 The changing economic environment

Over the past 50 years, the need to provide better methods to value forest resources has been increasingly recognized. This recognition has come from the growing awareness that deforestation is partly the result of the failure to assign adequate economic value to the many non-timber products associated with forests (Gordon *et al.*, 1999). These products include both tangible products, such as rattan or honey, and intangible benefits, such as the conservation of biodiversity and the maintenance of spiritual values associated with forests. Fortunately, improvements are occurring in the way that such benefits are valued. These include better valuation of services such as the supply of fresh water (Echavarria, 2002; Pagiola, 2002), the conservation of biodiversity (May *et al.*, 2002; Pagiola and Ruthenberg, 2002; Veisten, 2003), the sequestration of carbon (Niesten *et al.*, 2002), and the flow of non-timber forest products (Campbell and Luckert, 2002), amongst others.

Partly as a result of the failure to value forests adequately, there remain huge pressures to convert forests to other forms of land use. This is particularly apparent in the tropics, even though the resulting agricultural land is often only capable of producing crops for a few years. There are many reasons why such deforestation occurs, and these are examined in Chapter 5. Of particular importance is the relationship between agriculture and the general economy of the country. In areas where other forms of wealth are being generated, there is often (but not always) reduced pressure to convert forests to agricultural production (Wunder, 2003), a point that is worth considering when trying to devise solutions to the conversion problem.

The increasing attention that is being paid to tangible non-timber forest products (NTFPs) is having a significant effect on the way that forests are viewed.

This is not necessarily reflected in forest regulation: very few countries have paid attention to the sustainable harvesting of NTFPs, with Brazil and Peru being notable exceptions (Broekhoven, 1996). In examining the value of NTFPs, it is important to recognize that they have both a subsistence value and a market value. The subsistence values are particularly important for forest-dependent peoples with no or few other sources of income. However, largely for this reason, the economic value that is attached to NTFPs tends to be less than their true worth (Gillis, 1992).

The use of forest plantations has a long history, but in recent decades there has been an acceleration in the establishment of new plantations and a growing proportion of the global wood supply coming from plantations. The Global Forest Resource Assessment 2000 estimated that the area of global plantations in 1980 was 17.8 million hectares, a figure that had expanded to 187 million hectares by 2000 (FAO, 2001). Traditionally, fast-growing plantations have been used for pulpwood and charcoal (such as the 1 million hectares of *Acacia mangium* plantations in Indonesia) and slow-growing ones for sawn- and veneer wood. For example, Brazil has 2 million hectares of eucalypt plantations devoted to charcoal production (Cossalter and Pye-Smith, 2003). However, the use of plantation-derived wood is changing, reflecting the benefits of the investment in research and development that this part of the forest sector has enjoyed. This not only applies to plantations established for one purpose and now used for other products, as has happened for example with rubberwood (*Hevea brasiliensis*). While pulpwood and charcoal remain important uses, a range of engineered wood products, including oriented strand board (OSB), laminate strand lumber (LSL), medium-density fibreboard (MDF), hardboard and particleboard, are being produced. Some fast-growing plantation species, such as *Paulownia tomentosa* in China, are grown specifically for timber, and advances in the utilization of small logs have enabled more plantation wood to be used in solid timber, particularly that derived from eucalypts (Cossalter and Pye-Smith, 2003). As advances are made in composite products (e.g. wood-cement composites), even more uses for plantation wood are likely to be found.

Plantations are generally associated with industrial forestry, and many of the plantations established in the 1980s and 1990s were done so by industrial concerns. Private-sector investment in plantations in the 1980s was dominated by companies from North America, Europe, southern South America, South Africa, New Zealand and Australia, although the 1990s also saw significant investments being made by Asian companies (Cossalter and Pye-Smith, 2003). For many people, these plantations are often associated with very intensive methods, with single-species single-age crops of trees grown through the frequent use of pesticides, herbicides and fertilizers. While such plantations exist and continue to be established (e.g. in China), increasingly there has been a tendency towards what Kanowski (2001) terms complex plantations. These may have multiple species of trees and a more diverse structure, and may have greater involvement of local people.

The significance of the changes to the global forest economy being brought about by plantations has not always been noticed. South Africa provides a good example of the trend. Within the last 20 years, South Africa has become a significant source of fibre, with 19 million m^3 of roundwood being harvested annually. Half of this is pulpwood, so that the country is now the 12th largest producer of pulp in the world (Mayers *et al.*, 2001). The pulp and paper industry in the country is dominated by four groups: Mondi Business Paper, Sappi Ltd., Nampak Ltd. and Kimberly-Lark of South Africa (Pty) Ltd., and similar patterns are emerging in other countries, with the plantations of Brazil's Aracruz Cellulose SA now being responsible for the production of 31% of the global supply of bleached eucalyptus pulp.

Although there are uncertainties associated with the future of forestry in some countries, it remains an attractive investment, and private funds continue to be invested in forests (e.g. Mertz, 2001). Continued purchases of forest land by investment groups such as the Hancock Natural Resource Group Inc. are having an impact on forest land prices, although the long-term implications of such purchases are unclear. As foresters move towards sustainable forest management, they face a number of challenges; not least amongst these is the problem of applying traditional economic methods, such as discounting. This is examined in Chapters 3 and 4.

1.4 The changing social environment

Forests are a major resource in many areas, but there has been a recent history of local people failing to benefit from them (see, for example, Peluso, 1992). The social origins of many environmental problems are now much better understood as a result of the use of political ecology to examine the diverse issues that are usually involved. This approach incorporates a number of different methods that are not normally considered in traditional forest research, such as the analysis of local social relations and the contextualization of local situations within the broader context of their political-economic, geographic and environmental settings (Peluso, 1999). In particular, there is now better understanding of the underlying causes of deforestation and the methods available to reduce the current rate of forest loss (e.g. Angelsen and Kaimowitz, 1999; see also Chapter 5).

There have been steadily increasing pressures to adopt a more participatory approach to the development of forest policies. Such a move is seen as an important step towards more ecologically based forest management, although the connection is not always clear. Environmental groups have strongly advocated the use of more participatory techniques in forest management (e.g. Jackson and Ingles, 1998), and this is also reflected in the scientific literature (e.g. Pierce Colfer and Resosudarno, 2002). Increased participation of local communities in forest management is often viewed with considerable caution by the forest industry, and not all attempts to build such participation have been successful (Mayers *et al.*, 2001), but there are numerous examples of successful partnerships (Anon., 2000; Mayers and Vermeulen, 2002; Mayers and Bass, 2004). In particular, there has been a spirited defence of private property rights in countries such as the USA, where any attempt to allow a more consultative approach to the management of forests on privately owned lands has been considered unconstitutional and vigorously (and successfully) opposed.

In addition to the increased involvement of local people in the management of publicly owned forests, a number of other attitudinal changes are occurring. Illustrative of the change in attitudes towards forests is the recent switch in emphasis from forest quantity to forest quality, particularly in relation to the boreal and temperate forests of the world. Whereas previously concerns focused on the loss of forest area, increasingly concerns are being expressed about the loss of quality of the remaining forests. This loss of quality relates to issues such as the loss of species diversity, the loss of structural diversity and the breakdown of natural processes. These changes are examined in detail in Chapter 2.

There continue to be calls to see large areas of forest set aside as reserves. This could have considerable impacts on global wood supply and markets, a possibility that is examined in Chapter 9. There has also been a tendency to set aside forests traditionally viewed as a source of timber, such as the National Forests of the USA. As shown in Chapter 10, such changes require the identification of alternative supplies of timber if markets are not to be unduly disrupted. One possibility is to intensify the production of fibre in existing forests; another is to develop new forest lands through afforestation.

Some of these changes can be related to the emergence of a conservation ethic as a key component in sustainable forest management. While such an ethic has long been advocated (Leopold, 1949), it is only quite recently that it has risen to prominence within forest management. It is frequently associated with indigenous peoples, but attempts to quantify the relationship between indigenous peoples and a conservation ethic have not always been very successful (e.g. Pierce Colfer *et al.*, 2001). Instead, it has permeated much of the thinking about sustainable forest management in the developed world, especially in North America. In the tropics, it is evident from the emphasis that is being placed on techniques such as reduced-impact logging (e.g. Grieser Johns, 1997).

1.5 The changing policy environment

One of the most critical aspects of the policy environment as it relates to forests is the tenure arrangements for those forests. This is nowhere more apparent than in the countries that comprised the former Soviet Union and Eastern bloc. In many of these countries, major privatization programmes are under way, and these have resulted in far-reaching social and economic impacts, discussed in Chapter 7. The rates of reform vary between countries, as does the scope of the reforms, and there has been a tendency to focus on institutional issues, with environmental management only recently becoming more important. Social issues have not been considered in depth, but are likely to assume a greater role in the future (Simula, 2003). While the changes are most apparent in former communist bloc countries, there are also many changes in governance occurring elsewhere as control of forests is devolved to local people (Edmunds and Wollenberg, 2003). Such changes have not always been successful, often because the interests of state officials allowing the devolution differ markedly from those who in theory should benefit.

There are many different types of tenure, and the way in which these are allocated has a profound influence on the forest sector. How land use is allocated within these tenures is also extremely important. In the 1960s and 1970s, there was a belief that forests could be managed for multiple values, although this has been much more difficult to operationalize than was initially anticipated (Peluso, 1999). This in part relates to the difficulties associated with assigning appropriate economic values to the many facets of multiple-use forestry (see Chapter 3), but is also related to problems associated with planning (see Chapter 11) and with ownership patterns (see Chapter 7). More recently, the practicality of land-use zoning has been examined, and Chapter 6 of this book looks at examples of the application of the 'Triad' zoning approach.

In many areas, the relative advantages and disadvantages of the management of forests by the state have been examined. State ownership of forests is often advocated as the only way that public values and rights of access can be maintained. For example, in British Columbia, Canada, the Crown owns 95% of the forest area. Management is undertaken by licensees under a range of different tenure options. Any suggestion that this arrangement be changed in favour of more private ownership has always been rejected by the public, usually amid fears that a transfer of ownership would reduce the quality of forest management and access by the public to those forests. However, such fears are based largely on the property rights associated with common law, particularly as applied in the USA, where a landowner generally has extensive rights. In areas dominated by civil law, these rights are much more restricted and, for example, laws of trespass may be much less stringent, enabling more widespread access to forests. For example, in Switzerland, a private landowner may not prevent public access to privately owned forests, provided that no damage is being done. In the Nordic countries, the right of access guarantees the public open access to non-agricultural land (such as forests), and the law protects this right.

Spears (1999) lists a number of policy and institutional reforms that influence the sustainable management of forests. These are:

Regulatory instruments

- Land-tenure arrangements and land-use zoning
- Legislation to protect traditional property rights and access to forestlands
- Legislation to protect intellectual property rights to germplasm
- Trade regulations
- Timber concessions, licence agreements and penalties for non-compliance
- Legislation to protect biodiversity reserves and national parks
- Global conventions with implications for forest resource management

Financial instruments

- Removal of taxes and subsidies that encourage deforestation
- Revenue-sharing contracts with local communities
- Stumpage fees and timber harvesting or export tax levels that capture economic rent

The influence of changing institutional arrangements on the traditional rights of owners is examined in Chapter 13.

1.6 The changing cultural environment

A major change that is occurring in many parts of the world, and one that is not covered in this volume, is the emergence of the recognition that indigenous peoples have an important part to play in the management of forests. In some countries, such as the USA, the treaties that were settled in the 19th century largely separated the indigenous peoples from their traditional lands. In other areas, such as Canada, New Zealand and Australia, indigenous peoples are now increasingly calling for a greater role in the management of forests on their traditional territories.

Concerns about indigenous peoples' rights are often traced back to the UN Universal Declaration of Human Rights in 1948. This recognized the 'equal and inalienable rights of all members of the human family', but did little to further the plight of some forest-dependent peoples affected by development in their traditional territories. Progress occurred with the 1989 International Labour Organization Convention on Indigenous and Tribal Peoples, and led to the inclusion in Agenda 21 (1992) of a special section dealing with indigenous peoples, and inclusion of indigenous rights in the Convention on Biological Diversity (Barsh, 1994).

The cultural changes associated with changing perceptions of the rights of forest-dependent indigenous peoples have been considered to be so important and so far-reaching as to warrant separate treatment, and a special group is being considered within IUFRO to do this. Any such group will have to consider changes in the entire scientific philosophy that has dominated IUFRO for the past 100 years: the approaches to science that are being adopted by indigenous scholars are very different to those adopted by Western-trained scientists (e.g. Cajete, 1999, 2000; Smith, 1999). Similarly the traditional approach taken by Western scientists that indigenous peoples are an object of study needs to be revised to see indigenous peoples as equal partners in scientific investigations (Mills, 1994).

Governance is an issue that transcends policy, social and cultural issues. The nature of governance is often determined by the culture of those in authority. When that culture differs from the culture(s) of those being governed, problems arise (as in the case of indigenous peoples). A particular governance issue that is of major concern is corruption, and its relationship to illegal logging. Illegal logging is generally considered to be a response to market forces (Dudley, 2002), but the responses to it are largely a cultural issue. However, the extent of illegal logging has been noted by society, and methods to address it are being examined. For example, most certification schemes specifically exclude companies that utilize illegally obtained wood, although the degree to which this is audited and enforced appears to be rather variable. This may do little to deter illegal logging that is conducted in connivance with corrupt officials, but again some progress is being made in identifying the processes that lead to such corruption, and the ways in which it can be tackled (e.g. Elliott, 1997; Kartodihardjo, 2002).

Another change in cultural values is the increasing recognition of the importance of gender-related issues in forestry. This in part can be related to the increasing value that is being placed on non-timber forest products (Brown and Lapuyade, 2001). In some areas, women are increasingly earning more than men because of the way in which they utilize forest resources, with Brown and Lapuyade (2001) citing the example of the village of Komassi, Cameroon. Gender-related issues cover a range of different fields (McDougall, 2001), but in the context of a changing cultural environment, it is the social relationships between men and women (particularly those related to the empowerment of women in forestry) that are often receiving the greatest interest.

1.7 Bringing it all together

There is a growing realization that many forest policies have been less than successful. Too often, development strategies have failed to consider forests within a broader social and economic context, with development programmes neither focusing sufficiently on people who live near or in forests nor considering forests within the context of the non-forest characteristics of the country (Gordon *et al.*, 1999).

During the 1990s, various movements in Europe and North America and, to a lesser extent, elsewhere, advocated a major change in the way that forestry is practised. Prominent amongst these are the concepts of near-to-nature and continuous-cover forestry in Europe (e.g. Bode, 1997; Egli, 1999) and ecosystem management (and the rather different ecosystem-based management) in North America. The approach of continuous-cover forestry has been gaining ground in many European countries, and the case of Turkey is presented in Chapter 11. Gordon (1994) identified four major themes of ecosystem management. These were that forests should be managed with people in mind, that management should be based on local conditions, that management should use site-specific information and that management should be undertaken without externalities. Many of these have since been taken up in practice, and are evident in the increasing number of guides that take a more ecologically sensitive approach to forest management, such as *Positive Impact Forestry: A Sustainable Approach to Managing Woodlands* (McEvoy, 2004) and *Low-Impact Forestry: Forestry as if the Future Mattered* (Lansky, 2002).

These changes have become apparent in both hard and soft law. Of the various soft law mechanisms, it is certification that has received the greatest attention. In Chapter 8, the development of certification is examined. Certification is particularly interesting as it is an example of a market-driven process, although the process has not exactly worked in the way that was intended. For example, there is rarely a premium for wood products from certified forests except where these are able to fill small niche markets. However, it is certainly a factor that is driving changes in the way that forests are managed, moving traditional forest management towards a more sustainable approach (Mayers *et al.*, 2001). Unfortunately, some of the greatest certification-induced changes in forest management appear to be occurring in developed countries, whereas the original idea was to help improve the management of tropical forests (Poore, 2003).

The changes that are occurring in forestry require the development and application of suitable planning tools. These are examined in Chapters 11 (for Turkey) and 12 (for Malaysia), and in Chapter 14. Forest planning is becoming increasingly sophisticated, and an increasing number of tools have become available in recent years (see, for example, Schmoldt *et al.*, 2001). The need for such tools has never been greater than today. Dawkins and Philip (1998) state:

> In the past, it was the task of the professional forester to decide the limits of technical feasibility, propose the strategies and implement specific forest programmes to attain the objects. Now that forest products and services are insufficient to meet all demands, the allocation decisions made by the professionals often fail to match what society perceives as optimal. (p. 305)

This is at the root of the widespread loss of confidence in the profession of forestry that has occurred in recent years. Foresters are no longer considered by many to be the most suitable people to manage forests, primarily because they have failed to meet the expectations of society. This is a problem that is going to get worse before it is resolved. Currently, forestry education institutions around the world are under pressure, and many have closed or have been merged into other institutions. Continuing education is also under threat, with the example of British Columbia being illustrative. Here, forestry education was served through a number of specialized institutions, including the universities, colleges, and three agencies dealing with continuing education in various forms, the Forest Management Institute of British Columbia (FMIBC), the Forestry Continuing Studies Network (FCSN), and FORREX, a natural resources extension partnership. By mid-2005, both the FCSN and the FMIBC had been closed down, because government was unwilling to invest in funds for the continuing education of the managers responsible for maintaining the public resource and the licensees were unwilling to give up human resources for or invest in further training. Such changes are likely to have far-reaching implications for the future, and the future sustainable management of the world's forests will occur only if there are suitably trained people to do so (Poore, 2003).

Another unifying theme has been the development of criteria and indicators of sustainable forest management. This idea was first introduced in relation to tropical forest management, and an example of their application in Sabah, Malaysia, is provided in Chapter 12. However, the use of criteria and indicators has become highly politicized, especially in developed countries, and they have become a focal point for repeated discussions, rather than a means of improved forest management through comprehensive reporting on critical issues. As Poore (2003) states, they have become 'a good idea that has lost its way' (p. 255). Criteria and indicators should be an indication of whether or not a forest is well managed. The extent to which indicators should be determined locally is debatable – Rasmussen *et al.* (2000) indicate that their study in Thailand has little applicability because it was developed with minimal input from local stakeholders. However, local stakeholders may be unaware of larger issues, particularly in relation to global processes such as loss of biodiversity or carbon conservation. Consequently, it seems that a balance between local and outside expertise is needed when developing local indicators of sustainable forest management.

1.8 Conclusion

While there has been a great deal of attention paid to the impact of phenomena such as air pollution and climate change on forests, the impacts of other forms of change have not received the same type of attention. Many studies have been undertaken, but they have rarely attempted to place these socioeconomic changes within the broader context of all the changes that are occurring. This book represents an attempt to approach the issue of the changing sociocultural context of forestry. It is not, and could not be, comprehensive, but it will hopefully point to areas where further work is needed.

References

Abrahamsen, G., Horntvedt, R. and Venn, K. (1984) Air pollution – the risk of forest injuries in Norway. *Norsk skogbruk* 30(11), 15–17. (In Norwegian.)

Angelsen, A. and Kaimowitz, D. (eds) (1999) *Agricultural Technologies and Tropical Deforestation.* CABI Publishing, Wallingford.

Anon. (1991) *National Acidic Precipitation Assessment Program. 1990 Integrated Assessment Report.* National Acidic Precipitation Assessment Program, Washington, DC.

Anon. (2000) Participatory forest management: a strategy for sustainable forest management in Africa. In: *Proceedings of the International Workshop on Community Forestry in Africa.* FAO, Rome.

Barsh, R. (1994) Indigenous peoples in the 1990s: from object to subject of international law. *Harvard Human Rights Journal* 7, 33–86.

Berkes, F. and Folke, C. (eds) (1998) *Linking Social and Ecological Systems. Management Practices and Social Mechanisms for Building Resilience.* Cambridge University Press, Cambridge.

Bode, W. (ed.) (1997) *Naturnahe Waldwirtschaft. Prozeßschutz oder biologische Nachhaltigkeit?* Deukalion, Hamburg.

Bowles, I.A. and Prickett, G.T. (eds) (2001) *Footprints in the Jungle. Natural Resource Industries, Infrastructure, and Biodiversity Conservation.* Oxford University Press, Oxford.

Broekhoven, G. (1996) *Non-timber Forest Products. Ecological and Economic Aspects of Exploitation in Colombia, Ecuador and Bolivia.* IUCN, Gland.

Brown, K. and Lapuyade, S. (2001) Changing gender relationships and forest use. A case study from Komassi, Cameroon. In: Pierce Colfer, C.J. and Byron, Y. (eds) *People Managing Forests. The Links Between Human Well-being and Sustainability.* Resources for the Future, Washington, DC, pp. 90–110.

Cajete, G. (1999) *Ignite the Sparkle: An Indigenous Science Model.* Kivaki Press, Skyland, North Carolina.

Cajete, G. (2000) Indigenous knowledge: the Pueblo metaphor of indigenous education. In: Battiste, M. (ed.) *Reclaiming Indigenous Voice and Vision.* University of British Columbia Press, Vancouver, pp. 181–191.

Campbell, B.M. and Luckert, M.K. (eds) (2002) *Uncovering the Hidden Harvest. Valuation Methods for Woodlands and Forest Resources.* Earthscan, London.

Cossalter, C. and Pye-Smith, C. (2003) *Fast-wood Forestry. Myths and Realities.* Centre for International Forestry Research, Bogor.

Dawkins, H.C. and Philip, M.S. (1998) *Tropical Moist Forest Silviculture and Management. A History of Success and Failure.* CABI Publishing, Wallingford.

Dudley, R.G. (2002) Dynamics of illegal logging. In: Pierce Colfer, C.J. and Resosudarno I.A.P. (eds) *Which Way Forward? People, Forests and Policymaking in Indonesia.* Resources for the Future, Washington, DC, 358–382.

Echavarria, M. (2002) Financing watershed conservation: the FONAG water fund in Quito, Ecuador. In: Pagiola, S., Bishop, J. and Landell-Mills, N. (eds) *Selling Forest Environmental Services. Market-based Mechanisms for Conservation and Development.* Earthscan, London, pp. 91–101.

Edmunds, D. and Wollenberg, E. (2003) Whose devolution is it anyway? Divergent constructs, interests and capacities between the poorest forest users and states. In: Edmunds, D. and Wollenberg, E. (eds) *Local Forest Management. The Impacts of Devolution Policies.* Earthscan, London, pp. 150–165.

Egli, B. (1999) *Naturgemässe Waldwirtschaft.* Neujahrsblatt der Naturfoschenden Gesellschaft Schaffhausen, N. 51/1999. Naturfoschenden Gesellschaft Schaffhausen, Schaffhausen.

Elliott, K.A. (1997) Corruption as an international policy problem: overview and recommendations. In: Elliott, K.A. (ed.) *Corruption and the Global Economy.* Institute for International Economics, Washington, DC, pp. 175–233.

FAO (2001) *Global Forest Resource Assessment 2000: Main Report.* FAO Forestry Paper 140. FAO, Rome.

Gillis, M. (1992) Economic policies and tropical deforestation. *Advances in Economic Botany* 9, 129–142.

Gordon, J.C. (1994) Vision to policy: a rule for foresters. *Journal of Forestry* 92(7), 16–19.

Gordon, J.C., Berry, J.K. and Schmidt, R. (1999) Forests, poverty, and this book. In: Schmidt, R., Berry, J.K. and Gordon, J.C. (eds) *Forests to Fight Poverty. Creating National Strategies.* Yale University Press, New Haven, pp. 8–22.

Goudie, A. (1983) *Environmental change.* 2nd edition. Oxford University Press, Oxford.

Grieser Johns, A. (1997) *Timber Production and Biodiversity Conservation in Tropical Rain Forests.* Cambridge University Press, Cambridge.

Horntvedt, R., Aamlid, D., Rørå, A. and Joranger, E. (1992) Monitoring programme for forest damage. An overview of the Norwegian programme. *Norwegian Journal of Agricultural Sciences* 6, 1–17.

Innes, J.L. and Haron A.H. (2000) *Air Pollution and the Forests of Developing and Rapidly Industrializing Countries.* CABI Publishing, Wallingford.

Innes, J.L. and Oleksyn, J. (2000) *Forest Dynamics in Heavily Polluted Regions.* CABI Publishing, Wallingford.

Innes, J.L. and Skelly, J.M. (2002) Forest decline and air pollution: an assessment of 'forest health' in the forests of Europe, the Northeastern United States, and Southeastern Canada, In: Bell, J.N.B. and Treshow, M. (eds) *Air Pollution and Plant Life*, 2nd edition. John Wiley and Sons, Chichester, pp. 273–293.

Jackson, W.J. and Ingles, A.W. (1998) *Participatory Techniques for Community Forestry. A Field Manual.* IUCN, Gland, Switzerland, and World Wide Fund for Nature, Gland, Switzerland.

Kanowski, P. (2001) Plantation forestry at the millennium. In: Woodwell, G.M. (ed.) *Forests in a Full World.* Yale University Press, New Haven, pp. 97–109.

Karnosky, D.F., Ceulemans, R., Scarascia-Mugnozza, G. and Innes, J.L. *The Impact of Carbon Dioxide and Other Greenhouse Gases on Forest Ecosystems.* CABI Publishing, Wallingford.

Kartodihardjo, H. (2002) Structural problems in implementing new forestry policies. In: Pierce Colfer, C.J. and Resosudarno, I.A.P. (eds) *Which Way Forward? People, Forests and Policy Making in Indonesia.* Resources for the Future, Washington, DC, pp. 144–160.

Kauppi, P., Anttila, P. and Kenttämies, K. (eds) (1990) *Acidification in Finland.* Springer Verlag, Berlin.

Landmann, G. (1992) Research into forest decline and air pollution in France. Major findings and relevance for policy applications. In: Schneider, T. (ed.) *Acidification Research: Evaluation and Policy Implications.* Elsevier, Amsterdam, pp. 383–395.

Lansky, M. (ed.) (2002). *Low-Impact Forestry: Forestry as if the Future Mattered.* Maine Environmental Policy Institute, Hallowell, Maine.

Leopold, A. (1949) *A Sand County Almanac.* Oxford University Press, Oxford.

May, P.H., Veiga Neto, F., Denardin, V. and Loureiro, W. (2002) Using fiscal instruments to encourage conservation: Municipal responses to the 'ecological' value-added tax in Paraná and Minas Gerais, Brazil. In: Pagiola, S., Bishop, J. and Landell-Mills, N. (eds) *Selling Forest Environmental Services. Market-based Mechanisms for Conservation and Development.* Earthscan, London, pp. 173–199.

Mayers, J. and Bass, S. (2004) *Policy that Works for Forests and People. Real Prospects for Governance and Livelihoods.* Earthscan, London.

Mayers, J., Evans, J. and Foy, T. (2001) *Raising the Stakes. Impacts of Privatisation, Certification and Partnerships in South African Forestry.* International Institute for Environment and Development, London.

Mayers, J. and Vermeulen, S. (2002) *Company–Community Forestry Partnerships. From Raw Deals to Mutual Gains?* Instruments for Sustainable Private Sector Forestry series. International Institute for Environment and Development, London.

McCarthy, J.J., Canziani, O.F., Leary, N.A., Dokken, D.J. and White, K.S. (eds) (2001) *Climate Change 2001: Impacts, Adaptation, and Vulnerability.* Cambridge University Press, Cambridge.

McDougall, C.L. (2001) Gender and diversity in assessing sustainable forest management and human well-being. Reflections on assessment methods tests conducted in Bulungan, East Kalimantan, Indonesia. In: Pierce Colfer, C.J. and Byron, Y. (eds) *People Managing Forests. The Links between Human Well-being and Sustainability.* Resources for the Future, Washington, DC, pp. 50–71.

McEvoy, T.J. (2004) *Positive Impact Forestry: A Sustainable Approach to Managing Woodlands.* Island Press, Washington, DC.

Mertz, P. (2001) Investing in sustainable forestry. In: Chipeta, M.E. and Joshi, M. (eds) *The Private Sector Speaks: Investing in Sustainable Forest Management.* Center for International Forestry Research, Bogor, pp. 111–119.

Mills, A. (1994) *Eagle Down is our Law. Witsuwit'en Law, Forests and Land Claims.* UBC Press, Vancouver.

Niesten, E., Frumhoff, P.C., Manion, M.M. and Hardner, J.J. (2002) Designing a carbon market that protects forests in developing countries. In: Swingland, I.R. (ed.) *Capturing Carbon and Conserving Biodiversity. The Market Approach.* Earthscan, London, pp. 332–345.

Pagiola, S. (2002) Paying for water services in Central America: learning from Costa Rica. In: Pagiola, S., Bishop, J. and Landell-Mills, N. (eds) *Selling Forest Environmental Services. Market-based Mechanisms for Conservation and Development.* Earthscan, London, pp. 37–61.

Pagiola, S. and Ruthenburg, I.-M. (2002) Selling biodiversity in a coffee cup: shade-grown coffee and conservation in Mesoamerica. In: Pagiola, S., Bishop, J. and Landell-Mills, N. (eds) *Selling Forest Environmental Services. Market-based Mechanisms for Conservation and Development.* Earthscan, London, pp. 103–126.

Peluso, N. (1992) *Rich Forests, Poor People: Resource Control and Resistance in Java.* University of California Press, Berkeley.

Peluso, N. (1999) The role of forests in sustaining smallholders. In: Schmidt, R., Berry, J.K. and Gordon, J.C. (eds) *Forests to Fight Poverty. Creating National Strategies.* Yale University Press, New Haven, pp. 38–64.

Pierce Colfer, C.J. and Byron, Y. (eds) (2001) *People Managing Forests. The Links between Human Well-being and Sustainability.* Resources for the Future, Washington, DC.

Pierce Colfer, C.J., Woelfel, J., Wadley, R.L. and Harwell, E. (2001) Assessing people's perceptions of forests. Research in West Kalimantan, Indonesia. In: Pierce Colfer, C.J. and Byron, Y. (eds) *People Managing Forests. The Links between Human Well-being and Sustainability.* Resources for the Future, Washington, DC, pp. 135–154.

Pierce Colfer, C.J. and Resosudarno I.A.P. (eds) (2002) *Which Way Forward? People, Forests and Policymaking in Indonesia.* Resources for the Future, Washington, DC.

Poore, D. (2003) *Changing Landscapes.* Earthscan, London.

Rasmussen, J.N., Kaosa-ard, A., Boon, T.E., Diaw, M.C., Edwards, K., Kadyschuk, S., Kaosa-ard, M., Lang, T., Preechapanya, P., Rerkasem. K. and Rune, F. (2000) *For Whom and for What? Principles, Criteria and Indicators for Sustainable Forest Resources Management in Thailand.* Skov and Landskab Report 6, Danish Centre for Forest, Landscape and Planning, Hørsholm.

Schmoldt, D.L., Kangas, J., Mendoza, G.A. and Pesonen, M. (eds) (2001) *The Analytic Hierarchy Process in Natural Resource and Environmental Decision Making.* Kluwer Academic, Dordrecht.

Schulze, E.-D., Lange, O.L. and Oren R. (1989) *Forest Decline and Air Pollution. A Study of Spruce (Picea abies) on Acid Soils.* Springer Verlag, Berlin.

Sidle, R.C. (2003) *Environmental Change and Geomorphic Hazards in Forests.* CABI Publishing, Wallingford.

Simula, M. (2003) Forest sector reforms in Eastern European countries – overview and lessons learnt. In: Anon. (ed.) *Experiences of Countries with Transition Economies: Problems and Solutions.* Program on Forests, World Bank, Washington, DC, pp. 5–12.

Smith, L.T. (1999) *Decolonizing Methodologies: Research and Indigenous Peoples.* Zed Books, London.

Smouts, M.-C. (2003) *Tropical Forests, International Jungle. The Underside of Global Ecopolitics.* Palgrave Macmillan, New York.

Spears, J. (1999) Creation of country-specific strategies. In: Schmidt, R., Berry, J.K. and Gordon, J.C. (eds) *Forests to Fight Poverty. Creating National Strategies.* Yale University Press, New Haven, pp. 146–167.

Stewart, D.I. (1994) *After the Trees: Living on the Transamazon Highway.* Texas University Press, Austin.

Veisten, K. (2003) Valuation of non-market forest products – methodological and empirical studies. Doctor Scientarium thesis 2003:39, Agricultural University of Norway, Ås.

Watson, R.T., Zinyowera, M.C., Moss, R.H., Dokken, D.J. (eds) (1996) *Climate Change 1995. Impacts, Adaptations and Mitigation of Climate Change: Scientific-technical Analyses.* Cambridge University Press, Cambridge.

Watson, R.T., Zinyowera, M.C., Moss, R.H. and Dokken, D.J. (eds) (1998) *The Regional Impacts of Climate Change. An Assessment of Vulnerability.* Cambridge University Press, Cambridge.

Watson, R.T., Noble, I.R., Bolin, B., Ravindranath, N.H., Verardo, D.J. and Dokken, D.J. (eds) (2000) *Land Use, Land-use Change, and Forestry.* Cambridge University Press, Cambridge.

Wunder, S. (2003) *Oil Wealth and the Fate of the Forest. A Comparative Study of Eight Tropical Countries.* Routledge, London.

The Emerging Role of Forest Quality in Setting Perspectives in Forest Management

2

N. Dudley

Equilibrium Ltd., UK

Over the last 15 to 20 years there has been a profound cultural shift within the worlds of forestry and conservation, from a narrow focus on the area of trees to an increasing emphasis on the quality of forest ecosystems. These changes have taken place in the temperate and boreal forests of the richer world, and have been driven by wider recognition of the breadth of values that forests contain and a renewed focus on their conservation. A similar shift in values can be seen emerging in the tropics. The new perspective has profound practical implications for the way in which forests will be managed, restored and maintained in the future. It has challenged what had become a rather comfortable view of the state of forests in the richer countries. All over the world, dissident natural and social scientists are starting to challenge the official views about forest status.

This new focus on forest quality did not come in one leap, but instead filtered in gradually, developing from simple notions like 'old growth forest' into an increasingly sophisticated recognition of the range of different goods and services that can accrue from a forest mosaic. These concerns have prompted an increasing focus on optimal management at a landscape level.

2.1 Introduction

Comparing a natural forest with a monoculture plantation of exotic trees is a bit like comparing a wildflower meadow with a football pitch. Football and wildflowers both have their uses and their supporters, but they are definitely not the same thing. Yet, until recently, the word 'forest' was used to describe everything from intensively managed tree crops to ancient, near primary, forests with their vast associated ecology.

© CAB International 2005.
Forestry and Environmental Change: Socioeconomic and Political Dimensions (eds J.L. Innes, G.M. Hickey and H.F. Hoen)

Over the last 15 to 20 years there has been a profound cultural shift within the worlds of forestry and conservation, from a narrow focus on the area of trees to an increasing emphasis on the quality of forest ecosystems. These changes have taken place first in the temperate and boreal forests of the richer world and have been driven by wider recognition of the breadth of values that forests contain and a renewed focus on their conservation. A similar shift in values can be seen emerging in the tropics. The new perspective will have profound practical implications for the way in which forests are to be managed, restored and maintained in the future.

This shift has also challenged what had become a rather comfortable view of the state of forests in the richer countries. Should Western Europe really be smug about the status of its forests when most countries have less than 1% (Anon., 2000) of their forest estate remaining in anything like a natural state? Can New Zealand claim to have an effective network of protected forests when Taylor and Smith (1997) have argued that a good proportion of the native wildlife that they contain is being destroyed by invasive species? A switch from natural forest to oil palm or acacia plantation has been good for foreign exchange in many tropical countries, but does this compensate for the losses in non-timber forest products suffered by the poorest members of society? All over the world, dissident natural and social scientists are starting to challenge the official views about forest status.

The increasing focus on forest quality did not come about in one leap, but instead filtered in gradually, developing from simple notions like 'old growth forest' into an increasingly sophisticated recognition of the range of different goods and services that can accrue from a forest mosaic. Those of us interested in these issues – both within conservation and industry – are now at the stage of grappling with models and approaches to help achieve optimal management at a landscape level.

This chapter attempts to provide a quick overview of how ideas related to forest quality have developed. It looks at the constituents of forest quality (from one particular experience in their measurement) and examines how others have tackled the issue. Finally, it explores what we may glean about global forest quality from the data that are currently available.

2.2 A growing understanding about forest quality

In 1977, US President Jimmy Carter directed the Council on Environmental Quality to review the likely state of the environment in the year 2000. The resulting *Global 2000 Report to the President* (Barney, 1980) painted a fairly gloomy picture of overcrowding, pollution and destabilized ecosystems. While the report was notable in being one of the first to point out the consequences of tropical forest loss, it presented a relatively benign picture of conditions in the temperate world, noting that forest area in Europe and North America was stable and in most countries increasing. In doing so, it reflected the commonly held view that forests in the developed world were in good shape. Over the next decade, the debate about forest management centred almost entirely on the tropics. Attempts to track forest cover, notably through the Food and Agriculture Organization

(FAO)/United Nations Economic Commission for Europe (UNECE) Forest Resource Assessments, counted all types of forest as one, so that, while the tropical forest estate continued to shrink, the temperate forests were seen as rebounding after historical deforestation.

During the 1990s, three factors helped to change this perception. The failure of the United Nations Conference on Environment and Development to agree on a global forest convention, settling instead for a set of forest principles (Anon., 1993a), caused such a backlash that it spurred many governments to address forest issues in a way that they had not done before. This move came at a time when there was growing concern about threats to native, particularly old-growth, forest in North America, Scandinavia, New Zealand and Australia and later across most of Western Europe. The debate grew into one of the longest and most bitter disputes in the short history of modern environmental concern, and still shows little sign of abating (Dietrich, 1992). Scientists in Europe began to be concerned about issues of tree health and whether long-range air pollutants were damaging trees in the same way that they had impacted on freshwater ecosystems. Subsequently, there was a growing perception that global forest quality was declining as a result of human activities. This decline has been linked to a number of changes in management and resources. For example:

- Replacement of biologically rich old-growth forest with species-poor young forests, intensively-managed forests or plantations;
- Decline in the health of trees and other forest species as a result of anthropogenic changes (e.g. air pollution, climate change and the impacts of introduced pests and diseases and invasive species, etc.);
- Breakdown of ecological support systems associated with forests, including hydrological systems, soil structure and fire ecology;
- Threats to social rights in forests, including issues related to tenure, access and management practices that have resulted in declining non-wood goods and services; and
- Loss of intangible values (e.g. aesthetic, cultural and spiritual)

Each of these issues has its own champions and detractors. The public debate concerning the role of both plantations and air pollution, for example, has frequently been bitter. In those countries where forest cover has stabilized, the debate has shifted to what kind of forests remain or could be recreated.

2.3 Defining and measuring forest quality

In response to these issues, the WWF International expanded its work on forests from a focus on the tropics to incorporate temperate and boreal issues. In the temperate and boreal regions, it became clear that conservation concerns were less about the total quantity of forests than about the quality of the forest that remained. Subsequently, a definition of forest quality was proposed: 'The significance and value of all ecological, social and economic components of the forest landscape.' In an attempt to capture what this meant, WWF proposed four criteria related to quality (Dudley, 1992): authenticity, forest health (this was later conflated into authenticity under indicators of resilience), environmental benefits and socioeconomic benefits. The remaining criteria are outlined in Table 2.1. The criteria were tested theoretically in workshops and meetings in a number of countries and later developed into landscape scale indicators suitable for use in field assessments. This was done in partnership with the World Conservation Union (IUCN) and the École Polytechnique Fédérale de Lausanne in Switzerland (Dudley and Rae, 1998).

Table 2.1. Three criteria of forest quality

Criteria	Brief description
Authenticity	Authenticity is a measure of ecosystem integrity and health in the broadest sense. It concentrates on current ecosystem functioning, regardless of the forest's history, and thus also has relevance to disturbed forests. One definition of an 'authentic' forest is a forest in which: **all the expected ecosystem functions can continue to operate indefinitely** (Dudley, 1996).
Environmental benefits	Environmental benefits **encompass a range of issues that have direct relevance to both ecosystem health and to the health of human societies**. Important elements include the extent to which forests interact with soil and water systems, the impacts on climate and the forests' role in harbouring biodiversity.
Social and economic benefits	The last criterion is exclusively focused **on interactions between forests and human societies.** Benefits range from products such as wood and game to the use of forests for living, recreation and other 'hard-to-measure' values such as the cultural, aesthetic and spiritual values of particular forest types or locations.

A methodology was then applied through field tests and further development in the UK, Switzerland, Cameroon, Gabon, Central America and Vietnam (Dudley *et al.*, 2003). From these experiences a methodology and a core set of indicators emerged (Dudley *et al.*, in press) (see Table 2.2), although variations were encouraged to address local conditions. Assessment was then based around a seven-stage process that was either expert driven or participatory (see Fig. 2.1). Along with the practical problems associated with identifying measurable indicators that will capture information about different aspects of forest quality (a task that is by no means complete), the work also identified a number of broader lessons. It became clear that 'quality', which everyone agreed was important, meant completely different things to different people. Commercial timber producers do not look at a forest in the same way as local villagers, holidaymakers or indigenous people. Yet their views are all valid, and experience shows that, in most cases, stakeholders can understand other ideas of quality.

Forests provide society with a large range of goods and services, and reconciling these within sustainable forest management policy presents major challenges. Some countries, such as Germany, have attempted to do this by managing forests so that each particular forest stand supplies a wide range of economic, social and environmental benefits, while countries like New Zealand have made a sharp distinction between commercial timber and fibre plantations and 'natural' forests for biodiversity and social values. The recent work on forest quality has suggested that representing the entire range of different 'qualities' is difficult in one forest stand, but possible at the forest landscape level (made up of a range of forest stands serving different needs). However, given the interconnectedness of the forest systems, it is not feasible to divorce elements of the forest estate completely from one another. Therefore, management decisions in one stand need to take account of impacts beyond the stand's borders. The ideal situation is a forest mosaic planned and managed as an integrated whole: perhaps midway between the multipurpose model and the New Zealand approach. However, this is extremely difficult to achieve in the 'real' world.

Table 2.2. Some suggested indicators of forest quality

Indicators of authenticity	
Composition	Resilience
Pattern	Continuity
Functioning	Development patterns
Process	
Indicators of environmental benefits	
Biodiversity conservation	Impacts on other ecosystems
Soil and watershed protection	Climate stabilization
Indicators of other social and economic benefits	
Wood products	Spiritual, cultural and artistic values
Non-wood products	Management and land use
Employment and subsistence	Rights and legal issues
Recreation	Knowledge
Homeland	Nature of incentives
Historical values	Local distinctiveness

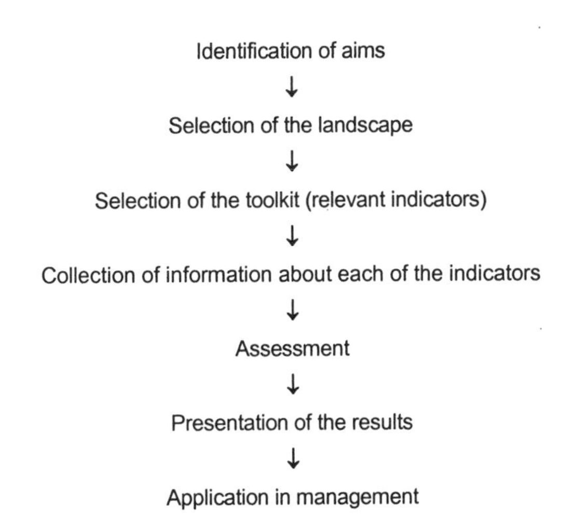

Fig. 2.1. Seven-stage process for the identification of indicators of forest quality.

More recently, forest quality methodology has been developed into wider work that deals with landscape approaches to conservation management, where assessment plays one part in a broader planning and negotiation process. Issues of quality have focused on methods for the identification of High Conservation Value Forests, which take both social and environmental issues into consideration (Jennings, 2003).

2.4 A plethora of attempts to define forest quality

In the early 1990s, there were suddenly a plethora of attempts to define good forest management and high quality forest estates. In late 1992, European governments met in Helsinki and agreed to develop criteria and indicators of sustainable forest management. One year later other temperate forest countries did the same in Montreal, Canada. The Helsinki Process (now the Ministerial Conference on the Protection of Forests in Europe (MCPFE)) and the Montreal Process started a trend and at least six regional criteria and indicator processes emerged. Some of these continue to be influential while others have since lost funding and appear to have been abandoned. Most extend well beyond traditional concerns about growing stock and tree health, and include issues relating to the environment and human society. At a global level, criteria and indicators have been developed for tropical forests (by the International Tropical Timber Organization (ITTO)) and for forest biodiversity (by the Convention on Biological Diversity (CBD)). Issues of quality were also addressed by the temperate and boreal forests components of the United Nations Forest Resource Assessment 2000, organized jointly by FAO and the UN Economic Commission for Europe (Anon., 2000). The work at the international and regional levels has been complemented by the development of national systems. There has also been a range of non-governmental initiatives, many of which are based around stand-level assessments (e.g. forest management certification). There are also some attempts by non-governmental organizations (NGOs) to reflect regional forest quality criteria, such as the WWF European forest scorecards (Sollander, 2000).

A selection of forestry-related initiatives is summarized in Table 2.3. In general, there are a number of similarities between regional systems, to the extent that there was an unsuccessful attempt to combine them in 1996 (Anttila, 1996). There have been a number of comparisons of the various forestry-related schemes, most notably by the Center for International Forestry Research (CIFOR) (Prabhu *et al.*, 1996) and the Tropenbos Foundation (Lammerts van Bueren and Blom 1996). CIFOR has also developed a methodology for users to select a portfolio of indicators suitable to their particular situation and has tested this in many parts of the world (Prabhu *et al.*, 1999).

There has undoubtedly been some overlap in these efforts, although this has ensured that many different groups of people around the world have been forced to wrestle with the issues. It also indicates that there is an emerging consensus on the constituents of a high quality forest estate, which have remained remarkably consistent through numerous discussions. These include maintaining sufficient proportions of near-natural forest to maintain biodiversity, balancing different social and economic needs from particular forest estates and making the best use of forests in terms of environmental management.

The precise details of the criteria and indicator schemes vary between the initiatives and are unlikely to remain constant. For instance, aesthetic appreciation of forests has changed dramatically over the last century in many countries, with an increasing recognition of the appeal of natural forests as compared with more heavily managed and manicured parkland, and this perception could well switch again in the future. Similar changes can be expected to occur in the demands for industrial timber and in the ways that indigenous people wish to manage their resources.

2.5 Can we measure the status of global forest quality?

Most attempts to measure forest quality (whether or not they use this precise term) focus on individual forests at the stand level, rather than the landscape level. Over the last decade there have also been some efforts, mainly through the United Nations Forest Resource Assessments and the regional criteria and indicator processes, to identify indicators of forest quality at the national level. While these currently represent our best available information, none of the assessment methodologies take into account the full range of forest quality objectives, and all have been plagued by absent or uncertain information. There is, therefore, no survey of forest quality at the global level that encapsulates all of the elements discussed above. There is, however, some very interesting information being collected. In the following section a preliminary attempt has been made to see what data are available regarding forest quality at the global scale. Drawing on the very large amount of time that has gone into indicators of good management and forest condition, I have identified the following as covering the majority of issues:

- Conservation value of forests
- Area of native forest
- Biodiversity (including, in particular, species under threat)
- Connectivity and fragmentation
- Age of native forests
- Forest health (particularly unnatural health impacts related to pollution, invasive species, etc.)
- Susceptibility to climate change
- Wood products
- Non-timber forest products
- Environmental services
- Use of forests by traditional and indigenous peoples
- Management effectiveness

There are a number of other important issues, such as the recreational or spiritual values of forests, which were omitted because no global databases have been developed.

Table 2.3. Global and regional attempts to define forest quality

Criteria and indicators (C&I)	Details
(a) Global-level processes measuring forest quality on a country scale.	
Convention on Biological Diversity	The CBD drew up draft C&I for forest biodiversity at an experts meeting in Helsinki in 1996 although these have never been applied.
International Tropical Timber Organization	ITTO has drawn up C&I for tropical forests, including C&I for forest management (Anon., 1997b), planted forests (Anon., 1993b) and forest restoration (Anon., 2003a).
UN Forest Resources Assessment	Development of the Forest Resource Assessment 2000, by FAO and the UNECE, included aspects of biodiversity, naturalness and non-timber forest products. An expert meeting in Kotka, Finland, in June 1996 identified new criteria including 'naturalness' of forests.
(b) Regional-level criteria and indicator processes measuring forest quality on a country scale	
Ministerial Conference for the Protection of Forests in Europe (MCPFE)	Previously known as the Helsinki Process. An initiative of the Finnish and Portuguese governments, launched in 1993 in Helsinki (Liaison Unit in Lisbon, 1998). The Helsinki process has drawn up indicators of good forest management at a national level (Anon., 1995a).
Montreal Process	Launched in October 1993 and has drawn up criteria and indicators of sustainable forest management with 10 non-European temperate and boreal countries and produces regular reports (Anon., undated) including a definition of sustainable forest management (Canadian Council of Forest Ministers, 1995).
Tarapoto Process	Launched by the Amazon Cooperation Treaty at a *Regional Workshop on the Definition of Criteria and Indicators for Sustainability of Amazonian Forests* in Peru in February 1995 (Anon., 1995b).
Dry-Zone Africa Process	Launched at a *United Nations Environment Programme/United Nations Food and Agriculture Organization Expert Meeting on Criteria and Indicators for Sustainable Management in Dry-Zone Africa* at Nairobi in November 1995 (Anon., 1996).
Central American Process	Draft criteria and indicators for sustainable forest management (SFM) were developed at Tegucigalpa, Honduras, in 1997. C&I are set at regional and national level (Anon., 1997a).
North Africa and the Middle East	United Nations Food and Agriculture Organization process – draft C&I were produced in 1997.
African Timber Organization	Has drawn up C&I for sustainable management of African tropical forests in association with ITTO.
(c) National-level criteria and indicator processes measuring forest quality	
France	Detailed indicators for French forests for the pan-European process (Anon., 1994).
Finland	C&I for the pan-European process with a report published in 1997 (Eeronheimo *et al.*, 1997).
(d) Stand-level attempts to set criteria of forest quality	
Forest Stewardship Council	Established as an accreditation body for independent, stand-level assessment of sustainable forest management. The FSC principles and criteria provide guidance for certification bodies, which then draw up their own standards.

Table 2.3 (cont). Global and regional attempts to define forest quality

ISO 14000	The International Organization for Standardization has developed a certification scheme for timber, although this is not based on independent certification at stand level.
Programme for the Endorsement of Forest Certification (PEFC)	Certification scheme running as an umbrella, with 30 member national schemes in five continents, of which 18 have so far been endorsed (2005).
Soil Association	The organization launched a Responsible Forestry Programme in 1994, associated with the Woodmark label. Accredited to the FSC.
Scientific Certification Systems	SCS launched a similar certification scheme in 1995.
Center for International Forestry Research	Series of toolkits developed for choosing and testing criteria and indicators for stand-level forest management.
World Resources Institute	Carried out work on environmental indicators, including indicators of quality (Hammond *et al.*, 1995), sometimes in partnership with IUCN (Reid *et al.*, 1993).
WWF and ProForest	Developing indicators for High Conservation Value Forest.

Table 2.4. Identifying areas of high conservation value

Methodology	Principle values covered	Source
Global 200 ecoregions	▪ Species richness ▪ Endemism ▪ Higher taxonomic uniqueness (e.g. unique genera or families, relict species or communities, primitive lineages) ▪ Extraordinary ecological or evolutionary phenomena (e.g. adaptive radiations, intact large vertebrate assemblages) ▪ Global rarity of the major habitat type	WWF
Biodiversity hot spots	The key criteria for determining a hot spot are endemism and degree of threat. Plant endemism is the primary criterion for hot spot status because plants support most other forms of life through their ability to harness energy from sunlight. The degree of threat is measured in terms of habitat loss. Hot spots have lost at least 70% and in some cases more than 90% of their original natural vegetation.	Conservation International
Centres of plant diversity	Key criteria are biodiversity and endemism. The analysis selects a large number of sites but picks out 234 for special treatment through description in data sheets. Mainland sites defined by data sheets are all thought to have at least 1000 vascular plant species, of which at least 10% are endemic to the site or region: this rule was relaxed for island sites with high endemism but fewer overall species.	IUCN/WWF
Endemic bird areas of the world	The distribution patterns of 2609 land bird species with restricted ranges (i.e. which historically have had a total global breeding area below 50,000 km^2) have been recorded and some 51,000 separate locality records mapped. As a result a series of 'Endemic Bird Areas' (EBAs) have been identified, and classified with respect to biological importance, immediacy and type of threat, degree of existing protection and priority for conservation.	Birdlife International

Conservation value of forests

The methodology to identify High Conservation Value Forests could, in theory, provide a global picture of the most valuable forests but it has yet to be applied on anything but a test basis. It will also be too expensive to use in all forests, so cruder measures are needed at the larger scales of management. Over the last decade, several major conservation organizations have developed prioritization methodologies for identifying 'important' habitats. While these inevitably contain a great deal of overlap, they all use slightly different criteria, such as biodiversity, endemism, uniqueness, integrity and degree of threat. Therefore, a cumulative analysis of several of these initiatives provides an indicative picture of overall value from a range of different perspectives. Four prioritization methodologies are particularly relevant to forest quality (Table 2.4).

Area of native forest

No global statistics currently exist for native forest extent and indeed there is a vigorous debate concerning the meaning of the term 'native' in this context. If we assume it to mean forest made up predominantly of natural species, then, while some minority of countries have very good data, most do not. The best we can come up with on a global level is figures from the FAO *Global Forest Resources Assessment 2000*, updated in some respects by the biannual *State of the World's Forests* (SOF) reports, which distinguish plantations from the rest of the forest estate. FAO itself recognizes that these data are inadequate for today's uses and notes in the 2003 SOF report that 'future assessments will have to delve into aspects related to function, impact and potential, providing much more information than in the past' (Anon., 2003b).

Forest fragmentation

The World Resources Institute (WRI) has developed a methodology for identifying what it terms 'frontier forests', which are defined as large tracts of relatively undisturbed forest, with the following key characteristics:

- Free from substantial anthropogenic fragmentation (settlements, roads, clear-cuts, pipelines, mines, etc.);
- Free from detectable human influence for long enough periods to ensure that it is formed by naturally occurring ecological processes (including fires, wind and pest species);
- Large enough to be resilient to edge effects and to survive most natural disturbance events;
- Contains only naturally seeded indigenous plant species, and supports viable populations of most native species associated with the ecosystem (Lee *et al.*, 2003).

WRI has completed a global survey of frontier forests, which provides percentage figures for the remaining frontier forests (Bryant *et al.*, 1997), and detailed studies of several other regions, including Canada, the Russian Federation (Aksenov *et al.*, 2002), Chile (Neira *et al.*, 2002), the Congo Basin (Minnemeyer *et al.*, 2002) and others. While none of these data are precise (particularly concerning the global survey, which is also out of date) they provide the best available source of global data on fragmentation that allow broad conclusions about the degree of fragmentation in different countries.

Age

Despite the intense debate surrounding old-growth forests there had, until recently, been few attempts to identify old forests within countries or regions, except on a strictly sub-national basis. The UNECE Temperate and Boreal Forest Resources Assessment (covering Europe, North America, the CIS, Australia, New Zealand and Japan) requested that countries report on the proportion of their forests that were 'untouched by man', defined as unmanaged for a period of at least 200 years. Whilst this is still far from a definition of old growth as it relates to trees of their maximum age for most forest ecosystems, it does give a rough indication of areas likely to contain older forests (Anon., 2002).

Forest health

Data for Western Europe are good, but are largely lacking for the rest of the world in any systematic sense (although there have been local, detailed monitoring exercises in, for example, parts of North America and there are plans to draw together an Acid Deposition Monitoring Network in East Asia). Forest health in Western Europe has been regularly monitored for almost 20 years. This has been done in line with the United Nations Convention on Long-Range Transboundary Air Pollution with results based on 6000 permanent sample plots scattered across the continent (measuring defoliation/crown transparency; discoloration; stand and site characteristics). Causal relationships have been investigated (UNECE/EC, 2002) by intensively monitoring 860 plots to assess crown condition, air pollution levels, impacts of pests, diseases and invasive species and other, more detailed questions (see Table 2.5).

Susceptibility to climate change

The Intergovernmental Panel on Climate Change (McCarthy *et al.*, 2001) remains the most authoritative source of information on the impacts of climate change on forests. From the perspective of assessment, complications have arisen through the use of several different scenarios, all of which have widely differing impacts. The *IPCC Special Report on Regional Impacts of Climate Change* provides some good data (Watson *et al.*, 1998) and there have also been numerous scenarios described by academic bodies, NGOs and industry.

MALASPINA UNIVERSITY-COLLEGE LIBRARY

Wood products

Both the FAO and the UNECE maintain detailed and regular monitoring of the world's timber supply. This is done through reports from both exporter and importer countries. Whilst these reports are not perfect (particularly in countries with an active trade in illegal products), they do give an indication of the total volume and total value of trade, stretching back over several decades. This allows comparisons of trends and developments.

Table 2.5. Intensive surveys of forest health in Europe

Survey	Frequency	Intensity
Crown condition	At least annually	All plots
Soil (solid phase)	Every 10 years	All plots
Soil solution	Continuous	Part of the plots
Foliage	Every 2 years	All plots
Deposition	Continuous	Part of the plots
Ambient air quality	Continuous	Part of the plots
Meteorology	Continuous	Part of the plots
Forest growth	Every 5 years	All plots
Ground vegetation	Every 5 years	All plots
Phenology	Several times a year	All plots
Remote sensing	Preferably at plot installation	Optional

Non-timber forest products

Recognition of the importance of other forest products has increased rapidly over the past two decades, starting with qualitative assessments in the 1990 Temperate and Boreal Forest Resource Assessment and growing into figures that relate to the quantity and value of non-timber forest products in both MCPFE and Montreal Process reports (albeit rather incomplete). There are also individual studies and estimates from a number of other countries, although not enough to make generalized conclusions.

Environmental services

Forests can serve as protection against avalanches, to reduce soil erosion, as windbreaks and to maintain water quality for drinking and other purposes. In particular circumstances, forests can also help increase the quantity of water in catchments, while in other conditions they have the reverse effect. While a minority of countries categorized some of their forests by protection function (e.g. India), or worked out the proportion of the forest that fulfilled these uses (e.g. Switzerland), there were few attempts to pull this information together at a regional level. The MCPFE countries have since tried to address this by identifying forests that have a protective function (see Table 2.6). Whilst this is probably the most comprehensive attempt to define these values to date, it is unlikely to be an underestimate because other research has shown that many official protected areas (national parks, nature reserves and others defined as protected areas by IUCN), which are separated from protective forests in the MCPFE analysis, also serve a protective function, particularly towards water sources (Dudley and Stolton 2003).

Management effectiveness

In theory, interest in certification under a range of different, and sometimes competing, schemes should have provided excellent data sources on management effectiveness. This is because each certification scheme runs detailed monitoring systems for their own certified bodies (Raunetsalo *et al.*, 2002). However, whilst this certainly provides a good place to start, many well-managed forests remain uncertified.

Table 2.6. Different protection functions defined by MCPFE, compared with IUCN protected area categories

MCPFE classes		IUCN protected area categories
1. Objective: 'biodiversity'	*1.1 No active intervention*	I
	1.2 Minimum intervention	II
	1.3 Conservation through active management	IV
2. Objective: 'protection of landscapes and specific natural elements'		III, V, VI
3. Objective: 'protective functions'		n/a

2.6 Conclusion

This chapter has shown that data are indeed starting to emerge and that we can expect a growing picture of forest quality to emerge over the coming years. The 2000 Forest Resource Assessment and, to an even greater extent, the *State of the World's Forests* reports from the FAO have been a clear attempt to look beyond traditional statistics. Despite this progress, they still have a long way to go in defining and collecting information on the world's forest quality. Forest management paradigms have changed a lot in 10 years, from a time when, during the launch of the Montreal Process, one participant queried whether any indicator of forest quality other than biomass would ever be necessary. The ideas have been assimilated, sometimes almost unconsciously, and relevant information is starting to emerge. Collecting and reporting on these issues almost inevitably means that governments and companies have to rethink their own attitudes and their own practices. The willingness, and some of the expertise, is already present in many cases. Nevertheless, applying this knowledge in a complicated world of diminishing space and competing interests will pose a challenge for some time to come.

References

Aksenov, D., Dobrynin, D., Dubinin, M., Egorov, A., Isaev, A., Karpachevskiy, M., Laestadius, L., Potapov, P., Purekhovskiy, A., Turubanova, S. and Yaroshenko, A. (2002) *Atlas of Russia's Intact Forest Landscapes*. Greenpeace Russia, Moscow.

Anon. (1993a) *Agenda 21 and the Forest Principles*. United Nations, New York and Geneva.

Anon. (1993b) *ITTO Guidelines for the Establishment and Sustainable Management of Planted Tropical Forests*. ITTO Policy Development Series 4, Yokahama.

Anon. (1994) *Indicators for the Sustainable Management of French Forests*. Ministère de l'Agriculture et de la Pêche, Paris.

Anon. (1995a) *European Criteria and Indicators for Sustainable Forest Management: Adopted by the Expert Level Follow-Up Meetings of the Helsinki Conference in Geneva, June 24 1994 and in Antalya, January 23 1995*. Ministerial Conference for the Protection of Forests in Europe, Vienna.

Anon. (1995b) *Regional Workshop on the Definition of Criteria and Indicators for Sustainability of Amazonian Forests: Final Document – Tarapoto, Peru, February 25 1995*. Ministry of Foreign Affairs of Peru, Lima.

Anon. (1996) *Criteria and Indicators for Sustainable Forest Management in Dry-zone Africa*. UNEP/FAO Expert Meeting, Nairobi, Kenya, 21–24 November 1995, FAO, Rome.

Anon. (1997a) *Criteria and Indicators for Sustainable Forest Management in Central America*. CCAD, FAO and CCAB-AP, Tegucigalpa, Honduras.

Anon. (1997b) *Report of the Expert Panel on Criteria and Indicators for the Measurement of Sustainable Management in Natural Tropical Forests*. Yokohama, 8–12 September 1997, International Tropical Timber Council, Document ITTC(XXIII)/9, 8 October 1997.

Anon. (2000) *Forest Resources of Europe, CIS, North America, Australia, Japan and New Zealand*. UNECE and FAO, Geneva and Rome.

Anon. (2002) *New Zealand Country Report 2003*. MAF Technical Paper number 2002/21, [http://www.mpci.org/rep-pub/2003/2003newzealand.pdf].

Anon. (2003a) *ITTO Guidelines for the Restoration, Management and Rehabilitation of Degraded and Secondary Tropical Forests*. ITTO Policy Development Series 13, Yokahama, Japan.

Anon. (2003b) *State of the World's Forests 2003*. UN Food and Agriculture Organization, Rome.

Anon. (undated) *Criteria and Indicators for the Conservation and Sustainable Management of Temperate and Boreal Forests – The Montreal Process*. Canadian Forest Service, Hull.

Anttila, E. (1996) *National Experiences on Criteria and Indicators, Intergovernmental Seminar on Criteria and Indicators for Sustainable Forest Management*. August 19–22 1996, Ministry of Agriculture and Forestry, Helsinki.

Barney, G.O. (1980) *The Global 2000 Report to the President: Entering the 21st Century*. Volume 1, US Government Printing Office, Washington, DC.

Bryant, D., Nielsen, D. and Tangley, L. (1997) *The Last Frontier Forests*, World Resources Institute, Washington, DC.

Canadian Council of Forest Ministers (1995) *Defining Sustainable Forest Management: Canadian Approach to Criteria and Indicators*. Natural Resources Canada, Ottawa.

Dietrich, W. (1992) *The Final Forest: The Battle for the Last Great Trees of the Pacific Northwest*. Simon and Schuster, New York.

Dudley, N. (1992) *Forests in Trouble*. WWF International, Gland.

Dudley, N. (1996) Authenticity as a means of measuring forest quality. *Biodiversity Letters* 3, 6–9.

Dudley, N. and Rae, M. (1998) Criteria and indicators of forest quality. In: *Proceedings of International Conference on Indicators for Sustainable Forest Management,* IUFRO, CIFOR and FAO, Melbourne, Australia.

Dudley, N. and Stolton, S. (2003) *Running Pure: The Importance of Forest Protected Areas to Drinking Water.* WWF and the World Bank, Gland and Washington, DC.

Dudley, N., Cu, N. and Manh, V.T. (2003) *A Monitoring and Evaluation System for Forest Landscape Restoration in the Central Truong Son Landscape.* WWF Indochina and the Government of Vietnam.

Dudley, N., Schlaepfer, R., Stolton, S., Jackson, W. and Jeanrenaud, J.-P. (eds) (in press) *A Manual on Forest Quality.* WWF, IUCN and EPFL, Gland, Switzerland.

Eeronheimo, O., Ahti, A. and Sahlber, S. (1997) *Criteria and Indicators for Sustainable Forest Management in Finland.* Ministry of Agriculture and Forestry, Helsinki.

Hammond, A., Adriaanse, A., Rodenburg, E., Bryant, D. and Woodward, R. (1995) *Environmental Indicators: A Systematic Approach to Measuring and Reporting on Environmental Policy Performance in the Context of Sustainable Development.* World Resources Institute, Washington, DC.

Jennings, S. (2003) *HCVF for Conservation Practitioners.* Proforest, Oxford.

Lammerts van Bueren, E.M. and Blom, E.M. (1997) *Hierarchical Framework for the Formulation of Sustainable Forest Management Standards: Principles and Criteria.* Tropenbos Foundation, Leiden.

Lee, P., Askenov, D., Laestidius, L., Nogueron, R. and Smith, W. (2003) *Canada's Large Intact Forest Landscapes.* Global Forest Watch, Washington, DC.

Liaison Unit in Lisbon (1998) *Follow-up Reports on the Ministerial Conferences on the Protection of Forests in Europe – Volume 1: Report of the Follow-up of the Strasbourg and Helsinki Ministerial Conferences on the Protection of Forests in Europe.* Third Ministerial Conference on the Protection of Forests in Europe, Lisbon, June 1998.

McCarthy, J.J., Canziani, O.F., Leary, N.A., Dokken, D.J. and White, K.S. (eds) (2001) *Climate Change 2001: Impacts, Adaptation, and Vulnerability.* Cambridge University Press, Cambridge.

Minnemeyer, S., Walker, T., Collomb, J.-G., Cotton, L. and Bryant, D. (2002) *An Analysis of Access to Central Africa's Forests.* World Resources Institute, Washington, DC.

Neira, E., Verscheure, H. and Revenga, C. (2002) *Chile's Frontier Forests: Preserving a Treasure.* World Resources Institute, Washington, DC.

Prabhu, R., Colfer, C.J.P., Venkateswarlu, P., Tan, L.C., Soekmadi, R. and Wollenberg, E. (1996) *Testing Criteria and Indicators for the Sustainable Management of Forests: Phase 1 Final Report.* Center for International Forestry Research, Bogor.

Prabhu, R., Colfer, C.J.P. and Dudley, R.G. (1999) *Guidelines for Developing, Testing and Selecting Criteria and Indicators for Sustainable Forest Management.* Criteria and Indicator Toolbox Series number 1, Center for International Forestry Research, Bogor.

Raunetsalo, J., Juslin, H., Hansen, E. and Forsyth, K. (2002) *Forest Certification Update for the UNECE Region Summer 2002.* UNECE and FAO, United Nations, New York and Geneva.

Reid, W.V., McNeely, J.A., Tunstall, D.B., Bryant, D.A. and Winograd, M. (1993) *Biodiversity Indicators for Policy-Makers.* WRI and IUCN, Washington, DC, and Gland.

Sollander, E. (2000) *European Forest Scorecards.* WWF International, Gland.

Taylor, R. and Smith, I. (1997) *The State of New Zealand's Environment 1997.* Ministry of the Environment, Wellington.

UNECE/EC (2002) *Condition of Forests in Europe: 2003 Executive Report.* United Nations, Geneva, and European Commission, Brussels.

Watson, R.T., Zinyowera, M.C., Moss R.H. and Dokken D.J. (eds) (1998) *The Regional Impacts of Climate Change: An Assessment of Vulnerability.* Cambridge University Press, Cambridge.

Making Value Compromises in an Efficient Economy: Efficiency as Value Dialogue

3

M.C. Farmer[1] and A. Randall[2]

[1] *School of Public Policy, Georgia Institute of Technology, USA;* [2] *Department of Agricultural and Resource Economics, Ohio State University, USA*

Decades of controversy surround the deployment of discount rates in benefit-cost analyses (BCAs) to establish optimal forest rotation and to evaluate the appropriate mix of benefits in a multiple-use forest. Complaints remain nearly unchanged: efficiency demands discount rates in forestry decisions to prevent needless waste, while fairness implies that discounting inherently undervalues benefits from forest services over the very long run. We charge that this impasse results from conflation of welfare economic analysis with social choice, thus committing the sin of comparing the 'Pareto non-comparable' via an instrument derived to detect Pareto gains: BCA. Confusion emerges from traditional practices that overly attach market failure with the default corrective practice of BCA. Given the incomplete markets that exist over the long term, public policy can close only a few select market failures that long-term investors face.

If baseline social choice deliberations, or precedents, affecting a given decision are in place and clear, then a proper BCA can proceed as a neoclassical welfare economics exercise that quite properly includes discount rates. Yet, without reasonably clear signals regarding how society intends to treat a set of issues related to a long-term forest plan, BCA is premature. Ignoring the grey distinctions that divide welfare economics and social choice analysis is dangerous as it discards valuable economic information from the public value debate over long-term sustainable forestry.

© CAB International 2005.
Forestry and Environmental Change: Socioeconomic and Political Dimensions
(eds J.L. Innes, G.M. Hickey and H.F. Hoen)

3.1 Introduction

This chapter advocates a powerful role for standard economic welfare analysis (or benefit–cost analysis) in the regular course of forestry operations, especially multiple-use forestry.

At stake in multiple-use forestry are numerous public interests: recreation, hunting, biodiversity preservation and traditional silviculture objectives. One way to make decisions about how to manage a public forest is to maximize economic welfare through an economic benefit-cost analysis. That recommendation clearly worries some critics, especially since long-rotation forestry is driven so powerfully by economic discount rates - the process by which future forestry benefits are valued less than current forestry benefits. In economic analysis, environmental benefits a generation away can be discounted severely, perhaps all the way to zero under the efficiency criteria, so critics worry that economic benefit-cost analyses injure the future.

Our defence of economics and of discounting in welfare economics (benefit-cost analysis) is at once technically driven and ethically motivated. The debate over efficiency (and discounting), we charge, is largely misdirected, and the misdirection stems from a fundamental misunderstanding of what welfare economics is and what it measures. This misunderstanding requires benefit-cost analysis to do much more than it really can, or rightly ought to, do.

Critiques embed a latent ethical assumption about welfare economics that judges the instrument against the standard of how well it functions as a complete value theory. Given this standard, welfare economics is seen to reduce future persons to mere objects of the myopic whims and ephemeral preferences of those who live today. We reject, immediately, the absolutism of this extreme position and we expect most readers will accept this premise; but the extreme position persists. Anti-efficiency contentions nest many latent appeals that do, indeed, equate efficiency-premised practice with a very narrow range of moral values. Intentionally or not, participants repeatedly fall victim to an odd standard that judges welfare economics as a value theory that must be evaluated as a complete moral theory, a premise that founding scholars in economics rejected decades ago (Arrow, 1954). This fetish to expand welfare economics beyond its scope and then to reject it for its limitations is a straw man that diverts public attention from the real value challenges that societies face when they establish long-term forestry plans. Unfortunately, the lightning rod in the debate is often the discount rate and this focus, we argue, is a poor forum to explore the value content of economic efficiency.

We think that economists are unwittingly drawn into a debate over welfare economics as a complete moral theory when the discounting question is on the table; so we adopt a new language to circumvent miscommunication on both sides. We assert forthrightly that: 'Economics is not a complete and coherent value theory!' Rather, welfare analysis constitutes 'an economic way of valuing' that well serves a non-trivial number of policy contexts in both the short and the long term. Efficiency is best understood as a baseline practice that captures many shared values and resolves many day-to-day questions through a flexible, adaptable tool that contributes to the achievement of many things that people value. While it does not deliver a unique moral target, efficiency does adhere, imperfectly, to the goals of many different value concerns and delivers a decision tool that is adaptable and pliable to many venues.

For societies that recognize many different definitions of moral good, the search for an adequate balance among diverse views supersedes the duty to search for a singular resolution among competing views; yet this 'freedom' from one singular value hardly makes things easier for the forest manager. The duty to locate moral consensus and compromise among many competing social values and integrate them into a forest plan asks the forest manager to broker a highly complex value debate, while at the same time administering to a complex ecology. The beleaguered forest manager is simply overwhelmed by a duty to engage in deep value dialogue over every issue and still perform management tasks with the due diligence society expects. We argue that benefit-cost analysis aids the forest manager by allowing a large number of public forestry decisions to be resolved through efficiency analysis, and without committing irreconcilable violence to plausible ethical positions held by citizens. With a baseline default, the forest manager is free to address, more surgically, those values that fare poorly in the process and then focus on a much smaller set of heavily scrutinized decisions to level the field for values treated improperly.

Our technical defences are very pragmatic. First, efficiency clearly expands the 'size of the game' because adherence to it directly makes society wealthier under existing rules. Society obtains more flexibility to achieve several moral objectives at once simply because it has more resources to accomplish more things; but this indirect 'wealth adds flexibility' defence is only part of the contribution an efficient forest plan can make. Welfare maximization is an instrumental, morally relevant objective on its own because it is a compelling way to 'add up' the things people care about and then to record, however imperfectly, the intensity of those values. It is this adding up process that is most controversial, and it is here where the seductive but misdirected 'efficiency must be a stand alone moral theory or it is valueless' critique enters to confuse and to muddy the policy process.

3.2 Technical summary

Our first technical question centres on the stability of a welfare economics analysis over time. By stability, we simply mean the ability of a technical analysis to compare costs and benefits of roughly equivalent scope and scale when discount rates enter the analysis. One concern of critics is that efficiency analysis easily or routinely compares the very general – future global ecological collapse – with the particular – the decision to fell or replant a given forest stand. The confusion is natural enough, especially to the non-economist, because the stability of a welfare economics exercise centres on a technical distinction between general equilibrium and partial equilibrium analysis that gets lost too easily in the discounting debate. Many of the more horrific and graphic discounting scenarios that compare future catastrophe with trivial current concerns rely, explicitly or implicitly, on fixity in asset prices over time (mixing general equilibrium outcomes with partial equilibrium assumptions). As an example, the presumption that today's efficient prices (and discount rates) that predict one future can be used to evaluate a very different future is a non-starter that is equivalent to demanding that the price of water be the same in a drought as in a flood: it is not fair to use old prices to discount a new event.[1] Unfortunately, the way economists explain discounting to non-economists – as a measure of foregone opportunity – makes the welfare economics exercise appear much more unstable than it really is.

Our second concern relates to the first. To evaluate a forestry project by benefit-cost criteria presumes that the parts of the economy related to the plan, such as timber, construction and energy markets, are relatively efficient; or, at least, that these related sectors are inefficient in a very tractable, straightforward way – an import tax for instance. To conduct efficiency analysis can impose a rather demanding set of regularity conditions on social institutions: property rights have to be well assigned; costs of exploiting non-marketed commodities such as clean water and air have to be treated; production from factories and consumption of automobile trips have to be recognized in the pursuit of these activities. Our concern is that there are times when massive market failure occludes the sort of project analysis that society thinks it is getting when it licenses a benefit-cost analysis to resolve a certain question. This economic problem is much deeper than any adjustment to discount rates in benefit-cost analysis is likely to address properly, and it is important to separate the two types of economic questions.

Market incompleteness has to do with the management of highly inefficient markets where a society has to choose a strategy to address the rules and the rights and the procedures to make economic progress. Some economists and many non-economists alike are seduced into looking at the discount rate as a simple, one source target to understand the economy: to some a symbol of corruption within efficiency and to others the technical vehicle to 'fix' the welfare economics exercise and to manage market failure. We assert that the premise of resolving massive market failure, if that is indeed the condition an analyst faces, by fixing the discount rate or by discarding economic information is a poorly constructed question. Such a question eclipses what welfare economics contributes to the policy process and traps public value-making in a circular debate that diverts attention from authentic value challenges of the day.

In both technical defences above, efficiency looms large in the value debate. If markets are complete and tractable enough for a standard benefit-cost estimation to occur, then standard welfare economics and statistical analyses serve a valuable function. If market failure is very widespread in the sectors related to a given issue, especially if market failure persists through time, then an institutional analysis sustains active efficiency mindfulness in the orchestration of the laws, rules and public investments that help to deliver goods and services to citizens

[1] In the extreme, prices are never entirely fixed for any welfare economics exercise. Moving from inefficiency to efficiency are two states of the world that will alter prices. A reasonable debate in welfare economics centres on how far one can extend the efficiency calculus before the impacts of the change are so large that a simple welfare economics exercise constitutes a noticeable change in the way society conducts its business. A change too large suggests that the welfare economist ought to work in consultation with policymakers. While economists quibble and argue over how big is too big, it would be very hard to find a dedicated welfare economist who accepts the constant price assumption to evaluate a change that introduces global cataclysm.

today and through time. The vocal debate over discounting, we argue, frequently centres on a failure to recognize the technical distinction between these two types of economic efficiency appeals. The conversion of a standard benefit-cost estimation to inform and to resolve a deep structural question over-extends one efficiency practice as it mutes another; yet that is just what reductions in discount rates are often designed to do. What gets lost in this future friendly 'fix' is more than the role of economists: society sacrifices considerable clarity, precision and focus in a value debate that centres on how to respond to market failures that are often difficult to see.

3.2.1 Economic discounting challenge

Positive interest rates remain a special source of confusion in the ethics and economics debate. We assert that to use myopia attributed to the discount rate as a foil to economics – a foundational Achilles heel – is far too crude an assertion to affect the wide claims that critics make. Economies with very high interest rates have been quite sustainable while other economies with very low interest rates have exhibited severe welfare declines, even ecological collapse. A blanket policy to remove discounting for very distant time periods from long-term planning can convert a sustainable society into an unsustainable society and can accelerate the pace of severe ecological collapse. Simply, our very attempt to help the future injures it.

We charge that far too much structural information about the economy is lost in a simplistic reduction that equates lower discount rates with a better future. Precision is key. Societies able to define jointly a role for efficiency and for clear exceptions to efficiency enjoy a much sharper dialogue over environmental values. Efficiency can bring into presence, succinctly, value conflicts at well-defined friction points among citizens.

3.3 Strange discounting arithmetic

In one rendition, discounting allows society to evaluate the total collapse of a life-supporting activity in the distant future as worth today something less than the gate receipts at a premier sporting event. Attention to this strange arithmetic of discounting consumes a lot of academic print. Although we view the discounting controversy as a true insider's technical squabble among economists, the enormous ethical and policy implications of the outcome predictably draw attention from all corners of academia.

Oddly, environmental economists seem to view the discounting debate either as a non-issue (Howarth and Norgaard, 1992) or as the issue (Page, 1978) to guide multiple use forestry; so we review the early discounting calculus for the optimal extraction of a resource that resource economists inherit from Hotelling (1931). The immediate economic price, P_t, for a resource used today versus the price of a unit conserved for use tomorrow, P_{t+1}, is:

$$P_t = \frac{P_{t+1}}{r_{t+1}}.$$

The price (value) of a resource conserved for use tomorrow, P_{t+1}, is valued less than a resource used today; or, tomorrow's resource use by, say, a future generation, is discounted by the market rate of interest, r_{t+1} (for convenience, think of the market rate of interest as the lending rate to borrow money from a bank, or $r_{t+1} = 1.05$ for a 5% loan). In other words, resources conserved for use tomorrow are less valuable than those used today by this interest price.

Extrapolating to the far distant future, this peculiar economic value equivalence between resource use a century from now and its value today (or its present value) is:

$$P_t = \frac{P_{t+100}}{\Pi_t^{100} r_{t+1}}.$$

Or, a full forest stand that matures a century from now is almost worthless, discounted by the product of all intervening lending rates over the entire century, $\Pi_\tau^T r_{t+1}$, where T=100. Even with lending rates close to 2% over the period, the denominator is still large enough to assign to a forest stand available a century from now almost no economic value.

We can amend the Hotelling equation to accommodate the renewal rate for forestry resources. In its simplest form, conservation pricing changes to:

$$P_t = \frac{P_{t+1}}{r_{t+1}}(1+g_t),$$

where $(1+g_t)$ represents the marginal growth rate for the renewable resource conserved at time period t, carried into time period t+1.[2] Resource growth helps a little to offset the discounting arithmetic because forests grow in value by virtue of their biological regeneration properties. Yet, to make much of a difference in the conservation equation, this typically requires a fast-growing resource. Over an infinite time period, a relatively high interest rate at the beginning, followed by a declining interest rate (Howarth and Norgaard, 1992; Weitzman, 2001) continues to make it very easy to ignore the distant future. Discounting arithmetic speaks to a myopia that casts efficient prices as intrinsically destabilizing. So what is going on here?

3.3.1 Partial vs. general equilibrium

A typical defence of discounting argues that discount rates merely account for the cumulative effect of the assets that could have been acquired instead of the proposed project. Over a century, if the revenues that could be generated from

[2] This does not have to be fixed. It can take on a sigmoidal form that maps renewal rates to the amount of resources saved.

harvesting timber today instead of conserving it were put in a bank, money could be lent out for a new investment that would be allowed to grow instead of the forest. Discounting, seen this way, becomes a fair comparison between two clear options: letting a forest continuously get larger and accumulate its stock of lumber, or letting an investment continuously grow over that same period.

Discounting asks the following question: if we continuously reinvest an opportunity for a new business loan and cash it in, say, a century from now, or let a forest stand accumulate, which of these opportunities will be worth more in the future? This is not myopia at all. In fact, the forgone opportunity view of discounting compares economic opportunities in terms of how the future is likely to react to the alternatives. That is, discounting appraises which set of choices is better for the future.

In the rare case that a resource delivered to the future really cannot be used for an entire century (has no intervening use at all), we might expect a productive economy to multiply, or to compound, many times over the revenues diverted from conservation today. As the future accumulates continuously a compounded investment over a century, the future appears to acquire great wealth from the investments freed up for new enterprise by choosing not to conserve the forest. That sounds reasonable for a couple of growing seasons; but continuous accumulation of capital over a century made possible by passing over a relatively small conservation project today begins to become incredibly large.

An implausibly high value of the forgone opportunity lost when a society forgoes a modest-sized project reignites the nightmare scenario that disturbs discounting detractors. The apparent ease with which a net benefit stream goes from the particular (conserve a forest stand in this region today) to the general (at the expense of forgone investments that may accumulate to 20% of gross national product (GNP)) startles the citizen who has accepted efficiency analysis as part of the social decision process. If welfare economic analyses function as incremental rules that accomplish good without risking horrendous bad, then the forgone opportunity story as told is clearly worrisome. In particular, if the effect of a single project can explode to impact economy-wide wealth and prices, then it can also threaten ecology-wide viability in the distant future. Each efficiency decision becomes capable of committing tremendous violence exponentially. We disagree entirely that discounting in a proper welfare analysis is unstable in this way and that it compares really big future changes with really small immediate actions for reasons explored in the next two sections; but the concern persists and muddies debate.

3.3.2 The 'Gamma discounting' fix to discounting

Weitzman (1998, 2001) is perhaps the most prominent figure dedicated to the concern that something is still wrong with discounting beyond a simple appeal to 'forgone opportunity'. Though he never articulates a concern that discounting too easily compares the general (incredible future abundance or ecological disaster) with the particular (save this timber stand or invest the proceeds in a new enterprise), Weitzman (2001) does successfully record a clear suspicion among the majority of economists that it is inappropriate to employ high interest rates to value benefits that accrue to the distant future, opting instead to evaluate benefits that accrue to the far distant future using a zero or near-zero interest rate.

Weitzman's recommendation to benefit-cost analysts in the *American Economics Review* (2001) is to employ a so-called 'Gamma discounting" formula, where interest rates fall overtime towards zero, eroding the impact of discounting on the distant future benefit in benefit-cost analysis. 'Gamma discounting' is defended by a survey that reports a consensus among many, even most, economists to eliminate discounting on very long-term projects; but the argument strikes us as less a solution that finally removes an old canker on the efficiency calculus than a newly swabbed Petri dish that breeds new viruses.[3] We concede that there are good reasons to expect that interest rates may fall in some economies, but we also accept that high interest rates may be sustained over time in others. For economies with increasing rates, it is still appropriate to employ those high interest prices in a benefit-cost report.

The consensus argument for 'gamma discounting' sacrifices a lot of precision in the pursuit of general acceptance because there are lots of different reasons why economists might want discount rates to fall to zero in a welfare economics exercise; and not all economists who endorse gamma discounting do so for the same reason. In fact, many adherents among themselves disagree with the basic economic assumptions that lead other economists to support the rule. The non-economist may be unimpressed with the differences as long as many agree; but there are reasons why non-economists interested in the policy sciences should care. While the distinctions among economists are technical in nature, the social implications of each technical position are relevant to the policy process. The public, we assert, ought to know why society ought to adopt a rule that assumes declining interest rates. Is it to improve fairness? Enhance efficiency? Achieve market reform? These are distinct moral questions.

One reason for reduced discounting may be to circumvent the more egregious impact of myopia; so gamma discounting is fair. Other economists may conclude simply that the economy is expanding and that an ever-lower interest rate reflects this enviable economic condition; so gamma discounting is efficient. Still other economists may hope to nudge society towards a more sustainable future over time by taking advantage of every public investment to inject a sustainable price into the benefit-cost formula in order to deliver more assets to the future; so gamma discounting assists market reform. Finally, another group may view the incomplete nature of markets in the distant future as so compromised that true efficiency gains are distorted by inefficiently high market interest rates.

Each view above may, at times, agree on the same interest trajectory; but each position accepts a very different assumption about how an individual project actually delivers benefits to the distant future. Members of each camp may disagree strongly with the economic assumptions the others make; and those

[3] Perhaps the most important contribution of the work is to expose a common straw man. Non-economists routinely assume what economists must suppose and think regardless of what economists actually do suppose and think. That there is lively and active debate over an issue presumed to be a bedrock default in economics is noteworthy on its own.

disagreements are not value neutral.[4] When analysts depart knowingly from efficiency to effect, say, a fairness objective, those citizens who license the economist to perform efficiency analysis may object that the analyst usurps public authority. Other economists may object to the argument on technical grounds to charge that any presumed fairness gain from an adoption of the gamma discounting formula is illusory. Why society adopts welfare economics to resolve certain policy questions seems, arguably, to be about what efficiency contributes to public goals. To amend or to fix a single efficiency calculation to embrace a broad public agenda seems to be about devising a formulaic instrument to resolve all policy questions, something governing authorities are likely to reject.

Loosely, in the practice of a given project analysis, society wants to know when a single application of a benefit-cost analysis drifts too far from welfare economics, which is relatively incremental in scope, to a social choice issue that launches society into a wholly different future. It is reasonable then for the worried ecologist to insist that an economic analysis be transparent about the expected future impacts that flow from a given project. If efficiency is to become a tool to clarify value challenges, it seems unlikely that society will allow welfare economics to mask tough ethical choices inside a single discount rate adjusted benefit-cost report. Society wants credible assurances that very general changes in future economic conditions are not hidden inside a single benefit-cost calculation, and it is the economist who is responsible to present a true picture of the scope of any project analysed.

Again, some public decisions do involve dramatic changes in future prospects and a simple welfare economics exercise may not be enough to decide those questions. Indeed, there may be many paths to efficiency, each with morally distinct content, among which society ought to adjudicate. That there can be noticeable, ethically distinct outcomes is well recognized in economics; but benefit-cost analysis by itself is not equipped to make those distinctions. As policy questions move out from the incremental issue onto a plain of highly differentiated social outcomes, to remedy the problem by adjusted welfare analyses is a very unwise practice – and we warn the economist and the policy analyst of these hazards. However, discount detractors are more hostile to welfare analysis than needed for a disputable technical reason. Many suggest that a standard neoclassical welfare economics exercise is intrinsically too unstable; that the decisive benefit-cost analysis itself creates a very different world in the regular course of operations – meaning that every welfare economics exercise over the very long term is heavily loaded with morals that introduce very distinct outcomes.

This charge is based on a misunderstanding of what welfare economics is and what it measures and has several socially undesirable results. This misunderstanding can lead to the development of a cottage industry of technicians who repair the instability of the day-to-day welfare analysis by proposing discount rate 'fixes'. This benefit-cost repair business can blur the lines between incremental welfare analytic contributions and deeper social choice questions by accepting the regular instability argument too quickly. The result can be the

[4] Over the short and intermediate term there is more consensus regarding the mechanics of asset accumulation.

demise of fluid value debate that commits society to an unintentional pervasive inefficiency that removes and offends the very social choice deliberations that originally licensed the benefit-cost analysis process.

3.4 Intertemporal substitution

We can think of two plausible scenarios where far distant interest rates are positive and the value of a good delivered in the distant future falls legitimately to zero with no unacceptable delegation or usurpation of authority by a single technical analysis.

3.4.1 Discounting as forgone opportunity

Our first target is to return to and clarify the story that every modest sized long-term project analysis seems unstable in that any goods delivered to the far distant future sacrifice an implausible fortune of forgone wealth. Or, conversely, an implausible bounty of future environmental benefits is worth no more than a modest-sized project today.

Interpreting forgone opportunities as infinite, continuous capital accumulation leaves an impression that almost any project with defined benefits delivered only in the far distant future necessarily sacrifices an astonishing future treasure. Over the years scholars who have commented on discounting recognized that some productive benefits will be consumed along the way (Arrow and Lind, 1970; or Pearce and Turner, 1990, for discussion), which to some justifies a lower interest rate in a benefit-cost analysis: discount only that portion of capital actually reinvested. Yet the interpretation commits a serious error in the full range of forgone substitutes.

Discounting captures all sorts of responses to the announcement of a forestry project. If a small marginal project delivers, say, a mature and diverse forest stand a century from now, what is forgone is some share of both consumption and investment alternatives of all kinds. So what is forgone is a trajectory of both investment and consumption opportunities that the future enjoys, opportunities that can be legitimately represented by the interest rate. To discard the consumed portions along the way by a downward-adjusted discount rate is to discard the value of future consumption in, say, years 50 or 60 or 99 on behalf of those alive at year 100. By extension, pick a project with a 101-year pay-off and consumption in year 100; by the same discount-adjusted analysis, it becomes a valueless forgone opportunity, hardly a consistent practice to curb injury to future persons. The lesson is instructive. The economically efficient rationale for endorsing 'gamma' discounting where interest rates fall is that it is unreasonable to expect a society to continuously grow at a very high rate. Eventually, the true cost of investment capital, the interest rate, should converge to the aggregate economic returns to investment, the economic growth rate; or it is unreasonable to expect a modern developed economy to continue to grow at, say, 5%–8% for ever. Things will eventually slow down, so the interest rate should fall as well. However, circumstances will vary from economy to economy and from sector to sector. The continuous compounding analogy, if taken too literally, can support a formulaic declining interest rate rule that 'kicks in' too early in wealthy economies, is entirely inaccurate in poor or growing societies, or is simply a crude fix to benefit-cost analyses when not enough economic regularity exists in an economy or in a sector to conduct a proper benefit-cost analysis. But big changes seldom go unnoticed by market actors.

The point here is that a single change, or project, is unlikely to reallocate resources so massively that interest prices (the value of a forest or of productive capital at a given period in time) will change, especially 20–50 years into the future. If a project truly alters future welfare conditions in such a measurable way, there ought to be a clear and discernible impact on prices; so if a project does leave a noticeable imprint on true opportunity costs of aggregate economic resource prices, say, 50 years from now, that is a pretty good signal that a benefit-cost calculation has strayed too far into the realm of value-making. It has introduced to society a widely different future before any political and ethical review takes place. Yet this sort of change is a pretty hefty precondition, far from a regular prospect emergent from single welfare economics applications.

Interest rates in efficient economies serve as a unified record of a host of forgone opportunities – a very broad indicator of the price of various assets that move between two time periods. Indeed, the closest substitute response to the forestry project is another forestry project. As time moves closer to the 100-year mark, society will recognize that there is a mature forest stand that meets many social objectives that will be 'coming on line' in perhaps three or four more decades; and the discount rate is a device to capture this substitute alternative.

Clearly we do not want to count the forestry stand conserved today as a true net benefit in 100 years if that forest conservation act comes at the expense of an equally large forest stand nearby. The zero present value of the 100-year pay-off simply recognizes that the effect of this one small project has no remaining discernible impact on the future welfare given the time frame in which many substitute activities can take place. To assume that society literally discounts future welfare because it discounts this project is to assume away all intervening reactions to the project over the next century. Discounting viewed in this more comprehensive context is not unstable at all. There is no explosive impact at a scale tomorrow that far exceeds the modest scale of the decision today. Rather, discounting records for the analyst the limits and reach of the modest project under review, and that reach is finite. It is the presumed confidence of foresight openly claimed by discount rate detractors that this forest stand unambiguously delivers a net gain 100 years from now that strikes us as the heroic assertion.

The story of continuously accumulating capital is a pretty good pedagogic device for the very short run, but in the long run the story must include the wider array of real reactions that do occur. Discounting simply recognizes that there are lots of alternatives available in complex economies, and, due to the flexibility and adaptability of market efficiency over a very long time, any project that delivers a valuable asset to the far distant future will likely fail to make any measurable impact on net future welfare. To assume otherwise begs the question as to how the critic presumes to know that this investment is so valuable and so unique that no reaction over the next century can possibly make up for the forestry project. A certain myopic arrogance, which attributes to the immediately seen the universe of social reaction, infects the argument.

Declining interest rates in benefit-cost analyses make sense in some economies but not all. In a growing economy, for instance, where investment capital becomes more abundant, the price of investment capital (the interest rate) falls. In this case, economists are well grounded in an endorsement of declining interest rates as interest rates align eventually with a more sustained long-term level of economic growth. Yet, if society is not accumulating capital (we are getting poorer) or there is considerable technological innovation (we are getting much richer from the same capital base), both scenarios can lead to increasing interest rates because investment capital is becoming scarcer. In the case of the impoverished future, failure to discount delivers a portfolio of assets less valuable than would appear otherwise – further depriving future generations by giving them assets that meet fewer of their needs. In the case of growing technological innovation or of expanding trade or integration with an economically wealthier society, interest rates may also climb for several more generations. It is inefficient to deprive the technically advanced future of assets that it can put to better use. In both cases interest rates can rise for long periods and in both cases failure to discount at the right level means that all generations suffer. Again, a legitimate benefit-cost analysis requires that the analyst use efficient prices to measure costs and benefits.

Social discount rate

Another reason to allow rates to remain high, even increasing, but perhaps adjusted below the prevailing market rate is that the market rate itself may not be efficient. Much of the hand wringing and highly nuanced mechanics (Weitzman, 1998) over discount rates seems to suggest that something is wrong with the market, not that something is wrong with the progress to an efficient market. For more opaque critiques, market failure seems to be the chief concern and this is a very different focus from the 'discounting is myopia' critique. Sometimes the market failure is manageable in a given efficiency analysis, but sometimes more policy review is needed.

A case of a manageable market failure related to the interest rate is found in a much older and more precise distinction between the social discount rate and market discount rate that makes adjustment to the market rate reasonable. Loosely, markets may fail to find a vehicle to satisfy our collective paternalistic desires to care for future persons. Markets may have no efficient venue to capture our desires to see the future better off; so it may be more efficient for the public sector to boost future welfare through public projects that are analysed by lower discount rates. In this case, the price actually used for analysis is a better representation of efficiency and forgone opportunities than the market interest rate.

The rule recognizes that, whenever markets fail to get interest prices right, adopting a reduced discount rate improves economic performance. Unfortunately, if this 'correction' takes place only in public project analysis and no true 'fix' ever really appears in the markets, then the adjustment can introduce a new distortion. To account for market failure and also to prevent undue distortion, Eckstein

(1958) outlines a complicated formula (simple to write down, complicated to implement) that would mitigate distortion – a so-called 'second-best solution' – that works within the existing inefficiency. The social discount rate is one application of corrections to market prices known generally as shadow price corrections, so named because there is some trail (or shadow) of the true cost to provide a certain good.

This correction is instructive as much for what it does as for what conditions are required to make it work. The lower interest rate in the social discount rate case presupposes a manageable structural failure in the economy that prevents society from organizing its institutions in an efficient way. The issue is not about the evils of discounting but a more technical exploration into economic structure that, in this case, distorts the value of assets passed between two generations. Analysts assume that the inefficiency will continue; so Eckstein's social discount rate is a second-best formula crafted to operate in a market failure that is never repaired. Gamma discounting (Weitzman, 2001) and the market irregularities that motivate it (Weitzman, 1998) are modernizations of this concept to a more complex set of economic arrangements.

Clearly, the shadow-price corrected analysis is still a proper efficiency account. The public gets the efficiency report it wants and the orderly 'shadow price' correction makes sense; but there are limits to these corrections. If nothing related to a given choice is efficient, the modest scope of a welfare economics exercise suggests that there may not be enough market regularity (or determinacy) to locate the singular efficient 'shadow' price for a resource allocation claim. So shadow-pricing must be done with care and with respect.

If economists hope to live up to the claim that they offer policymakers precision and clarity, then it is incumbent on economists to articulate precise reasons for altering discount rates in welfare economics exercises. Sometimes, as we have seen, this makes sense for efficiency reasons; but a carte blanche majority-rule 'gamma' discounting over all public projects fails to rise to the standard of deliberative and transparent policymaking.

Market incompleteness

Our concern flows from our original assertion that there is a misinterpretation over what an efficiency analysis or benefit-cost analysis is; namely, that critiques evaluate it as a free-standing moral theory, not a way of valuing, and this confusion can lead paradoxically to an overextended benefit-cost practice that undermines entry of other valuable economic information into public discourse.

The 'shadow price' that produced the social discount rate above enters from a very tractable inefficiency; but 'shadow-pricing' is a more general concern that can involve more than just discount rates. A good welfare analysis recognizes that any sector where markets send the wrong signal regarding the actual opportunities forgone to acquire a good requires an appropriate efficiency price. Distorted wages and capital prices from labour or import equipment taxes are examples.

If there are lots of inefficiencies, corrections become increasingly more difficult to manage. Since welfare economics works best inside a relatively stable consensus that sets down a coherent set of market rules, benefit-cost analysis may not be appropriate for some markets; so, as society searches to fit the right tool to the right problem, it is important not to confuse market incompleteness in the general economy with discounting applications in a single setting.

We charge that both economic detractors and interest rate reformers often worry about the wrong thing. It is perfectly reasonable in an economic analysis to adjust true labour costs when minimum wage laws exist. This shadow price assumes that the inefficiency in the labour market is a political institution that expresses a clear public sentiment to be sustained throughout the project life. In this case the public at large, or an effective minority, has spoken with enough authority for analysts to accept the minimum wage institution as a permanent feature on the economic landscape. But other cases of inefficiency are less well deliberated, and discount rate adjustments applied to these circumstances frequently substitute (poorly) for deliberation over a host of undifferentiated market failures. This outcome, we think, is risked in the evaluation approach Weitzman (1998) offers.

Market incompleteness is not so abstract a concept. It simply recognizes that many of the goods and services that affect the distant future are not provided by efficient market exchange. This concern is not about uncertainty, but about the absence of laws and rights and investment securities needed to raise the types of capital to finance future activity, securities readily available to shorter-term investors. Clearly, these laws and rights and securities are not free of moral content, which suggests that lots of markets and insurances that support investment are not in place because society has not yet resolved the ethical choices required to complete these markets.

Without the proper institutions in place to support private initiatives (and the values they represent), many investments that affect the future require direct public support. That, after all, is why welfare economics is frequently called upon. But it is unlikely that the entry of the public sector to support one issue, say education, pension reform or forest protection, will be comprehensive enough to 'complete' the market entirely (e.g. resolve for ever all value debate). If society has yet to do the work of outlining basic ethical compromises to guide certain reforms (as they have done, arguably, with minimum wage laws of the Earned Income Tax Credit (EITC)), then the case-by-case entry of public sponsored programmes leaves a very uneven public investment pattern. This creates a problem for the benefit-cost analyst searching for a uniform 'shadow-pricing' standard across all public investments. As the public picks and chooses which long-term services to provide, some classes of long-term investments will be favoured over others. For services treated well by the public sector, governing rules will be more complete and therefore more receptive to reliable benefit-cost appraisal, even if inefficiencies remain and require extensive shadow-pricing. Yet investments left behind are relatively vulnerable. For the long-term investments that society forgoes, there may be far too many loose ends for the analyst to complete a credible benefit-cost analysis to evaluate these public projects. If one attempts heroic corrections in such an environment, the analyst risks getting the efficiency calculus wrong or, by presuming certain shadow-price value calculations, treads too far into value-making itself.

Public investment selectivity introduces a highly uneven public investment pattern that belies the orderliness and tractability to establish uniform shadow-price taxonomy across long-term public investments. Though we suggest that this marks a natural division between welfare economics and social choice, the shadow-pricing tradition from social discount rate adjustments assumes that distortions that affect public investments are rather uniform or regular enough to leave a clear market shadow.[5] Eckstein, for example, tracked one class of distortion that affects all public investments uniformly. But the uneven distortions in the real world are arguably much more obstructive for certain public investments; so to suggest a 'one size fits all' gamma discount rate for all long-term public investments is far too crude a 'force fitting exercise' that is likely to visit injury on the future; and the risk is twofold.

First, economic benefit-cost analyses take place in the context of stable rules and institutions; so, in the face of great inefficiency, benefit-cost analysts must presume which markets will become efficient and which will not through a series of shadow-price corrections. Since each distinct shadow-price correction in an incomplete market presupposes one set of rules and one set of rights to the exclusion of all others, the whole price array will be inaccurate if society, in fact, resolves these missing market challenges in a fashion much different than that predicted by the analyst, so the resulting choice could be inefficient. That is, context matters.[6]

For those who aspire to raise the level of debate, there is a second concern. An overextended benefit-cost model that adopts a uniform shadow-price rule, imposed on all analysts to adopt the same correction, is a premature move that might make shadow-price corrections a self-creating force. If all large public investments across the economy are evaluated consistently by a uniform shadow-price taxonomy such as that explored by Eckstein and Weitzman, these technical corrections can make very difficult ethical choices inside economic analysis, prior to any public airing. The sheer weight of this 'across the board' shadow-pricing standard can move society quite far in a single direction that may be difficult to reverse – all this before any value hearing over the embedded ethical trade-offs takes place. The problem is that interest rate detractors and discount rate repairpersons share a focus of debate that we think is really about something else.

The scholar wed to the starting assumption that economics as benefit-cost analysis is a value theory, albeit a poor or a satisfactory one, will be blind to the opportunity for economics to participate in shaping the value debate by constructing rules and rights and institutions rather than to simply analyse prices. Just because many different economic solutions are equally efficient does not mean at all that all efficient solutions are morally equivalent. The tragic fallout of this false dichotomy, on both sides, necessarily overly attaches the scholar to welfare analysis generally and, we assert, to the discount rate specifically as the

[5] The degree of completeness or, in market language, the satisfaction of all possible derivatives, may be beyond the reach of an economy with finite wealth – there may be a limit. At least for small countries, some markets may have to remain incomplete, disadvantaging them and complicating their public investment choices considerably.

[6] On the subject of context, Eckstein addressed only one inefficiency – the inability for individuals to express their concern for the future in an individual transaction. If public investments, however, took into account altruistic or paternalistic concerns, then a more efficient set of assets would be delivered to the future. Significantly, no other long-term incomplete markets were admitted into that discussion; his intertemporal distortion was uniform and single dimensioned, and therefore relatively more tractable.

barometer of moral validity for economics. The result is a narrowness that irresponsibly undermines fruitful value dialogue about how to handle highly inefficient and incomplete markets. In the process of placing a square peg in a round hole, critics will note the unfitness of benefit-cost analysis as a stand-alone value theory, while the devout analysts dedicated to fixing the benefit-cost analysis will force-fit the analysis through tortuous, heroic grinding adjustments to make the benefit-cost analysis work. Much of the critique of economics generally, and from discounting particularly, originates from just such a confusion between straight-forward efficiency analyses and the wider public value-making process itself.

These warnings reinforce our original message. A true benefit-cost analysis, evaluated in a straightforward neoclassical fashion, solves many problems. If that way of valuing is inappropriate, it is better to adopt a social choice rule than try to 'fix' the welfare analysis – but that does not mean economic principles are not useful. A short historical example may be of help.

Second-best social choice example

Arguably, long before government gets all the rules and institutions in place, it will strike out incrementally into the unprotected future and cross a relatively narrow passage of incomplete markets without much protection. In some cases, these decisions can be tracked by competent benefit-cost analyses. Yet one does not have to look very far to locate bold institutions created by the public to stimulate long-term investment that appear long before society details every public value needed to complete the market, or even to conduct an adequate benefit-cost analysis.

The creation of the corporation is an excellent example. Incorporation is a legal shelter to investors from losses beyond the actual amounts invested. Limited liability laws exculpate investors from deficiencies that might be attached to an insolvent corporate entity. So, while the rule improves the transfer of long-term assets from generation to generation, it is also highly distorting. To shelter corporate investors from selective risks is a move that misallocates resources terribly; yet, as an imperfect innovation, corporate guarantees seem to improve economic performance in ways consistent with values that prevailed when they were adopted.

What is noteworthy is that this type of public decision has nothing to do with discounting and everything to do with inefficient market structure. When the path to efficiency is far from unique, society has to make hard choices and set priorities to approach a clear enough set of rules and rights to conduct its business. Under conditions when many rules and rights have not yet been established, economic thinking can help to frame the question in a way that does not rely principally on welfare economic analyses.

Clearly the creation of the corporation and the corollary shipping and trade insurances of the Bank of England created by Queen Elizabeth I opened many new forms of trade finance. The programme converted reluctant peasants previously taxed to build ships into eager investors who, on their own, built and launched ships and then carried goods and services across the planet, enriching both the monarchy and her subjects. It was this deliberate attention to economic incentives and the identification of particularly acute institutional hold-ups from incomplete markets that propelled a still somewhat backward England into world prominence.

A benefit-cost analysis that attempted to estimate the untapped trade potential of incorporation would not only have been quite difficult, but a report to the Crown of a benefit-cost ratio greater than one seems much less informative than an institutional evaluation of the transformative power of the programme. I suspect bankers' reports (and there were some) were less effective than a full explanation to the Court of the potential to liberate traders from a nagging dependence on the Crown to license and to secure long-distance trade, approvals that included ship-building and even cargo selection. More impressive is a review of new incentives for small communities to self-organize to build ships and to deliver tradable local goods on escrow credits without the initial attention of government. The possibility of back-end benefits from taxes out of the explosion of new trade must have been attractive.

Curiously, the vastly improved productivity that followed surely increased long-term interest rates; yet we would not view incorporation and the creation of the Bank as having mortgaged England's prospects over the next two centuries. If some other public project, say, a road project designed to facilitate intra-national land transport in England, had been evaluated by a reduced discount rate in this rapidly growing economy, at the margin, England would have diverted more valuable capital (that was pouring in to support her profitable overseas trade) to overbuild domestic infrastructure too soon, given the long-run opportunities that were opening up for global trade. It is hard to see how a future friendly interest rate reduction for a road project to safeguard future welfare accomplishes its putative goal. It is easier to infer that such a fix, economy-wide, would have injured both the England of the 17th century and of the 19th century.

It seems likely that policymakers in Elizabethan England thought just like this as they navigated their way through the morass of emerging global markets and the vast array of incomplete trades and transactions that were not even yet imagined. Again to discount or not to discount in the decision to create the Bank of England cannot be the question. It would be absurd to attempt to fix a benefit-cost report to capture the full range of social questions surrounding the creation of the Bank of England in a one-stop free-standing report; but it is equally absurd to insist that the question is really only a political value question about the power and stability of the monarchy and not an economic value question. Clearly, to treat the issue properly, it is economic institutional analysis that frames the decision environment for many big questions that affect the future, and this is a requisite discipline for legitimate policymaking to prevent harm to the future.

3.4.2 Ethical pluralism challenge

Our review of some of the challenges and issues surrounding discounting serves as a good defence of efficiency in multiple-use forestry decisions. We are purists who insist that economists either do a neoclassical economic benefit-cost analysis to inform social questions or perform a detailed institutional analysis to help resolve a social choice dilemma. We have no ambition to transform welfare economics into a comprehensive social choice instrument; but the exit of welfare economics from the policy field does not mark the exit of economic efficiency thinking from informed value discourse, passing the baton from economists to ethicists.

The distinction between ethics and applied economics is less like the forced consolidation among strange bedfellows than a squabble between jealous siblings as the two disciplines are inextricably bound. The rules and laws already in place to make a welfare economic analysis possible are not value-free. Clearly, whether polluter or pollutee pays for environmental clean-up embeds different ethical content, but both choices are equally efficient. To label one choice an 'economic' value and the other choice a 'non-economic' value is a meaningless distinction that can only confuse public debate and value discourse.

As societies search for values that guide how to care for the future, ethical discourse turns quite directly on an economics question of selective reform in missing and incomplete markets. A dead-end discussion over discounting invents an artificial division between 'economic values' and our care for future persons, which obfuscates the social choice crisis over how we construct institutions to administer to the future. Without the missing markets challenge to frame the value dialogue, society dangerously discards an effective and legitimate value discourse instrument.

The folly of separating care for the future from economic thinking, on the premise that care for the future is simply not about economics, is best illustrated by a historical example. The several thousand-year transition out of hunting and gathering into agriculture involves enormous social institutional challenges. At the dawn of agriculture as humans began rudimentary cultivation of wild seeds, which they replanted and harvested again, it would have seemed crude, disgusting and even taboo for a group or a person to lay claim to ownership over that harvest. Those societies would surely declare the value challenge as not really about economic institutions, something well beyond any mercantile rent seeking far outside their experience; yet clearly, in the end, the resolution to the eight thousand-year-old problem of agriculture did revolve around innovations that set down new economic rules and institutions that coincide nicely with current notions of economic efficiency, property rights, trading practices and eventually the rise of larger states to oversee and to secure financial economic transactions (lending, trade escrows etc.).

In the end, human communities resolved some of the most cumbersome missing markets that stalled progress in agricultural innovation for millennia. Fortunately for us, we do not have to wait that long. To surgically divide ethics about future care from discussions of economic rights and institutions, in the face of the widespread market failures that are largely responsible for placing this question on the public agenda, is a hopelessly irresponsible convenience that delays solution and permits policy to duck the really hard ethical choices.

While a full mapping of complex land ownership and tenure structures would be inconceivable to the hunter and gatherer tribes at the end of the ice age, our modern civilization clearly has the experience to adequately digest the problem of missing markets, ambiguous ownership, unreliable profit claims and structural risk as the critical institutions that affect future welfare. This institutional and market structure of economic activity can, over time, easily motivate and frame value dialogue in a clean and effective way. We argue that enough ethical cohesiveness exists in modern states and Western-style democracies to use the framework of functional economic institutions to anchor value-making without fracturing value dialogue.

One of the powerful ingredients in the economics approach to the missing markets framework is that society is not committed to a singular outcome: there are choices with distinct ethical content but there are also transparent boundaries. Unlike welfare economics where the majority of rules that determine efficiency are established a priori, this form of social choice marks out a relatively large core of options within the efficient system that extend from Rawlsian difference principles to cooperative fairness to utility maximization – all are obtainable.

The reality of setting down one rule, however, suggests that society may not reach complete efficiency (market completeness); but it will aspire to remain close. Remaining ethical contention may lead society near to an efficient outcome but will require compromises among ethical positions at the margin that are still inefficient. In the economists' language, consensus can move society closer, but not all the way, to the Pareto frontier (complete, efficient markets). In the end, process constraints and compromise lead modern states to a liberal respect for efficiency, but with accommodations to certain interests or moral positions that pragmatically occlude a perfectly efficient system (Castle, 1996); and deep, decades-old traditions in resource economics recognize this tension (Ciriacy-Wantrup, 1952; Bishop, 1978; Farmer and Randall, 1998).

Modern societies have the experience and the economic acumen to prevent ethical compromise from needlessly overlooking opportunities for consensus that can make at least one person better off without injuring some other. The ability to clarify new processes opens doors to ethical compromise that expand options and add precision to moral dialogue. Indeed, much of our value search seems irresponsibly under-informed by such an artificial separation between 'economic' and 'non-economic' values over the choices available to care for future persons.

Society's concern for the future is multidimensional: we want our progeny to carry forward our values and cultural traditions, we want them to be relatively free of want, and we hope that they enjoy the options to explore their own creative ambitions to the extent that we can affect that outcome. If ethical dialogue progresses by an incremental sequence of constant updating of provisional value statements (Lindblom, 1977; Taylor, 1992), then the decision over which market failures command the greatest attention is a powerful component of ethical dialogue, a dialogue consensus that is far too incomplete to permit the market to be complete. In this real world, economists help to locate the severity and vulnerability of those issues society wishes to treat, such as education, research, pollution protection and environmental conservation. At the same time, they remind society of the vulnerabilities of the issues left behind, or the unresolved ethical choices that leave many markets incomplete. Efficiency helps to make consensus, to keep consensus and to unveil the ethical regions of incomplete attention. To reduce the quality of this rich discussion to a niggling discussion over discount rates is so far off the mark as to be irresponsible and, arguably, ethically and politically illegitimate.

Our claim of potential consensus through the liberal instrumental reliance on the economic way of valuing and an exploration of the incentives nested inside various economic structures assumes no more than many modern ethicists require.

The same assumptions that license many forms of community dialogue that enjoy currency appeal directly to moral consensus foundations nearly identical to ours. The very premise in value pluralism voiced from large circles of modern ethics centres more and more around primitive value consolidation that enables responsible discourse on particular ethical tensions (Rorty, 1990; Rawls, 1996). That is, individual expressions of value aggregated across persons are unlikely to deviate too far from social norms if a minimal shared value core exists.

3.5 Conclusion

What makes economic analysis highly adaptable in moral practice is the very thing that eclipses its universality. Simply, no process can encompass the entire breadth of social enquiry.

The moral question of the day turns on the role of efficiency analysis in a pluralist moral society, especially in the historical context of Western-style democracies. The social courage to face this more complex moral question 'head-on' is still in its infancy, and the contribution of the economic way of valuing, we argue, is a compelling real world moral concern.

Value competition in the modern world requires society to be progressive, not static: vibrant, not stale. What makes our programme work is the precision nested in an efficiency account to make material allocation choices in society without undue disruption. When competently identified, adherence to the efficient outcome is a valid moral act that can decisively resolve many questions for multiple-use forestry even if the issue extends over a very long time period.

Efficiency is a good rule for making material choices because it preserves material wealth and complements many diverse value claims on the public sphere that involve material allocations. An efficiency default helps social choice to be more selective in the topics for open dialogue, thereby contributing focus and discipline to public value-making. In our assertion that economics is not everything, we hold fast to the notion that it is far from nothing. We defend efficiency as a stand-alone concept which ought to enter public debate uncorrupted, presented with open eyes. Our pragmatic moral concern is to maintain transparency in social choice to allow a society to be very clear about what it gains and what it sacrifices under different choices.

In this world, economics becomes valuable as a moral instrument that adds precision to value-making. Attempts to reform a welfare analysis into a fully formed value expression, by converting prices to 'future-friendly interest rates' or by injecting distributional corrections or quality of life indices into the efficiency account, obscure the core social value conflicts that a true efficiency analysis helps to isolate and bring to public view. To make welfare economics into a full value system takes what was transparent and perverts it into an opaqueness that blurs the valuable distinctions between welfare economics and social choice that assist ethical choice.

We defend efficiency and the entry of all of the prices (or shadow prices if they can be efficiently estimated) that prevail as an important moral accounting device that takes seriously most plausible moral positions modern societies entertain. At the same time, as society opts at times to forgo direct benefit-cost analysis to evaluate structural reforms for the benefit of the future, understanding the incentives inside any set of rules adopted is crucial to value-making and to the delivery of true net benefits to the future. It is the transparency of the efficiency account and of the efficiency-guided market reform that helps the moral agent to assess whether value concerns are treated decently well across the scope of public enquiry. This practical role for efficiency has the advantage of being pluralist, adaptable and politically liberal. To ignore these advantages to legitimate decision-making on the grounds that the economic way of valuing does not constitute a complete or universal moral theory is, in the end, to desire the opaque and to risk, indeed to assure, the unseen harm.

References

Arrow, K. (1954) Existence of an equilibrium for a competitive economy. *Econometrica* 22, 265–290.

Arrow, K. and Lind, R.C. (1970) Uncertainty and the evaluation of investment decisions. *American Economic Review* 60, 364–378.

Bishop, R. (1978) Endangered species and uncertainty: the economics of a safe minimum standard. *American Journal of Agricultural Economics* 60(1), 10–18.

Castle, E. (1996) Pluralism and pragmatism in the pursuit of sustainable development. In: Adamowicz, W., Boxall, P.; Luckert, M., Phillips, W. and White, W. (eds) *Forestry, Economics and the Environment.* CAB International, Wallingford, UK.

Ciriacy-Wantrup, S.V. (1952) *Resource Conservation: Economics and Policies*, 1st edn. University of California Divisions of Agricultural Science, Berkeley, CA.

Eckstein, O. (1958) *Water Resource Development: The Economics of Project Evaluation.* Harvard University Press, Cambridge, MA.

Farmer, M. and Randall, A. (1998) The rationality of a safe minimum standard. *Land Economics* 74, 287–302.

Hotelling, H. (1931) The economics of exhaustible resources. *Journal of Political Economy* 39, 137–175.

Howarth, R.B. and Norgaard, R.B. (1992) Environmental valuation under sustainable development. *American Economic Review* 82, 473–477.

Lindblom, C.E. (1977) *Politics and Markets: the World's Political Economic Systems.* Basic Books. New York.

Page, T. (1978) *Conservation and Economic Efficiency.* Johns Hopkins University Press, Baltimore, MD.

Pearce, D.W. and Turner, R.K. (1990) *Economics of Natural Resources and the Environment.* Johns Hopkins University Press, Baltimore, MD.

Rawls, J. (1996) *Political Liberalism.* Columbia University Press, New York.

Rorty, R. (1990) The priority of democracy to philosophy. In: Masachowski, A. (ed.) *Reading Rorty.* Blackwell Publishing, Cambridge, MA.

Taylor, C. (1992) *The Ethics of Authenticity.* Harvard University Press, Cambridge, MA.

Weitzman, M.L. (1998) Why the far-distant future should be discounted at its lowest possible rate. *Journal of Environmental Economics and Management* 36(3), 201–208.

Weitzman, M.L. (2001) Gamma discounting. *American Economics Review* 91(1), 260–271.

4 An Intergenerational Perspective on Effects of Environmental Changes: Discounting the Future's Viewpoint

C. Price

School of Agricultural and Forest Sciences, University of Wales, Bangor, UK

In economic appraisal, the level of significance given to future environmental changes depends on the discounting procedure. The most common justification for discounting is that early cash revenues can grow at compound interest: thus compensation for distant-future environmental costs can be provided at small present cost. However, this argument requires that early benefits are in an investable form, and are actually completely invested. Conventional negative exponential discounting does not correctly reflect the form of subjective preferences for early consumption, causing inconsistencies of choice. In any case, time preference is better explained as a preference for immediate consumption over consumption at all other times, with no premium at all for consumption sooner rather than later. Neither physical changes through time nor risk in its various manifestations are reasonably incorporated by discounting, which may indeed have perverse effects by favouring risky projects. Although increasing consumption per head would be an ethically appealing argument to put less weight on extra future consumption, such an increase cannot be confidently predicted for all people, products and scenarios, and cannot apply to the totality of future consumption. Techniques are available for allocating investment funds rationally under low discount rates. Adopting sustainability constraints is an unsatisfactory alternative to rethinking discounting and applying the process only when it can be rationally and ethically defended.

© CAB International 2005.
Forestry and Environmental Change: Socioeconomic and Political Dimensions
(eds J.L. Innes, G.M. Hickey and H.F. Hoen)

4.1 Introduction

Forestry has traditionally concerned itself with continuity of production and, given the growth rates common in Europe and North America, silviculture has necessarily entailed investment for future generations. Even where growth rates are faster, concern about site quality maintenance through several rotations has kept time horizons long.

How such long-term considerations enter economic appraisals is largely mediated by the process of discounting.[1] Discounting has been widely, if sometimes grudgingly, accepted in assessing relative profitability of silvicultural treatments, harvesting and research. With the development of techniques to monetize the value of environmental changes, discounting is also increasingly applied to effects like degradation of visual environment, depletion of biodiversity and impact of land use on watercourses. Usually, the process has been transplanted unmodified from the world of commerce: 'environmental costs and benefits should be discounted, just like other costs and benefits' (Department of the Environment, 1991, p. 31).

The discounting procedure dramatically affects the importance ascribed to long-term environmental changes. For example, with the 6% discount rate endorsed by the UK Department of the Environment, global warming severe enough to eliminate the entire present gross domestic product (GDP) of the UK in 500 years' time would be valued at only a fraction of £1. Even over a shorter period, discounting crucially affects not only the value ascribed to forestry programmes for environmental purposes (Thomas and Price, 1999) but also the price assigned to such functions as carbon fixing (Healey *et al*., 2000).

This chapter reviews arguments for and against discounting, with particular emphasis on environmental changes that may affect future generations. The arguments are numerous, and some are complex, so this review is simplified. The literature on discounting is voluminous; therefore, reference is made only to seminal and recent contributions. The literature up to 1990 has been reviewed in Price (1993) and Arrow *et al.* (1996), which explore in more detail many of the arguments outlined below.

4.2 Rate of return and compensation

The most commonly proposed basis for a discount rate has been the market rate of return on investment. The argument is theoretically persuasive in the context of provision for future generations. A project that has adverse long-term environmental consequences but offers early revenues could endow an investment fund that will pay compensation to the affected future generations.

The further into the future that the compensation must be paid, the longer the period of compound growth, and so the smaller the sum needed now to provide that compensation (see the heavy curve in Fig. 4.1). The argument is particularly

[1] Discounting is the process of giving a present equivalent for a future value by dividing it by a factor $(1 + r)$ once for each year between the present and the time when the future value occurs. r is the discount rate.

appealing in that it requires no trade-offs between generations. If net present value (NPV)[2] is positive even after provision for compensation, the project would be accepted by the present generation, while future generations are no worse off, given that compensation is fully paid.

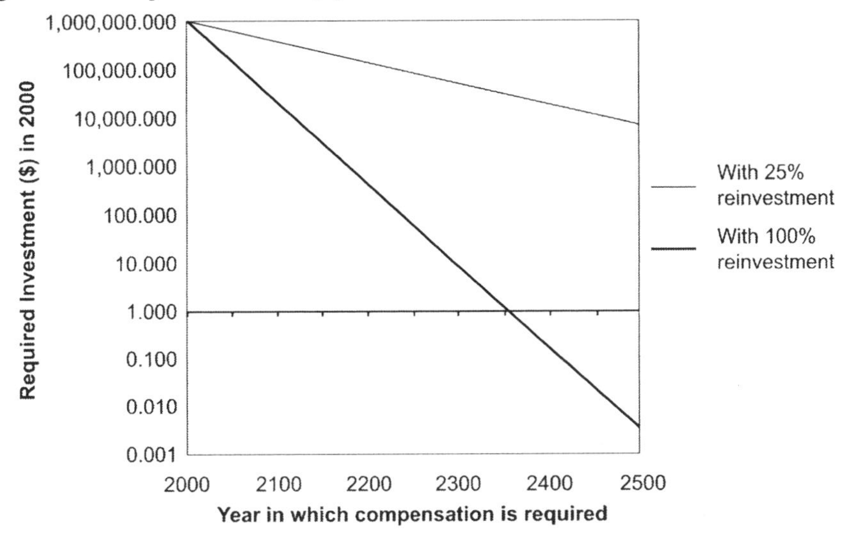

Fig. 4.1. Present investment needed to pay future compensation of $1,000,000. Rate of return on investment is 4%. Note the logarithmic scale for required investment.

The weaknesses of the argument are, however, rather evident. In the first place, the discount rate used is normally the real private pre-tax rate of return on financial investment. But, if the compensation fund is lodged in investments, which themselves have adverse environmental effects, part of the fund's returns should be reserved to compensate for these secondary damages. This reduces the surplus available to compensate for the damages of the original project. Hence the appropriate rate of return for discounting would be the rate of return net of externalities. Very generally, Howarth and Norgaard (1992) argue that the rate of interest should be lower under policies of sustainability, which preclude some high-return investments, and oblige others to pay compensation directly.

Consider, secondly, that many environmental changes are externalities, providing neither revenues nor expenditures. Thus, for example, early beneficial externalities necessarily accrue to the present generation, and provide no financial resources to compensate future generations for harmful externalities. By contrast, if harmful environmental effects are priced at the cost of restoring or replacing damaged environments, discounting remains relevant. For example, it is better to delay the visual costs of a mining operation, because this reduces the sum to be invested now to pay for the costs of cosmetic tree planting.

[2] Net present value is the sum of all the individual benefits and costs of a project, each discounted from the time in the project's life when it arises.

Thirdly, changed flows, even of freely investable revenue or expenditure, may not actually induce commensurate additions to or withdrawals from a compensation fund. Conspicuously, exploitative logging has not routinely been followed by the establishment of such a fund. It is also notable that carbon taxes have been welcomed for their 'double dividend' (Fankhauser, 1995); that is, their ability to benefit the present generation by replacing efficiency-distorting income taxation. Clearly, taxes used in this way cannot compensate future generations for adverse climate change. Recently, many contributors to Portney and Weyant (1999) have repeated long-expressed doubts about whether profits from projects that cause environmental degradation will be totally reinvested. It is still more problematic to enforce such a reinvestment policy upon immediate future generations for later generations' benefit.

Of course, it is technically possible for all changes in cash flow of a particular project to cause only changes of investment funds, leaving consumption wholly unaffected. However, it has long been explicit in formulations of the opportunity cost of investment funds (Eckstein 1957, and many later authors) that only a fraction (typically 10–25%) of project revenues will be reinvested – and that assumption seems to have gone unchallenged. The consequence of partial reinvestment is that the compensation fund grows at a slower rate than the rate of return. The light line in Fig. 4.1 shows how much would have to be invested presently to provide future compensation of $1,000,000 if only 25% of investment income is reinvested. For compensation to be paid in 500 years, the difference of required investment is huge: $6907 with 25% reinvestment, 0.3 cents with 100% reinvestment.

More recently, published valuations of investment funds have included no reinvestment at all (e.g. Livingstone and Tribe, 1995; Pearce and Ulph, 1995; Cline, 1999): instead, investments are assumed to generate a stream of consumption over a short period only, typically 15 years. After this time period nothing further changes: no scope exists to compensate future generations. Although this assumption is just as unrealistic as that of total reinvestment, its adoption by well-known environmental economists shows how casually the reinvestment assumption is treated in practice.

Revealingly, El Serafy (1989, p. 16) asserts that 'the setting aside of part of the proceeds [of natural resource exploitation] in reinvestment is only a metaphor'. Such metaphorical compensation might suit the needs of metaphorical future generations well enough. But to compensate real future people requires real compensation, paid from a fund really established and maintained, from the time when it is established, to the time when compensation is actually due. As d'Arge *et al.* (1982, p. 255) put it:

> Economists often use the notion of 'hypothetical' compensation to justify discounting. In an ethical context such arguments play no role whatsoever. Rather, if no actual compensation occurs, the market rate of return has no relevance for discount rates.

This does not mean that discounting cannot ever be related to rates of return: the investment growth rate, equal to:

[rate of return on investment] × [percentage of investment income reinvested] can be used to determine the present investment required to yield given future compensation. The circumstances in which investment growth rate might be used as a discount rate are alluded to later.

4.2.1 Discounting as a floodgate

There is one further argument for protecting future interests by 'high' – though not necessarily market-based – discount rates: high rates decrease the number of projects deemed worth undertaking, and hence reduce concomitant long-term environmental damage. However, this procedure preserves the environment haphazardly. Environment is favoured, not because it is itself valued, but because medium-term benefits of some projects may be discounted to a smaller sum than the short-term investment outlays. However, on this basis, high-return logging projects that destroy significant biodiversity resources in the long term might still be deemed acceptable, while low-return multipurpose forest development would probably be rejected. If environmental values are important despite being distant in time, the evaluation procedure should reflect that explicitly by using a suitably low or even zero discount rate.

4.3 Preferences for early, propitious circumstances or now?

Typically, though not invariably, when people are offered the choice between good things now and good things later, they prefer them now. They usually require some incentive to postpone a given act of consumption. By contrast, unpleasant experiences are delayed if possible. Within a market democracy, such evident and widespread time preferences should, it is argued, be respected by decision makers by discounting on behalf of collections of citizens, consumers or shareholders.

However, numerous investigations have shown that the form of the time preference function differs from the uniform negative exponential of conventional discounting. A lower subjective discount rate seems to be used:

- for longer as opposed to shorter periods (Cropper and Laibson, 1999);
- for large and negative sums of money rather than small and positive ones (Benzion *et al*., 1989);
- for 'public' as opposed to 'private' matters, which, of their nature, tend to embrace environmental as opposed to material products (Luckert and Adamowicz, 1993);
- for advance in timing of good events as opposed to delay (Loewenstein, 1988).

The first point is particularly significant, implying a relationship that is not exponential at all: the ratio of values between one date and another changes according to the position in time from which it is viewed. Evidence from several disciplines, (i.e. psychology, sociology, biology and economics) suggest a hyperbolic relationship of form [discounted value] = [future value] / (1 + [discount rate] × [time of future value])

(Ainslie, 1991; Henderson and Bateman, 1995). Unlike negative exponential discounting, hyperbolic discounting has no clear underlying theory: it is just empirically observed to fit researchers' data-sets. Similarly, much research into different discount rates used in different circumstances simply records perceptions without explaining them.

I know of no empirical or theoretical research that questions these findings, and yet time preference continues to be routinely invoked as a 'democratic' justification for conventional discounting. It is surely unjustifiable to invoke democratic sanction for the existence of time preference, and then undemocratically ignore the form of that preference.

4.3.1 The inconsistency problem - different rates

One consequence of deviating from standard discounting is inconsistent preferences as time moves on, with plans for the future subsequently revised (Horowitz, 1996). Consider landscape restoration to be undertaken by a mining company after a 50-year lease. The private financial costs, £100,000, are discounted at 6%, but the public environmental benefits, assessed as £3000 per year in perpetuity, are discounted at 4%. Seen from the present, this project has positive NPV:

$$£3000 / 0.04 / 1.04^{50} - £100,000 / 1.06^{50} = £5125$$

But a subsequent generation, reconsidering the project at the time of restoration, would re-evaluate it as:

$$£3000 / 0.04 - £100,000 = -£25,000$$

In this case, the scheme is not worth undertaking.

4.3.2 The inconsistency problem - hyperbolic format

Hyperbolic discounting results in a rapid decline of value during initial periods, but only a slow decline in value later on. Fig. 4.2 illustrates cumulative discounted benefits from a carbon-fixing plantation programme, in which 1 ha is planted each year for 10 years at a cost of £1000 per year. Each hectare's carbon-fixing yields £40 of benefit per year for 100 years, after which net carbon flux is zero. Costs and benefits are discounted according to the hyperbolic equation:

[discounted value] = [future value] / (1 + 20% × [time of future value])

To the present generation, a proposal for planting after 30 years looks attractive: very distant benefits are only modestly discounted compared with moderately distant costs, and the cumulative discounted value flow eventually shows a substantial surplus. However, when the next generation reconsiders the proposal as an immediate investment, moderately distant benefits are discounted heavily compared with costs; immediate investment never breaks even. Unless contractually constrained, the next generation will not implement the planned plantation – although they too will consider it a good idea that the following generation should undertake it.

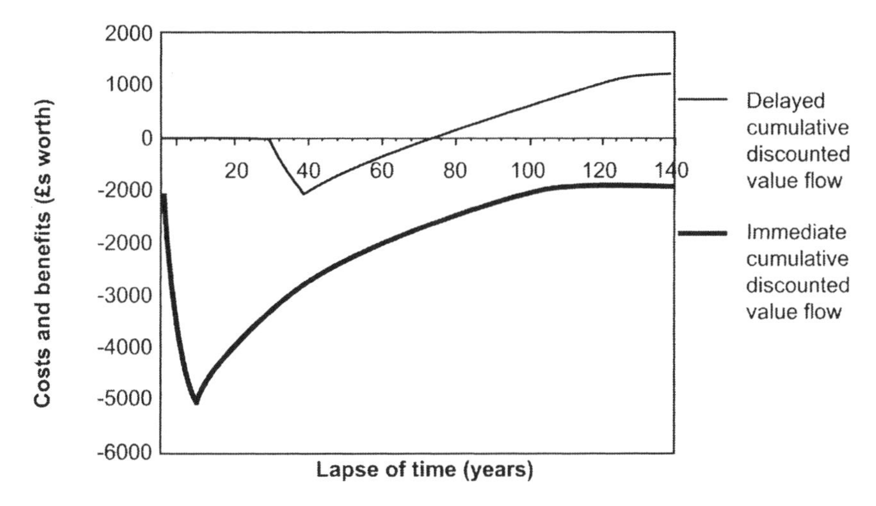

Fig. 4.2. Inconsistency under hyperbolic discounting.

4.3.3 The inconsistency problem: intragenerational and intergenerational discounting

There are some reasonable grounds for discounting whatever future consumption may accrue to oneself – the declining probability of one's being alive, for example, or life cycle effects that reduce capacity to enjoy a particular experience. (Whether a negative exponential is the appropriate functional form for such relationships is another matter.) These are not, however, reasonable grounds for discounting consumption by future generations, who outlive us and who pass through the same life cycles as us, but later in earth time.

Personal causes of discounting should be filtered out in intergenerational decisions. And, actually, we already take account of future generations in long-term discounting. We do it through our 'benevolent affections' (Rae, 1834) towards future generations, by which we derive pleasure from the prospect of their pleasure. Recently Kopp and Portney (1999) have proposed a 'mock referendum' on programmes with long-term implications, to reveal willingness to sanction a lower intergenerational discount rate. Several contributors to Portney and Weyant (1999) also believe that the present generation gives some weight to the interests of future generations, and that 'market rate of return' discounting should be gradually displaced by discounting at a lower rate for the distant future.

However, the value that the present generation places on intergenerational equity should not be confused with intergenerational equity itself. Despite its title, the recent compilation *Discounting and Intergenerational Equity* (Portney and

Weyant, 1999) generally treats time preference as a matter of intragenerational efficiency: the trade-off the present generation wants between the utility of its own direct consumption and that of its own pleasure in protecting the future's interests. But, if future generations count, they should count in their own right, not only valued according to the caprice of the present generation (Tenenbaum, 1989). Rather than asking people how they feel about long-term programmes, researchers might ask what ethically relevant reasons there are to give less weight to future than to present generations (see Partridge (1981) for diverse views on this matter).

Adam Smith, whose *The Wealth of Nations* is assumed to provide moral underpinnings for a system based on self-interest, was very clear that people's instinctive relative valuations (for example, of minor injury to themselves compared with large-scale catastrophe on the other side of the world) did not represent their dispassionate judgements of real worth. For the utilitarian, Sidgwick (1874, p. 414) concluded that '[t]he time at which a man exists cannot affect the value of his happiness'. Rawls (1972) suggests that insight into difficult problems of equity is given by considering an 'initial position' in which representatives of all stakeholders meet to discuss rules for just allocation. The key condition is that no one knows which interest group they will ultimately be born into: no one will agree to discount the well-being of any future generation, because that may be the generation into which he or she is born.

It is problematic to envisage what future generations' preferences will be for environmental conditions and for intergenerational distribution. However, it is more plausible that they will prefer conditions which we regard as good, and give a premium to the generation in which they happen to exist, than that they will have no intertemporal preferences at all.

Table 4.1. Options for oak forest rotation

Event	Cash flow	Discounted value seen from time 0	Discounted value seen from time 80
Plant	−£2000	−£2000	
Fell at age 80	£6000	£6000 × 0.26619 = £1597	£6000
Fell at age 120	£12000	£12 000 × 0.26619 = £3194	
Fell at age 120	£12000		£12 000 × 0.29927 = £3591

Such thinking has led several authors (Kula, 1988; Bellinger, 1991; Bayer and Cansier, 1999) to conclude that, while the present generation is entitled to discount its own well-being as it pleases, the consumption of future generations should not be discounted according to the present generation's intuitive time preference. The problem of discounting in these circumstances is aggravated by the absence of a sharp cut-off point between 'the present generation' and 'future generations': generations overlap. Because different discounting formulas apply to present and future generations, the formulas for intragenerational and intergenerational discounting must be mixed in changing ratios as time moves on. The authors mentioned above address this problem by calculating discount factors for each future year, in which different proportions of the present generation survive, and in which the age distribution of recruits to the population also changes.

Whatever the reasoning behind a lower intergenerational discount rate, it provides another source of inconsistency, given that future generations are likely to view their future from their own perspective, rather than merely accepting discount factors handed down to them (Price, 1989). Take for example, an oak plantation with alternative rotations of 80 and 120 years (see Table 4.1). Discount factors, calculated according to Kula (1988) with an intragenerational time preference rate of 5%, are: for 0 years, 1; for 80 years, 0.26619; for 120 years, 0.26619.

From the perspective of an owner in AD 2000, the wood is worth planting for its timber benefits on a 120-year rotation, there being no differential discounting between AD 2080 and AD 2120. However, in AD 2080, a future owner would conclude that the revenue from felling immediately outweighs the heavily discounted revenue from prolonging the rotation. And yet the present owner would have decided that the crop was not worth planting, had the future owner been expected to fell at 80 years. There are clear consequences for environmental benefits, which are likely to increase with the age of the crop, and to become more important with lapse of earth time.

4.3.4 The 'nowness' argument

These conundrums can be resolved if the 'special case' is not discounting the present generation's well-being, but the peculiar way in which each generation views the moment in time at which it makes evaluations. Shackle (1958, p. 13) asserts that 'there is for us a moment in being [the present] which is the locus of ... every decision'. But that moment moves on with passing time.

Arguably, the value of consumption should be seen from the perspective of the moment in which it occurs. As Sen (1957, p. 746) puts it: 'We are interested in tomorrow's satisfaction as such, not in today's assessment of tomorrow's satisfaction'. If every moment in an individual's life eventually becomes uniquely important 'now' once and once only, there should be no intragenerational discount for the 'not-nowness' of future acts of consumption. If no generation's consumption is to be discounted merely on account of 'the time when it is born', there is no intergenerational discount for pure passing of time either.

Under this 'nowness' interpretation of time preference, the supposed disjunction between intra- and intertemporal discounting vanishes. For individual-specific values, such as cherished familiarity with a landscape as it was before clear-felling, intragenerational discounting is done according to factors relevant to the individual (mortality, adaptation of taste). (However, such discounting is unlikely to have an exponential format.) These factors have no relevance to intergenerational discounting. For values that are transferable between generations, only factors that affect society apply, such as the aesthetic impact of forest growth, whether the beneficiary is a currently living or a yet-to-be-born person. Pure time preference (preference for immediacy) has no relevance to discounting later time, for either present or future generations. There is no time inconsistency, and no conflict of view between generations.

4.4 Physical change

Although they are not usually treated explicitly by discounting, processes of change affecting future generations are sometimes represented in the same way. For example, reduction of atmospheric CO_2 concentrations, by oceanic and terrestrial uptake, mitigates the long-term impact of present CO_2 fluxes. But discounting does not actually mirror the profile of change. Fig. 4.3 shows:

- a simple negative exponential;
- the model of Nordhaus (1992), which misguidedly estimates an immediate uptake of 36% of emissions (Price, 1995);
- the sum of separate exponential processes representing five different oceanic carbon sinks (IPPC, 1990);
- a model also making an illustrative adjustment for terrestrial uptake.

It is noteworthy that the more realistic models leave a permanent CO_2 residue, presenting an ongoing problem for future generations.

4.5 Risk

Resource managers and actuaries interpret risk as the probability of bad outcomes, like devastation by forest fires, atmospheric pollution or insects. Forest economists have traditionally seen these bad outcomes in terms of lost timber and growth potential, but they may also diminish amenity and trigger CO_2 emissions. Risky situations are conventionally distinguished from situations of uncertainty by the existence of some known or estimated numerical probability of the conditions that produce bad outcomes. For clarity, such situations, where variation from a hoped-for norm means a less valuable outcome, will be termed 'threat'.

Financial analysts consider risk quite differently, as variability of outcome, both above and below some mean expectation.

4.5.1 Risk as unknown outcome

Discounting for risk, threat and time have often been discussed collectively (Lind, 1982), and risk and threat are sometimes treated simply by adding a risk premium to the discount rate. It is true that the further into the future an event is expected, the less certain it is to happen, and the less predictable are its consequences. But, while the intention of the risk premium might be to put less confidence in individual outcomes, the effect for future generations is that the riskier the project, the less weight is put on its outcomes collectively. In the absence of an explicit statement of what other outcomes should have more weight (so that the sum of probabilities is, as required, unity), implicitly an outcome having zero value is given the residual probability.

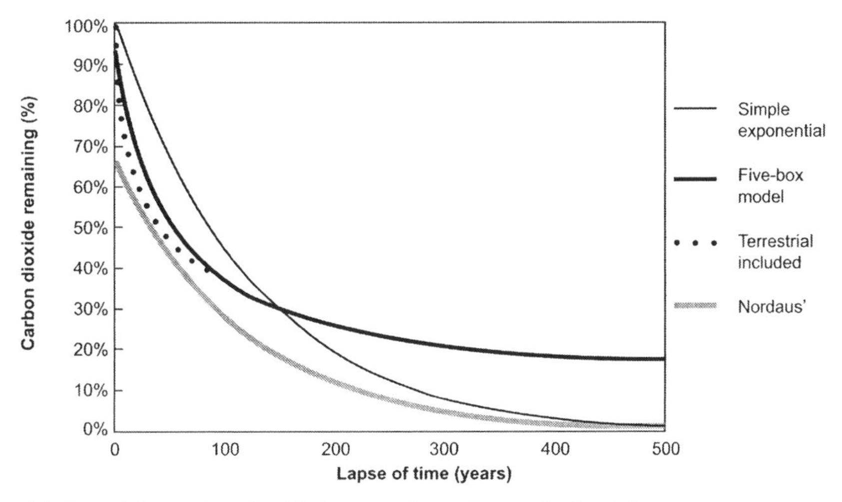

Fig. 4.3. Remaining carbon dioxide in atmosphere after uptake by sinks.

Table 4.2 shows a case with a single explicit outcome. It might represent the benefit of retaining a forest area for avalanche protection equal to the avoided annual damage cost, Fr.100,000. Climate change may reduce the required protection over time: for example, snowfall may decrease. But such an assessment neglects the possibility that climate change may mean increased storminess or more sudden snow instability, in which case one possible outcome is greater benefit from protection.

Table 4.2. Effect of risk premium on value of protection

Event	Time	Cash flow	Risk discount	Implied probability	Present value
Condition at present	25	Fr.100,000	6%	0.233	Fr.23,300
Implicit zero-valued outcome				0.767	Fr.0
Total				1.000	Fr.23,300

4.5.2 Risk as physical threat

A discount premium implies a particular time-path for threat and for its impact on utility. Consider damage by storms. Fig. 4.4 shows the cumulative probability of a tree crop not being hit by a storm sufficient to cause critical damage. The crop is coniferous, on an exposed site in the western UK (figures from Forest Research, 2000). Clearly discounting cannot represent the empirical data, for these reasons:

- The profile of survival probability within a rotation is not a negative exponential function;
- Events of this kind often cause reduction rather than destruction of crop value;

- Discounting gives relative value, not only within each crop's life, but between one cycle of growth and the next. Fig. 4.4 incorporates slowly increasing frequency of storms through earth time, but this is very different from the changes occurring within an individual crop cycle. Clearly discounting for an 80-year lapse of earth time does not sensibly estimate the probability of (say) the second cycle reaching the age of 20 without damage – the crop is almost immune up to that age.

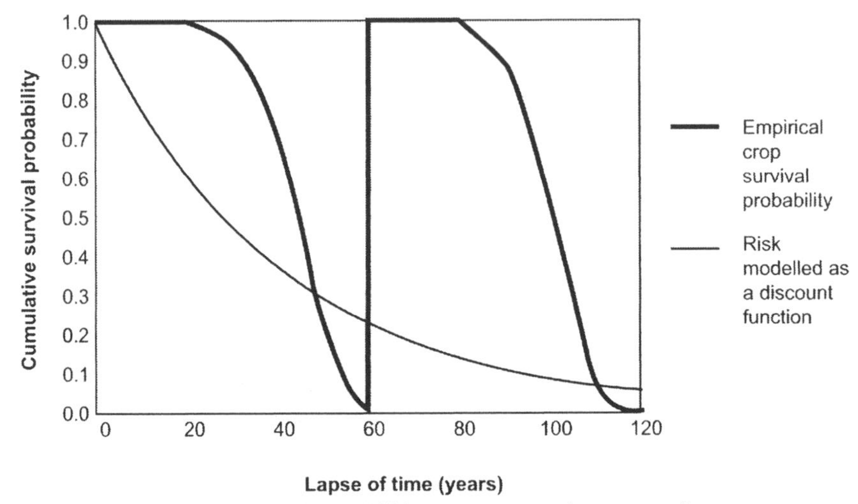

Fig. 4.4: Storm damage survival probabilities over two rotations: storm frequency increasing at 0.25% per year.

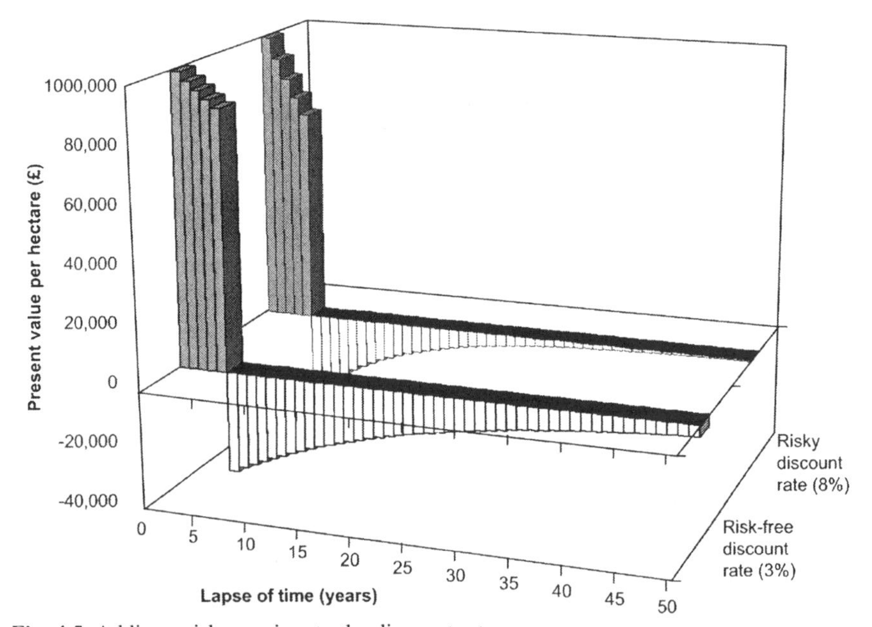

Fig. 4.5. Adding a risk premium to the discount rate.

4.5.3 Risk and the late-costs problem

Raising discount rates according to riskiness has systematically perverse outcomes when late costs dominate a cost–benefit flow, as frequently happens in exploitative projects. Consider the following: forest exploitation produces a 5-year revenue flow of Rs100,000 per hectare per year, after which deforestation causes environmental costs of Rs40,000 per hectare per year (impacts on watercourses, wildlife, etc.), though regeneration of the forest slowly mitigates these by 2% per year. The illustrative results are shown in Fig. 4.5. Calculation confirms the expectation and visual impression that adding a premium to the discount rate affects costs more than revenues, so that discounted (financial) revenues exceed discounted (environmental) costs. Thus an investment which would be unacceptable (NPV = –£1707 at 3%) becomes acceptable (NPV = +£1395 at 8%) because it is risky. Some authors (e.g. Celec and Pettway, 1979) have argued that risky costs uncorrelated with revenues should be discounted at a lower rate than risk-free costs, but this argument is too rarely recognized.

4.5.4 Risk as a threat to human existence

Unlike individual mortality, the possibility that environmental changes or other events will extinguish the human species affects both present and future generations. Resources saved for the future may never have use for any generation. Unlike threat within a project cycle, this threat is cumulative with the lapse of earth time. If we value future generations only through the benevolent affections of the present generation, there should be no discount on their possible non-existence, since it is a matter of indifference whether in the end they exist or not. On the other hand, if it is accepted both that it matters whether future generations exist and that, given that they exist, their well-being matters in itself, then the possibility of their non-existence should reduce the value ascribed to an uncertain future.

The problem is that by discounting on these grounds, the significance of devastating distant-future environmental changes might appear so trivial that we accept resource management strategies that result in such devastation. For example, Gordon stated that since '[resource] exhaustion will occur in the distant future - present generations may wisely ignore it' (Gordon, 1967, p. 276). Thereby, the probability of future generations not coming into existence increases. This seems dangerously self-fulfilling, especially at the level of policy analysis.

4.5.5 Risk as financial variability

The capital asset pricing model (CAPM) approaches risk more subtly: by considering the variation of a project's returns in relation to the variation in returns to investment generally. Projects whose returns vary inversely with those of general investment reduce the riskiness of the whole portfolio. They are thus even more desirable than risk-free projects, meriting a discount rate lower than the risk-free rate. For example, investment in greenhouse gas abatement measures would pay off well in a future where the returns from many industrial investments are reduced by environmental constraints and by the possible need to pay carbon taxes: but it would fare poorly in a prosperous future where these constraints are overcome.

There is some evidence that forestry returns vary inversely with market returns (McKillop and Hutchinson, 1990), so that a forestry component reduces risk in a portfolio. However, the increasing frequency of extreme events as a result of climate change would probably affect forestry returns in the same direction as returns in other markets. Besides, there are doubts about extending the CAPM to the multi-time-period framework of forestry. The model is derived from observing financial investments with short lead times and steady expected annual income. Although its results are usually interpreted in terms of an extra required annual rate of return, they may as effectively, and more simply, be explained as a multiplier on the required annual amount of return. Thus investment of $1000 in a risky project might be acceptable only if it offers $100 revenue per year, whereas $40 would be acceptable in a risk-free project. By such a reinterpretation, the 2.5 multiplier required on benefits to future generations is just the same as that required on benefits to the present generation, in so far as the riskiness (in the sense of variability) is not expected to increase through time. If a change in variability is predicted, then its nature and requirement for a different multiplier are better calculated explicitly.

4.6 Diminishing marginal utility of consumption

If future generations become much richer than we are, any addition to their abundant consumption will have comparatively low marginal value (Tullock, 1964). For future generations, any sacrifice of consumption (so that resources may be diverted to combat adverse environmental change) is less serious than the same sacrifice of consumption by the present generation (to avoid precipitating adverse environmental change). This is the most ethically appealing justification of discounting.

One recent recount of this argument for discounting combines it with the observation that US citizens will not support increased foreign aid. Schelling (1999) thus questions the case for restraint on greenhouse gas emissions, since this would aid future foreigners, who:

- are expected to be the chief beneficiaries of reduced global warming;
- by assumption will be richer and thus even less worthy than present foreigners.

But whether one group of people can correctly assess the worthiness of another group of people, relative to themselves, is questionable.

Additionally, not all nations share equally in the world's economic growth. Climate change may enhance output in some rich nations, while adversely affecting low-income countries. European forests are benefiting from CO_2 fertilization, but elsewhere desertification is the likely outcome of change. Even where GDP growth has been achieved, increased individual well-being may be compromised by:

- deductions from GDP to meet debt repayments;
- rapid population growth;
- the fact that some apparent growth results from drawing an existing subsistence sector into the money economy, and so amounts to mere reclassification, not gain, in well-being;
- allowances for depreciation of the natural resource base (see Repetto, 1995).

During the past 10 years, the GDP of Bangladesh has grown only fitfully, and the GDP per head hardly at all, despite them including, as a positive figure, remedial and protective works associated with increased fluvial and marine flooding. Thus apparent growth of well-being is doubly illusory: it reflects environmental changes that already compromise well-being, and which may be predicted to grow steadily more severe.

4.6.1 Averaging scenarios

If the mean expectation is for real growth, it remains possible that constraints imposed by environmental change – or the perceived need to avoid precipitating it – may significantly slow, halt or even reverse growth. Fig. 4.6 shows an optimistic scenario based on 5% economic growth and 0% population growth, while the equally likely pessimistic scenario envisages 2% and 1.5% growth respectively.

Damage resulting from climate change is predicted to increase in line with economic growth (Fankhauser, 1995), decreasing the effective discount rate for climate change effects. Marginal utility of income (which can be used to mitigate damage) declines with increasing income per head, the elasticity of marginal utility of income[3] here being –2. The overall effective discount rate is:

$$([\text{economic growth rate}] - [\text{population growth rate}]) \times$$
$$| [\text{elasticity of marginal utility of income}] | - [\text{economic growth rate}]$$

This yields a discount rate of (5% – 0%) × 2 – 5% = 5% for the optimistic scenario, and (2% – 1.5%) × 2 – 2% = –1% for the pessimistic scenario. The mean of optimistic and pessimistic discount rates, 2%, gives a discount factor for climate change costs asymptotic to zero, whereas the mean of optimistic and pessimistic discount factors eventually begins to rise. It is the mean discount factor (weighted according to the probability of each scenario coming to fruition), not the (weighted) mean discount rate that gives the correct mean expected value to future costs (Price, 1997; Weitzman, 1999).

[3] The percentage change in the value of one more unit of income, as income increases by 1%.

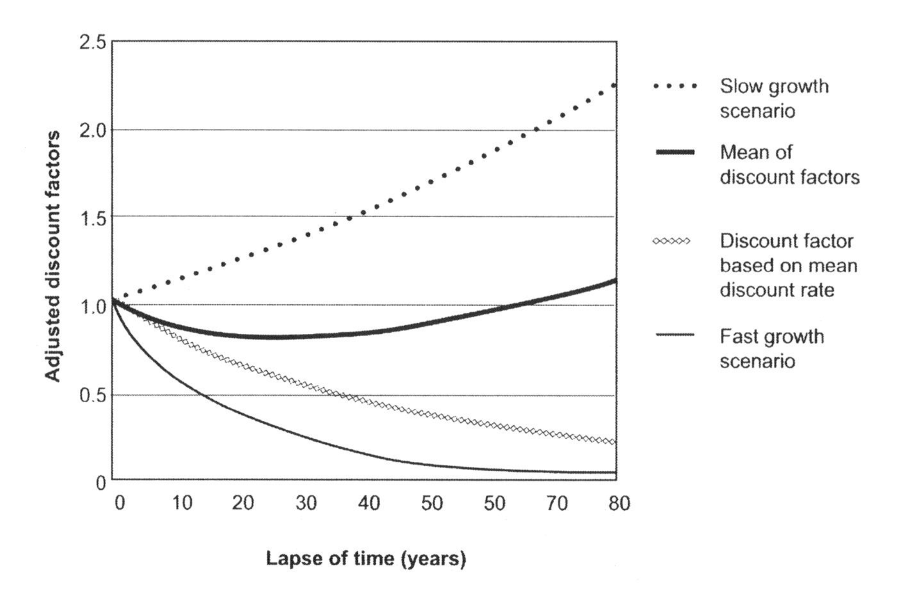

Fig.4.6. Averaging discount rates and discount factors for climate change damage.

4.6.2 Diminishing supply of environmental goods

Even for countries and social groups in which future income and supply of wood products would be assuredly higher, diminishing marginal utility does not apply equally to all products. The supply of many environmental products, especially those like real wilderness, which embody little modified natural capital, has been declining. This has led several environmental economists to propose a lower discount rate, or a relative price rise, for such products (Fisher and Krutilla, 1975; Price, 1978).

Advancing technology could improve access to the remaining products, but even this is restricted by capacity limits at the destination, and by resource and pollution limits on the transportation medium. Ecotourism should not only involve a light footprint on the ecosystem visited, but a preference for visitation that is less demanding on local resources.

Full technological substitution (virtual reality versions of natural environments) is little constrained by environmental changes and limitations of resources. But, unless they are complete substitutes, these technological possibilities cannot make up for shortages of natural prototypes (Price, 2000a).

It has been too readily supposed that differential discount rates favouring environmental goods lead to inefficiency (Nordhaus, 1999), as well as to inconsistencies of the sort discussed earlier. But the exact point of the differential is that uniform discounting understates the value of products that are expected to become relatively scarce in future. Under this condition the true inefficiency is to discount in a way that assumes constant relative scarcities. Alternatively and equivalently, a relative price rise might be predicted for relatively scarce goods and combined with a diminishing marginal utility discount rate. Price (1993, chapter 21) argues that this is a less satisfactory approach.

4.6.3 Diminishing marginal utility and compensation

Compensation to future generations may now be regarded in a new light. Broome (1994) and Price (2000b) argue that, even if a conventional compensation fund for future bodily injuries is established and grows at compound interest, it will not pay adequate compensation. Their argument can be exactly applied to environmental damage, particularly if it significantly affects people's lives. The seriousness of given environmental damage may remain constant through time, but the value of each dollar paid in compensation diminishes in a future of increasing real income. Consequently:

- more dollars are needed to compensate for given damage the further into the future that it occurs; therefore
- a bigger compensation fund is required than that calculated using rate of return on investment as a discount rate; therefore
- the discount rate to establish the required compensation fund must be less than the rate at which the fund is invested.

The discount rate may also decrease with an increase in the period of delay before damage is caused, and may be or become negative, indefinitely so at the point when no compensation is adequate. For example, in Fig. 4.7 required compensation becomes indefinitely large after 2055 (compare with Fig. 4.1). Mathematical proof of this surprising result is given in Price (2000b).

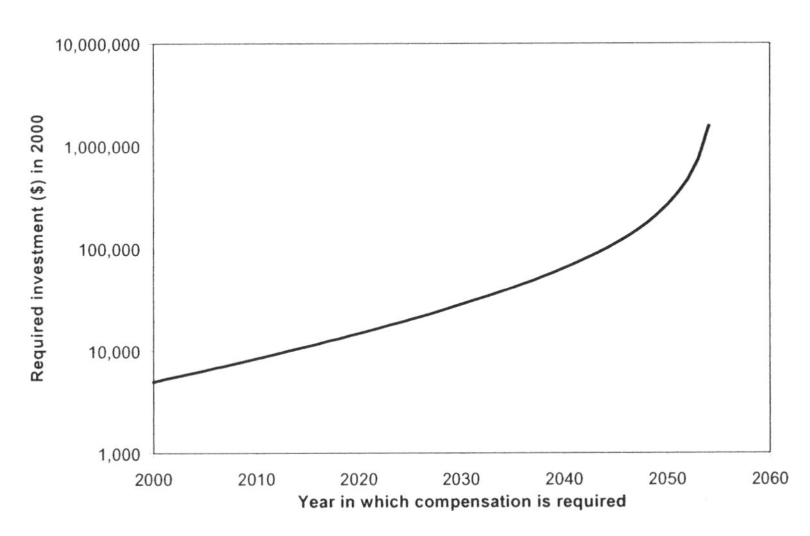

Fig. 4.7. Investment needed to compensate for future environmental damage = \$5000. Rate of return = 6%; rate of income growth = 2%; elasticity of marginal utility of income = –2.

4.6.4 Marginal discounting of totalities

Discounting based on expected diminishing utility of marginal units of consumption is, of course, appropriate only to utilities of marginal change. But, when public environmental goods are damaged, they are damaged not only for marginal consumers and marginal units of consumption, but for every consumer and for each unit of consumption. It is absurd and monstrous to discount, as though they were marginal, the entire value of uniquely cherished and irreplaceable things, of lives and livelihoods. And yet economists, usually without explanation or apology, routinely do it.

4.7 Allocating investment funds under a low discount rate

At low discount rates, more investments have positive net value (at least, before environmental costs are considered) than can be funded. Thus positive NPV does not guarantee that a project is worth funding. This problem is not new, and techniques to deal with it in an intellectually rigorous and realistic way have been developed and presented by Eckstein (1957), Marglin (1963), Feldstein (1964), Little and Mirrlees (1974), Squire and van der Tak (1975), Lind (1982), Price (1990) and others. The opportunity cost of a unit of investment funds is calculated as the discounted value of the stream of consumption, which arises from initial investment and partial reinvestment of revenues. Partial reinvestment leads to growth in the annual flow of consumption. The rate used in discounting is usually based on diminishing marginal utility of consumption and (sometimes, dubiously) on the pure time preference rate. This rate is commonly below the rate of return earned by the investment. The technical formulation of the opportunity cost is:

$$\frac{(1-[\text{reinvestment }\%])\times[\text{rate of return}]}{[\text{discount rate}]-[\text{reinvestment }\%]\times[\text{rate of return}]}$$

However, a problem arises if the point is reached where:

$$[\text{reinvestment }\%]\times[\text{rate of return}]\geq[\text{discount rate}]$$

Here, the discounted value of the consumption stream resulting from investment becomes indefinitely large. While this result is inconvenient, it is not impossible to accommodate it in conventional financial appraisal: the discount rate for actual cash flows is taken as the investment growth rate (= [reinvestment %] × [rate of return]) (Price, 1993; 1996). This does, however, mean that any non-investable benefits and costs have negligible significance, compared with those that do affect the investment stream: non-tradable environmental values, however large they are, cannot signify in a decision if they conflict with generation of investable monies. This unpalatable result does not arise if:

- investment and ensuing reinvestment have a finite duration; or
- environmental costs (or other factors) sufficiently reduce the overall rate of return on marginal investment, either immediately, or in due course (see Dasgupta *et al.*, 1999); or

- the rate of growth of consumption under continual reinvestment is such that a high diminishing marginal utility discount on the products of conventional investment is justified.

If none of these conditions hold, environmental values really may be trivial in relation to the indefinitely valuable future consumption stream that investment makes possible. Any contrary intuition needs to grapple with the problem. Reverting to a high discount rate, just because it gives some relative significance to environmental change, is not rationally defensible.

4.8 Sustainability as an alternative way to protect future generations' interests

It is often argued that adverse effects of discounting on future generations can be prevented by imposing sustainability constraints in project choice and policy analysis: environmental changes can be accepted, provided they do not reduce the well-being of future generations below the datum established by their predecessors. Yet sustainability has been redefined so often as to be almost meaningless except when heavily qualified: common usage does not clearly distinguish 'sustainable' from 'good'. Well-being can be defined so broadly that positive NPV suffices to pass the sustainability test (see the 'weak sustainability' of Pearce *et al.* (1989)), or so precisely that environmental change, of whatever character, is almost certain to infringe unacceptably some cherished aspect of well-being (Price, 2000a). Much hangs on how much substitution – among environmental values, or of other values for environmental ones – is permissible as a means of maintaining well-being.

And yet sustainable development – 'development which meets the needs of the present without compromising the ability of future generations to meet their own needs' (World Commission on the Environment and Development, 1987, p.43) – is only rational in a framework that rejects the discounting of future well-being. On the other hand, if such discounting were justifiable, then it should not matter if 'the ability of future generations to meet their own needs' were to be compromised. Sustainability's commitment to defending what exists could be paralleled in a protocol that does not discount existing intramarginal values, but does discount marginal increments to values. To that extent, these two approaches observe the same distinctions. The overt problem with sustainability is that it only protects existing environmental benefits and fends off detrimental changes of environment. (According to the definition of sustainability, it may not even do this if other forms of consumption are allowed as substitute means of maintaining well-being.) It does not – as not discounting would do – promote enhanced environmental benefits for distant future generations, no matter how efficiently the benefits are produced or how deserving the recipients. Nor does any strict principle of sustainability allow for 'worthwhile retreat', in which a trivial sacrifice of environmental quality might allow great benefits of other kinds to accrue to present or future generations.

Given that the arguments for indiscriminate discounting are in any case so defective, the problematic constraints of sustainability are an unnecessary complication. It could even be argued that by providing an ad hoc defence of future generations' interests, sustainability criteria reduce the perceived need for critically re-evaluating the case against conventional discounting, which so clearly jeopardizes those interests.

With discounting applied only selectively as appropriate, and with other mechanisms in place to allocate investments efficiently, something superior to mere sustainability might be provided to future generations: that is, a weighting of their well-being at parity with the well-being of the present generation.

References

Ainslie, G. (1991) Derivation of 'rational' economic behavior from hyperbolic discount curves. *American Economic Review Papers and Proceedings* 81, 334–340.

Arrow, K., Cline, W.R., Maler, K.-G., Munasinghe, M., Squitieri, R. and Stiglitz, J.E. (1996) Intertemporal equity, discounting and economic efficiency. In: IPCC (ed.) *Climate Change 1995: Economic and Social Dimensions of Climate Change.* Cambridge University Press, Cambridge, pp. 125–144.

Bayer, S. and Cansier, D. (1999) Intergenerational discounting: a new approach. *International Journal of Development Planning Literature* 14, 301–325.

Bellinger, W.K. (1991) Multigenerational value: modifying the modified discounting method. *Project Appraisal* 6, 101–108.

Benzion, U., Rapoport, A. and Yagil, J. (1989) Discount rates inferred from decisions - an experimental study. *Management Science* 35, 270–284.

Broome, J. (1994) Discounting the future. *Philosophy and Public Affairs* 23, 128–156.

Celec, S.E. and Pettway, R.H. (1979) Some observations on risk-adjusted discount rates: a comment. *Journal of Finance* 34, 1061–1063.

Cline, W.R. (1999) Discounting for the very long term. In: Portney, P. and Weyant, J. (eds) *Discounting and Intergenerational Equity.* Resources for the Future, Washington, DC, pp. 131–140.

Cropper, M. and Laibson, D. (1999) The implications of hyperbolic discounting for project evaluation. In: Portney, P. and Weyant, J. (eds), *Discounting and Intergenerational Equity.* Resources for the Future, Washington, DC, pp. 163–172.

d'Arge, R.C., Schulze, W.D. and Brookshire, D.S. (1982) Carbon dioxide and intergenerational choice. *American Economic Review Papers and Proceedings* 72, 251–256.

Dasgupta, P., Maler, K.-G. and Barrett, S. (1999) Intergenerational equity, social discount rates, and global warming. In: Portney, P. and Weyant, J. (eds) *Discounting and Intergenerational Equity.* Resources for the Future, Washington, DC, pp. 51–77.

Department of the Environment (1991) *Policy Appraisal and the Environment.* HMSO, London, 67 pp.

Eckstein, O. (1957) Investment criteria for economic development and the theory of intertemporal welfare economics. *Quarterly Journal of Economics* 71, 56–84.

El Serafy, S. (1989) The proper calculation of income from depletable natural resources. In: Ahmad, Y.J., El Serafy, S. and Lutz, E. (eds) *Environmental Accounting for Sustainable Development.* The World Bank, Washington, pp. 10–18.

Fankhauser, S. (1995) *Valuing Climate Change.* Earthscan, London.

Feldstein, M.S. (1964) Net social benefit calculation and the public investment decision. *Oxford Economic Papers* 16, 114–131.

Fisher, A.C. and Krutilla, J.V. (1975) Resource conservation, environmental preservation and the rate of discount. *Quarterly Journal of Economics* 89, 358–370.

Forest Research (2000) *ForestGALES*. [A computer based decision support tool]. Software from Forest Research, Roslin, Midlothian.

Gordon, R.L. (1967) A reinterpretation of the pure theory of exhaustion. *Journal of Political Economy* 75, 274–286.

Healey, J.R., Price, C. and Tay, J. (2000) The cost of carbon retention by reduced impact logging. *Forest Ecology and Management* 139, 237–255.

Henderson, N. and Bateman, I. (1995) Empirical and public choice evidence for hyperbolic social discount rates and the implications for intergenerational discounting. *Environmental and Resource Economics* 5, 413–423.

Horowitz, J.K. (1996) Environmental policy under a non-market discount rate. *Ecological Economics* 16, 73–78.

Howarth, R.B. and Norgaard, R.B. (1992) Environmental valuation under sustainable development. *American Economic Review* 82, 473–477.

Intergovernmental Panel on Climate Change (IPCC) (1990) *Climate Change: the IPCC Scientific Assessment*. Cambridge University Press, Cambridge.

Kopp, R.J. and Portney, P.R. (1999) Mock referenda for intergenerational decisionmaking. In: Portney, P. and Weyant, J. (eds) *Discounting and Intergenerational Equity*. Resources for the Future, Washington, DC, pp. 87–98.

Kula, E. (1988) Future generations: the modified discounting method. *Project Appraisal* 3, 85–88.

Lind, R.C. (ed.) (1982) *Discounting for Time and Risk in Energy Policy*. Johns Hopkins University Press, Baltimore.

Little, I.M.D. and Mirrlees, J.A. (1974) *Project Appraisal and Planning for Developing Countries*. Heinemann, London.

Livingstone, I. and Tribe, M. (1995) Projects with long time horizons: their economic appraisal and the discount rate. *Project Appraisal* 10, 66–76.

Loewenstein, G. (1988) Frames of mind in intertemporal choice. *Management Science* 34, 200–214.

Luckert, M.K. and Adamowicz, W.L. (1993) Empirical measures of factors affecting social rates of discount. *Environmental and Resource Economics* 3, 1–22.

Marglin, S.A. (1963) The opportunity costs of public investment. *Quarterly Journal of Economics* 77, 274–289.

McKillop, D.G. and Hutchinson, R.W. (1990) The determination of risk adjusted discount rates for private sector forestry investment. *Forestry* 63, 29–38.

Nordhaus, W.D. (1992) The optimal transition path for controlling greenhouse gases, *Science* 258, 1315–1319.

Nordhaus, W.D. (1999) Discounting and policies that affect the distant future. In: Portney, P. and Weyant, J. (eds) *Discounting and Intergenerational Equity*. Resources for the Future, Washington, DC, pp. 145–162.

Partridge, E. (1981) *Responsibilities to Future Generations*. Prometheus Books, Buffalo.

Pearce, D. and Ulph, D. (1995) *A Social Discount Rate for the United Kingdom*. CSERGE, Norwich.

Pearce, D.W., Markandya, A. and Barbier, E.B. (1989) *Blueprint for a Green Economy*. Earthscan, London.

Portney, P.R. and Weyant, J.P. (eds) (1999) *Discounting and Intergenerational Equity.* Resources for the Future, Washington.

Price, C. (1978) *Landscape Economics.* Macmillan, London.

Price, C. (1989) Equity, consistency, efficiency and new rules for discounting. *Project Appraisal* 4, 58–64.

Price, C. (1990) Deforestation and economic criteria. *Project Appraisal* 5, 159–166.

Price, C. (1993) *Time, Discounting and Value.* Blackwell, Oxford; also freely available electronically from c.price@bangor.ac.uk.

Price, C. (1995) Emissions, concentrations and disappearing CO_2, *Resource and Energy Economics* 17, 87–97.

Price, C. (1996) Long time horizons, low discount rates and moderate investment criteria. *Project Appraisal* 11, 157–168.

Price, C. (1997) Analysis of time profiles of climate change. In: Adger, W.N., Pettenella, D. and Whitby, M. (eds) *Climate Change Mitigation and European Land Use Policies.* CAB International, Wallingford, UK, pp. 71–87.

Price, C. (2000a) The landscape of sustainable economics. In: Benson, J.F. and Roe, M. (eds) *Landscape and Sustainability.* Spon, London, pp. 33–51.

Price, C. (2000b) Discounting compensation for injuries. *Risk Analysis* 20, 239–249.

Rae, J. (1834) *The Sociological Theory of Capital.* Macmillan, London.

Rawls, J. (1972) *A Theory of Justice.* Clarendon Press, Oxford.

Repetto, R. (1995) How to account for environmental degradation. In: Adamowicz, W.L., White, W. and Phillips, W.E. (eds) *Forestry and the Environment: Economic Perspectives.* CABI, Wallingford, pp. 3–18.

Sen, A.K. (1957) A note on Tinbergen on the optimum rate of saving. *Economic Journal* 67, 745–748.

Schelling, T.C. (1999) Intergenerational discounting. In: Portney, P. and Weyant, J. (eds) *Discounting and Intergenerational Equity.* Resources for the Future, Washington, DC, pp. 99–101.

Shackle, G.L.S. (1958) *Time in Economics.* North Holland, Amsterdam.

Sidgwick, H. (1874) *The Methods of Ethics.* Macmillan, London.

Squire, L. and van der Tak, H. (1975) *Economic Appraisal of Projects.* Johns Hopkins University Press, Baltimore.

Tenenbaum, S. (1989) Social discounting: retrieving the civic dimension. *Economics and Philosophy* 5, 33–46.

Thomas, A.L. and Price, C. (1999) Landscape valuation of farm woodlands. In: Burgess, P.J., Brierley, E.D.R., Morris, J. and Evans, J. (eds) *Farm Woodlands for the Future.* Bios Scientific, Oxford, pp. 69–79.

Tullock, G. (1964) The social rate of discount and the optimal rate of investment: comment. *Quarterly Journal of Economics* 78, 331–336.

Weitzman, M.L. (1999) Just keep discounting but.... In: Portney, P. and Weyant, J. (eds) *Discounting and Intergenerational Equity.* Resources for the Future, Washington, DC, pp. 23–29.

World Commission on Environment and Development (1987) *Our Common Future.* World Commission on Environment and Development, Oxford.

Deforestation Processes in Temperate and Tropical Forests: An Economic Perspective

5

O. Hofstad

Department of Forest Sciences, Agricultural University of Norway

This chapter addresses deforestation and forest degradation globally. Both economic and social explanations of deforestation are discussed. Deforestation is seen in the context of land allocation to competing uses. Economic externalities may explain why immediate actors do not consider the full range of effects resulting from deforestation activities. It is difficult to value properly the environmental effects of deforestation and forest degradation. The dynamic nature of both forest ecosystems and economic processes complicate the matter further. The irreversible nature of deforestation makes land allocation decisions complex. This chapter discusses the costs associated with intervention designed to regulate deforestation and forest degradation. Both market and intervention failures are quite common, leading to more deforestation and forest degradation in many places than what would be considered socially optimal. The purpose of the chapter is partly to illustrate the magnitude of the problem without dramatising the situation, but primarily to demonstrate how economic theory may contribute to explaining the reasons for deforestation, why deforestation is so difficult to control and reduce, but also to indicate some directions for further actions to contain deforestation and forest degradation where it causes most harm.

5.1 Introduction

It is generally recognized that economic theory has the potential to help explain why deforestation occurs and which processes can be seen to accelerate or retard it. Economic analyses may also indicate how economic conditions influence the rates of deforestation and woodland degradation and

© CAB International 2005.
Forestry and Environmental Change: Socioeconomic and Political Dimensions
(eds J.L. Innes, G.M. Hickey and H.F. Hoen)

which policy decisions are best suited to control deforestation in various situations. While the term 'deforestation' is widely used, it has not always been well defined. Deforestation is generally defined as the abrupt removal of all, or nearly all, of the tree cover (or woody biomass) from forests or woodlands spread over a fairly large and continuous area, either permanently or for an extended period of time. In comparison, forest and woodland degradation generally refers to either a quantitative or qualitative loss of vegetation cover over an extended period of time, thereby reducing the productive capacity of the forest (FAO, 1993). Forest and woodland degradation can be monitored in terms of crown cover, tree density, biomass density and changes in forest species.

This chapter aims to illustrate the magnitude of the problem and demonstrate where economic theory can contribute towards a better understanding of international deforestation and forest degradation issues.

5.1.1 The extent of deforestation

The extent of deforestation is not known with much precision. It has been known for centuries that many forested regions have been cleared for either agricultural or industrial purposes. The UN Food and Agriculture Organization (FAO) started publishing deforestation estimates around 1980 in response to the growing concern among conservationists, foresters and other stakeholders. While the quality of these estimates has often been criticized, their accuracy has undoubtedly improved over the years. The FAO's most recent estimates (FAO, 2001a) referred to forest cover change between 1990 and 2000. To demonstrate the likely magnitude of this change, a selection of summary figures is shown in Table 5.1.

It should be noted that the definition of 'forest' was different for developing and developed countries. In developed countries, the FAO defined 'forest' as land with crown cover greater than 20%, while in developing countries, crown cover was set at 10%. Furthermore, the FAO did not strictly define deforestation. Nevertheless, it is generally agreed that deforestation is a more widespread phenomenon in the tropics than in the temperate zones of the world. During the 1990s, the total loss of natural forests (i.e. deforestation plus the conversion of natural forests to forest plantations) was estimated to be 16.1 million hectares per year, of which 15.2 million hectares occurred in the tropics. There are, however, substantial variations in deforestation rates within and between different regions. For example, the annual deforestation rate in the worst performing African nation (Burundi) was 9%, compared to only 1.8% in the worst performing Asian nation (Nepal). The difference between Cuba and Haiti (both in Central America) provides another striking difference. Between 1990 and 2000, the forest area in Cuba increased by 1.3% per annum, while Haiti's decreased by 5.7% per annum.

Table 5.1 shows that deforestation is a serious problem facing nations in many parts of the world.

Table 5.1. Annual change in forest cover 1990–2000 (source: FAO, 2001a)

Region	1000 ha	%
Africa	- 5262	- 0.78
Asia	- 364	- 0.07
Europe	881	0.08
Oceania	- 365	- 0.18
North and Central America	- 570	- 0.10
South America	- 3711	- 0.41
Global balance	- 9391	- 0.22

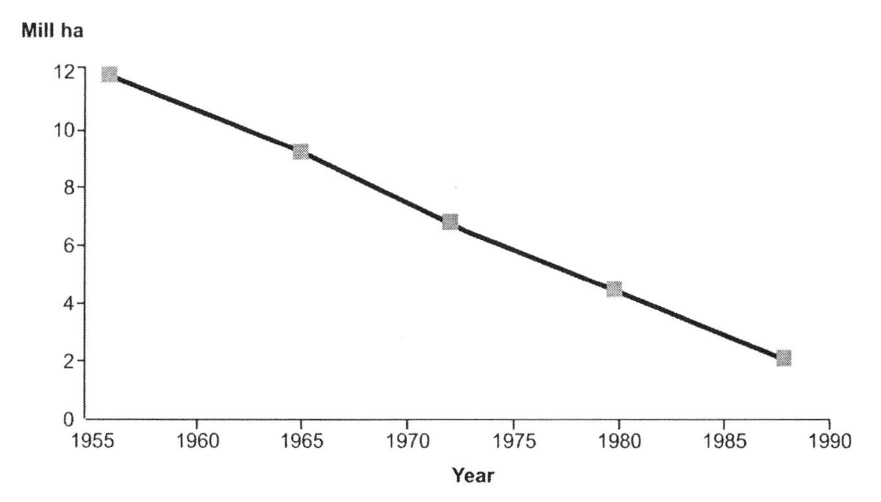

Fig. 5.1. Area of closed forest in Côte d'Ivoire from 1956 to 1988 (sources: FAO, 1981; Rietbergen, 1989).

In recent years, deforestation was seen as an entirely negative development that had the potential to lead to the complete destruction of forests over large regions. Fig. 5.1 shows the reduction in closed forest area in the African nation of Côte d'Ivoire between 1956 and 1988. Based on this dramatic diagram, it appears that the rate of reduction would soon lead to the complete deforestation of the country. However, several modifications to this simplistic picture are required:

1. Côte d'Ivoire is a small country on the fringe of the West African belt of rainforest. It may, therefore, be acceptable for most of the rainforest in Côte d'Ivoire to be cleared if a sufficient area of similar forest is conserved in other parts of its range.
2. The 1995 statistics on forest area in Côte d'Ivoire show that there are more than 5 million hectares of forest registered in the country by the FAO (1997). It therefore seems that we are dealing with forest degradation issues rather than deforestation.

3. The assumption that linear deforestation trends will continue until all forest is gone is seriously flawed.

Fig. 5.2 illustrates three deforestation scenarios. The first is a simple linear trend. The second is a trend based on the assumption that, if population grows exponentially, and agricultural productivity remains constant, forest area will be reduced exponentially. The third trend is sigmoid, assuming that there are economic feedback loops that will slow down the process of deforestation as the forest area decreases. The sigmoid curve is better suited for analysing developments over longer periods of time. In some countries, particularly in the temperate zone (e.g. USA, Britain and Denmark), forest area has increased after a long period of deforestation (Grainger, 1993).

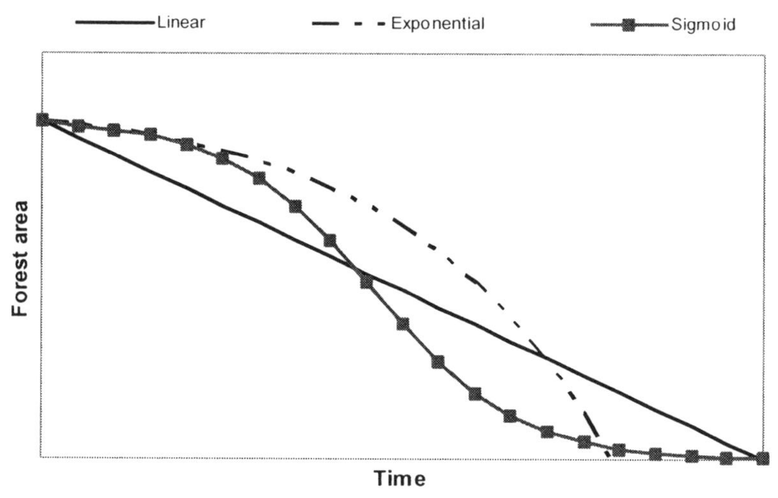

Fig. 5.2. Three scenarios of deforestation.

Table 5.2. Forest cover change matrix for tropical Africa (000 ha) (source: Singh, 1993)

	Area of classes in 1990										
Classes in 1980	Closed forest	Open forest	Shifting cultivation	Fragmented forest	Shrub	Short fallow	Other cover	Water	Planted forest	Total 1980	%
Closed forest	16,781	382	83	292	10	524	248	-	-	18,319	24.0
Open forest	24	10,049	48	371	13	118	397	-	1.4	11,022	14.0
Shifting cultivation	8	15	557	2	4	52	29	-	-	665	0.0
Fragmented forest	40	1	8,089	8	6	294	-	-		8,461	11.0
Shrub	1	11	-	1	3,878	-	164	-	-	4,055	5.0
Short fallow	8	11	10	2	-	2,255	53	-	-	2,339	3.0
Other cover	17	38	11	63	87	34	26,452	51	-	26,753	35.0
Water	1	-	-	1	-	3	82	2,960	-	3,046	4.0
Planted forest	-	-	-	-	-	-	-	-	4.6	5	0.0
Total 1990	16,863	10,546	709	8,820	3,999	2,992	27,718	3,012	6	74,665	
%	22.6	14.1	0.9	11.8	5.4	4.0	37.1	4.0	0.0		100

It was mentioned that the Côte d'Ivoire case is complicated by the fact that what had been referred to as deforestation was, in many cases, better described by the term degradation. Table 5.2 shows a summary of the land cover change in Africa between 1980 and 1990 (Singh, 1993). It shows that large areas of closed forest were lost, but a substantial proportion of this was due to forest conversion (to open forest) or fragmentation rather than deforestation. Table 5.2 also shows how the degradation process was continued through some of the open and fragmented forest being changed to fallow and 'other cover' land. In this case, there was a net loss of open forest, but a net gain in the area of fragmented forest.

The view that deforestation is, by necessity, a negative development can be nuanced by economic analysis. Based on economic theory, people can be seen to engage in deforesting activities to improve their welfare, at least in the short term. From a peasant's point of view, both the individual priorities and the actual situation will result in the best option being to clear the forest and even overexploit remaining woodlands. Deforestation and woodland degradation are important for individual welfare. To condemn peasants for being the agents of deforestation is both arrogant and a mistake. To look upon deforestation as a vicious process that, in all cases, should be combated is equally superficial. Deforestation and degradation can, in many situations, represent a small local cost compared to the benefits they offer poor people.

5.1.2 Explaining deforestation

What is perceived to be an explanation for a phenomenon like deforestation depends, to a large extent, on the scientific perspective. Furthermore, theories differ widely in the classification of immediate and underlying causes. Some may feel that a regression analysis that includes economic, biological and physical variables can 'explain' why large areas are deforested. Others may feel that this method offers no explanation as long as the social mechanisms are not explained in terms of economic or sociological theory. Economists may feel that market failure is an underlying cause of deforestation, while a psychologist may think that human greed (net present value (NPV) maximization at high interest rates) is a more fundamental underlying cause.

Palo and Lehto (1996) computed a number of regression models designed to explain the deforestation rate between 1980 and 1990, based on FAO (1993) data. The best model explained only 25% of the variation in deforestation rates in 70 tropical regions. The independent variables that they used were population density, national population growth rate, gross national product (GNP)/capita and 'wet' ecological zone area. Deforestation rates increased with all of the variables, except 'wet' area. The population growth rate at the national level had the highest deforestation elasticity (2.25).

This method of explaining deforestation is aggregated. It gives a general hint at some of the factors affecting deforestation across a wide variety of ecological and economic situations, but it can hardly improve our understanding of how deforestation happens and what the driving mechanisms are. Various authors have distinguished between the 'immediate' and 'underlying' causes of deforestation.

Palo (1987) attempted to describe a system of processes that contribute to increased or reduced deforestation rates. These processes span from general international relationships affecting the distribution of technology and purchasing power, to local conditions in tropical countries like poverty, public administration, cultural traditions, rainfall patterns and soils. According to Palo (1987), the most important immediate driving forces behind deforestation are:

- *agriculture*, where area productivity and the sustainability of farming systems are determining factors;
- *colonisation*, where public programmes to alleviate poverty by moving people to sparsely populated regions are central;
- *shifting cultivation*, an efficient farming system when population density was low, but one which induces deforestation when population density increases rapidly without an accompanying increase in area productivity in agriculture (e.g. by a shift to permanent agriculture);
- *grazing*, which may lead to degradation of vegetation when the number of animals surpasses carrying capacity of the land;
- *wood fuel supply*, which may lead to degradation when removals surpass increment of the forest for prolonged periods;
- *timber exploitation*, which may alter forest ecology dramatically, and often increases the accessibility of forest areas to shifting or permanent agriculture;
- development of *infrastructure*, which may, at least locally, result in flooding or clear-felling of fairly large forest areas.

Angelsen and Kaimowitz (1999) summarized the results of many modelling studies. As a result, they doubted some of the commonly held assumptions about the causes of deforestation:

- *The population thesis:* (Neo-)Malthusian explanations that population growth is a key driving force of deforestation find only weak support in the models.
- *The poverty thesis:* There is little empirical evidence on the link between deforestation and poverty (and associated phenomena such as high discount rates).
- *The win-win thesis:* The idea that economic growth and the removal of market distortions are good for people and forests finds limited support in the review.
- *The making-the-forest-valuable thesis:* The review suggests that lower timber prices should both reduce logging activities and also lessen agricultural encroachment stimulated by logging.
- *The tenure security thesis:* Increasing tenure security and giving land titles to farmers have contradictory effects. In many areas forest clearing gives farmers a claim to property rights over land, and increasing the security of such claims may stimulate a land race.
- *The intensification thesis:* How improvements in agricultural technology affect forest clearing cannot be determined, a priori, without information regarding the type of technology, and output and factor market elasticities.

The impact of different driving forces obviously varies from one region to another and from one political situation to another. Deforestation and woodland degradation are therefore the result of complex processes, and these processes, to a large extent, are social (i.e. they have to do with the organization of societies, rather than conditions determined by nature or geography).

5.1.3 A fundamental economic view of deforestation

Grøn (1931) distinguished between submarginal and intramarginal forest resources. He defined timber as submarginal when the cost of harvesting is higher than the market price. Simplified, one may express this as: $P < C$, where P is the price of timber, and C is the cost of harvesting. When $P > C$, we have a situation in which timber can be sold profitably after harvesting. Timber resources would then be termed intramarginal. Obviously, the marginal situation would only exist when $P = C$.

Similarly, one may analyse the profitability of agriculture or animal husbandry. If it is profitable to clear forests for different purposes, we may call them submarginal. On the other hand, when cultivation or the clearing of forests for pasture is profitable, the land may be termed intramarginal. This analysis leads to four different situations concerning the use of forest areas (see Fig. 5.3).

Fig. 5.3 shows that the forest will be left unexploited in places where it is not profitable to harvest timber or clear it for agriculture or pasture. In other places harvesting timber may be profitable while cultivating the land is not. The land may then be actively reforested (sowing, planting, etc.) or left to nature. In this latter case new forest may establish itself, but it may take a long time. In most cases the secondary forest will differ in some way from the pristine forest. As an example of such a situation, Grøn (1931) mentioned the large forest areas that were clear-cut and left for fallow in the USA at the beginning of last century. If agriculture is intramarginal and timber is submarginal, the forest will be cleared and cultivated without the timber being taken care of. This is a common situation across large areas within the tropics. Finally, a situation where both timber production and agriculture or animal husbandry are profitable (intramarginal) in the same area is possible. As an example of such a situation, Grøn (1931) mentioned the conflicts over forest clearing in Switzerland.

In the free market case, areas will be used in the most profitable way. If there are clear and significant externalities, regulating the markets (for timber, agricultural produce or land) may be optimal, although this also depends on the costs of regulation and the likelihood of intervention failures (Hofstad, 1997a).

5.2 Valuing the costs of deforestation

The benefits of deforestation, particularly in the short term, are evident and easily quantifiable to the immediate actors (peasant farmers, loggers, ranchers or plantation managers). The costs of deforestation, particularly the long-term costs, are not equally evident and, in many cases, are impossible to measure in monetary terms. Even the loss of future wood production can be difficult to measure economically, although many landowners are convinced that this is worth much less than what they would gain in future agricultural production. In practice, many forests are open access (although they may be owned pro forma by the state or some cooperative). This means that there is no market for standing trees in those areas, and the price of timber will not include the opportunity cost of forested land. Those who benefit from deforestation and, therefore, take action to bring about a reduction in forest area in many cases do not bear the costs of their actions. In such cases, stakeholders other than the decision maker must inform us about the costs of deforestation.

	Timber	
Land	P < C	P > C
P < C	Untouched	Cut timber, leave land fallow!
P > C	Clear forest, cultivate land!	Competition

Fig. 5.3. A matrix of land use in different situations of prices and costs.

According to Campbell *et al.* (1989), the Save valley in Zimbabwe is seriously silted so that it creates many problems for cropping and water supply. However, there are difficulties in determining to what extent this problem results from deforestation (Whitlow, 1988). Despite this, there is no doubt that the upper parts of the watershed are seriously deforested and that the remaining woodlands are also degraded. If geographers and hydrologists were able to determine the physical effect of deforestation, as opposed to the effect of general erosion of the old African continental shelf, economists might be able to value the reduction in crop yields. But how should the welfare effects of erratic water supply be valued? Several indirect and direct valuation methods (Mitchell and Carson, 1989; Hanemann, 1994; Bateman and Willis, 1999) could be employed in this case by attempting to estimate people's willingness to pay for remedial measures. In the Save valley, the inhabitants would have little problem understanding this issue and they would probably be able to compare the problems of water supply with some other element of their utility function (e.g. a school or health clinic).

Valuation becomes more difficult when the physical and cultural distance increases between the direct beneficiaries of deforestation and those who experience the losses. In central Argentina, the farmers that clear the Chaco forest for agriculture and cattle grazing (Fernández-Juricic, 1999) may have difficulties understanding that people in Germany are concerned about the habitat of blue-fronted Amazon parrots (*Amazona aestiva*) and the loss of biodiversity in general. Nevertheless, economists would employ contingent valuation to estimate the willingness to pay among the German public, so that the optimal resource use could be established by taking both the benefits to rural people in Argentina and the costs to German environmentalists into account. Certainly there would be more difficulties presenting the ecological effects of Latin American deforestation in an unbiased way to the German public than in explaining silting of dams to the people of Save. Despite this, the valuation exercise might still be valid and hint at the magnitude of the costs of deforestation. Another limitation relates to authority. In the example from Zimbabwe there is a principal in charge of both those who benefit and those who suffer (i.e. the Zimbabwean government). In the example from Argentina there is no such authority. Instead, there are weak mechanisms (international media, demonstrations, international negotiations and environmental conventions) to link the government of Argentina, which regulates the exploitation of the Chaco forest, to the concerned public in Germany and other affluent societies (see also Sandler, 1993).

Some arguments against deforestation, particularly in diverse rainforests, refer to the loss of uncertain future benefits such as useful chemicals, genes or organisms that have not yet been discovered. Attempts have been made to value such costs in monetary terms (Peters *et al.*, 1989; Loomis and White, 1996; Simpson *et al.*, 1996), but it seems that such estimates are very difficult to validate and to a large extent depend on arbitrarily chosen assumptions. The argument that such costs might exist seems valid, but the economic valuation appears quite hypothetical. There are, however, examples of 'debt for environment swaps' which demonstrate that a market can be established for some collective goods 'produced' in tropical forests and 'consumed' by people in other parts of the world (UNSO, 2000).

5.3 Optimal deforestation and degradation

We can use standard resource economics to demonstrate that there are optimal levels for both deforestation (Fig. 5.4) and woodland degradation (Fig. 5.5) in certain regions. If the decision-maker is a unified entity (i.e. a single farmer, a single family, the omnipotent state), the marginal benefits of alternative land uses can be balanced against the marginal costs of deforestation; and the marginal benefits of a new combination of multiple uses can be balanced against the marginal costs of woodland degradation.

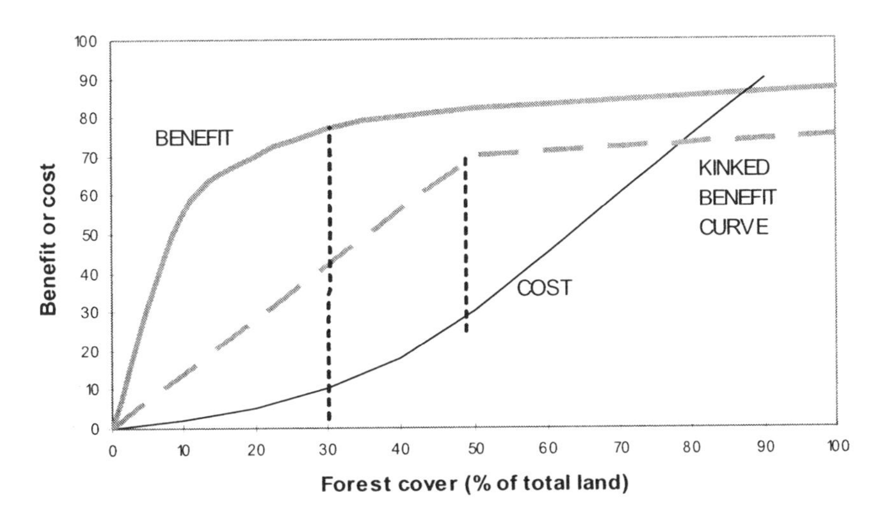

Fig. 5.4. Optimal deforestation with two benefit functions (source: Pearse *et al.*, 1990).

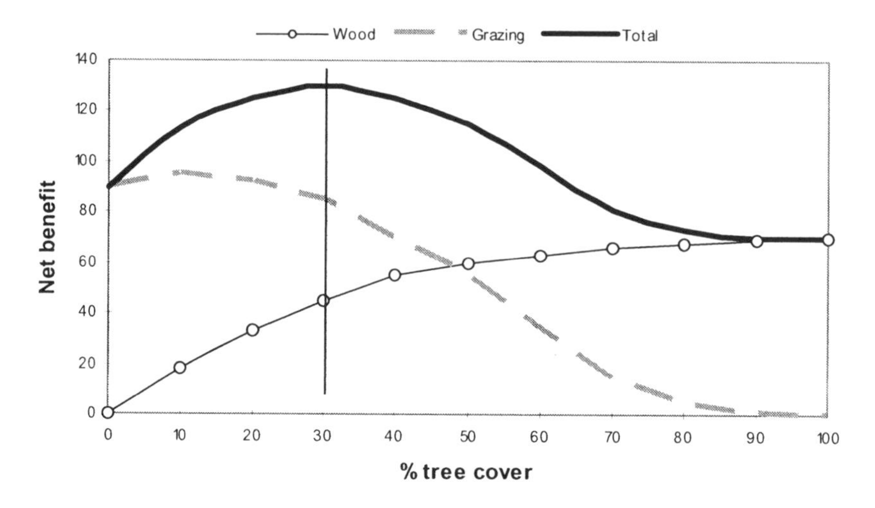

Fig. 5.5. Optimal tree cover in multiple uses of woodland.

5.3.1 Static optimum

The neoclassical explanation why the immediate agents find deforestation profitable while others consider deforestation a disaster is demonstrated by modelling the land-use decision as follows (Horne, 1999):

$$Q\,(t) = f\,[Z(t), S(t)] \qquad (1)$$

where Q(t) = agricultural production function, Z(t) = input package including labour and capital and S(t) = stock of woodland and soil resources. $\partial Q/\partial S > 0$, but $\partial^2 Q/\partial S^2 < 0$ since the first cleared woodland has the highest agricultural productivity. The stock effect can be interpreted as an increase in the cost of agricultural production as resources are depleted. The stock of woodland is reduced as production increases, therefore $\partial S/\partial Q < 0$.

The net benefit of a social manager would be:

$$NB_S = p\,Q(t) - w\,Z(t) + E[S(t)] \qquad (2)$$

where p = price of agricultural output, w = price of input package and E[S(t)] = environmental control provided by S(t). The on-site effect of erosion is incorporated in S(t), while E(t) denotes the off-site effect along with other environmental services. The ability of environmental control is an increasing function of the stock of woodland, $\partial E/\partial S > 0$.

An individual farmer may place zero value on environmental externalities. E[S(t)] is thereby omitted from the decision-making. The farmer would maximize:

$$NB_F = p\,Q(t) - w\,Z(t) \qquad (3)$$

When marginal net benefit of agricultural production becomes negative, the remaining land area would be conserved under forest cover. The net benefit to a social manager is lower than that to a farmer because of the additional cost of losing environmental services.

A more sophisticated treatment of the allocation of land between forest and non-forest use when the forest may deliver environmental services was described by Parks *et al.* (1998).

5.3.2 Dynamic optimum

We may also investigate the allocation of land by using a two-period model. The following symbols are used:

$A =$ total area, cultivable woodland plus cultivated fields,
$a_t =$ total cultivated area in period t,
$g'_t(\Delta a_t)$ = marginal utility of agriculture if Δa_t units of land are allocated for cultivation in period t,
$f'_t(-\Delta a_t)$ = marginal utility of forestry if $-\Delta a_t$ units of land are allocated for forestry in period t,
$r =$ discount rate.

Assume both $g'_t(\Delta a_t)$ and $f'_t(-\Delta a_t)$ to be decreasing. If a_0 and a_1 are cultivated in periods 0 and 1, respectively, total utility can be expressed by:

$$U = \int_0^{a_0} g'_0(\Delta a_0)\mathrm{d}\Delta a_0 + \int_0^{A-a_0} f'_0(-\Delta a_0)\mathrm{d}(-\Delta a_0) + (1+r)^{-1}\left[\int_0^{a_1} g'_1(\Delta a_1)\mathrm{d}\Delta a_1 + \int_0^{A-a_1} f'_1(-\Delta a_1)\mathrm{d}(-\Delta a_1)\right]$$

The conditions are:

$$0 \le a_0 \le a_1 \le A. \qquad (1)$$

$a_0 \le a_1$ is included since clearing of woodland is considered an irreversible process in this example. This is a reasonable assumption in the short run, and can be justified even in the long run if agricultural mining degrades the land seriously. There is also the assumption that arable land is a limiting factor of production, therefore no land is lies idle.

By defining the Lagrangian function,

$$L = \int_0^{a_0} g'_0(\Delta a_0)\mathrm{d}\Delta a_0 + \int_0^{A-a_0} f'_0(-\Delta a_0)\mathrm{d}(-\Delta a_0) + (1+r)^{-1}\left[\int_0^{a_1} g'_1(\Delta a_1)\mathrm{d}\Delta a_1 + \int_0^{A-a_1} f'_1(-\Delta a_1)d(-\Delta a_1)\right] + \lambda_0 a_0 + \lambda_1(a_1 - a_0) - \lambda_2(a_1 - A)$$

we can find the conditions that maximise U, given (1). If both a_0 and $a_1 > 0$, and therefore $\lambda_0 = \lambda_2 = 0$, these conditions are:

$$g'_0(a_0) - f'_0(A - a_0) = \lambda_1 \qquad (2)$$

$$g'_1(a_1) - f'_1(A - a_1) = -\lambda_1(1+r) \tag{3}$$

$$\lambda_1(a_1 - a_0) = 0 \tag{4}$$

If forestry's importance, compared with agriculture, does not increase from period 0 to 1, $\lambda_1 = 0$ (condition 4), the distribution of the area in both periods shall be such that marginal utility of agriculture equals marginal utility of forestry (Fig.5.6).

The optimal solution, presented in Fig. 5.6, shows an increase in the cultivated area from a_0^* to a_1^*. If, on the contrary, forestry's importance does increase compared to agriculture, $\lambda_1 > 0$ and consequently $a_1 = a_0$. The optimal distribution is given by conditions 2) and 3) such that the difference between marginal utility of agriculture and forestry in period 0 equals λ_1, the shadow price of the irreversibility restriction. In period 1 the same difference should be $-\lambda_1(1+r)$.

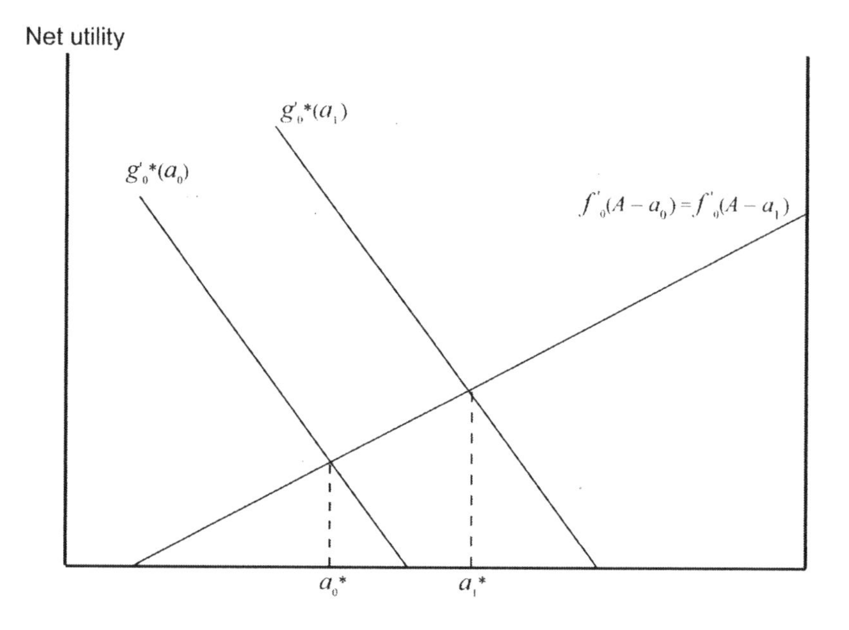

Fig. 5.6. Optimal land allocation with increasing relative utility of agriculture

The negative sign in period 1 shows that the allocated area for agriculture in this period is too large from a static point of view (see Fig. 5.7). Viewed in isolation for period 0, the optimal distribution would be the one that resulted in equal marginal utility of agriculture and forestry. In this case, a_0^+ should be used for agriculture, while $(A - a_0^+)$ should be kept for forestry. Since the reallocation of land from forestry to agriculture in this case is an irreversible process, it is necessary to avoid clearing land that might be optimal now but may lead to insufficient forest land later. If, in Fig. 5.7, more than a_0^- is used for agriculture, this would lead to future utility reductions. The benefit of further woodland clearing in period 0 must be balanced against the future costs. The balance is reached at a_0^* for agriculture and $(A - a_0^*)$ for forestry. Then the marginal benefits of clearing exactly correspond to the marginal costs in period 1. Ehui and Hertel (1989) have applied optimal control theory to compute an optimal path of deforestation in a two-sector (forestry and agriculture) model for Côte d'Ivoire where the same reasoning is fundamental.

As in other parts of the world, the productivity of agricultural land in Asia has increased remarkably as a result of the 'Green Revolution'. For example, in India the yield of food grains per hectare doubled over 30 years (DAC, 2000). According to Angelsen and Kaimowitz (2001) the rapid rise in rice yields associated with the Green Revolution in Asia reduced pressure on forests by pushing down rice prices, which discouraged upland rice production. From 1940 to 1990, the forest area in India was reduced from 75 to 65 million hectares (Raja, 1996). Since then, a further decline in total forest area has been avoided (FAO, 1997). This supports Swaminathan's (2000) argument that the Green Revolution has also saved forests.

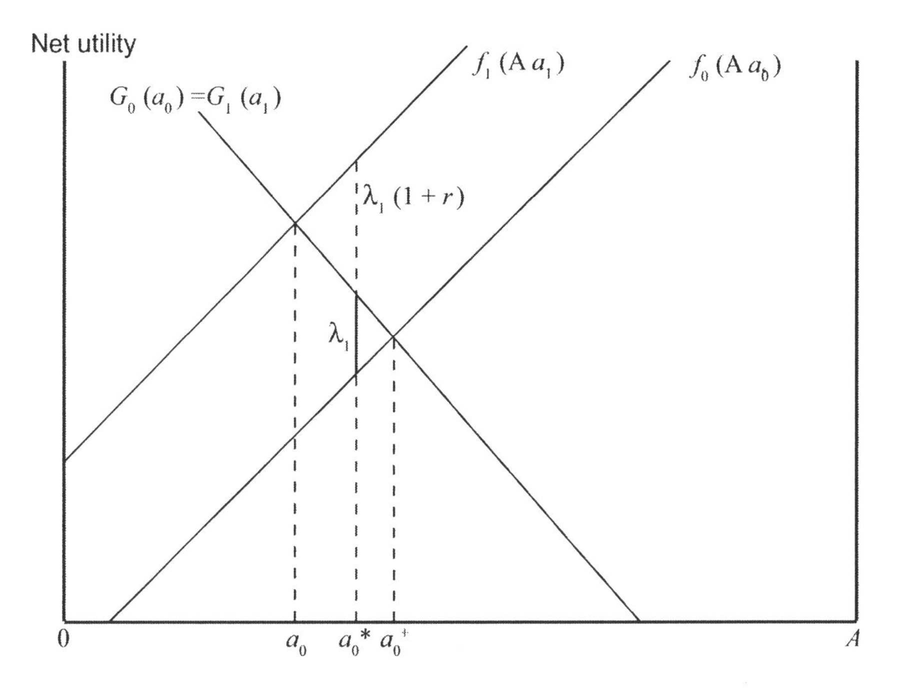

Fig. 5.7. Optimal land allocation with increasing relative utility of forestry.

However, anything that makes agriculture more attractive runs a risk of reducing forest area. It may encourage or permit existing farmers to clear additional land or attract new farmers. Technologies that improve the profitability of activities that do not require much labour, such as cattle ranching or mechanized soybean production, are especially problematic. So is the introduction of new export crops such as cocoa, bananas and rubber in areas undergoing rapid immigration (Angelsen and Kaimowitz, 2001).

5.3.3 Common forest resources, game theory and dynamic solutions

In many cases, particularly in developing countries, the decision maker is not in single command of forest resources. Actual tree cover in a region may, therefore, deviate considerably from what was previously defined as 'optimal'. In such cases, there is hardly any incentive to take the effect of present timber felling on potential future felling into consideration. Therefore, individual exploiters would tend to remove the quantity that maximizes the difference between the unit value of standing timber, h, and the costs of extraction, c. When we want to study the relationship between extraction and increment, and thereby the development of the standing volume, S, the number of exploiters, n, and the quantity removed by each exploiter, x, must also be considered.

Fig. 5.8 illustrates the development of such a system. When the standing volume is high, the number of exploiters low, and the value of wood high (phase B), removals are high per exploiter, but less than the increment in total. Consequently, standing volume increases, profits are high and new exploiters take up wood exploitation. After some time, removals increase beyond the increment (phase C). New exploiters continue entering business; there are still profits to be earned; however, the standing volume is now reduced. Costs increase, partly because of longer travelling distances, and some exploiters start leaving wood production (phase D). Removals then fall at the same time as standing volume falls. Exploitation implies losses to many exploiters who continue to leave business. After a while increment becomes higher than removals and standing volume increases again (phase A). Costs are reduced, and new exploiters are attracted.

The cycle is repeated until a stable situation, E = (n_s, S_s), is reached. Such a stable situation exists if $S_s > S^*$, but not necessarily if $S_s < S^*$. If the initial situation lies close to the stable point, E, the system may converge towards this point, but if the initial situation lies far from the stable point, E, we cannot exclude the possibility that the system may move further and further away from it and the forest eventually disappear. The last possibility is most likely when the forest is densely stocked, but will not happen at all, $c \rightarrow \infty$, when $S \rightarrow 0$. An increase in the price of fuel wood, and the value of standing timber, h, implies a negative shift of both $dS/dt = S = 0$ and $dn/dt = n = 0$, and will, therefore, shift the stable point, E, towards the origin.

Wirl (1999) points out that cyclical policies, (i.e. clearing of forests followed by reforestation and so on) may also be socially optimal. This may be due to the growth function of trees and the costs of clearing and reforestation.

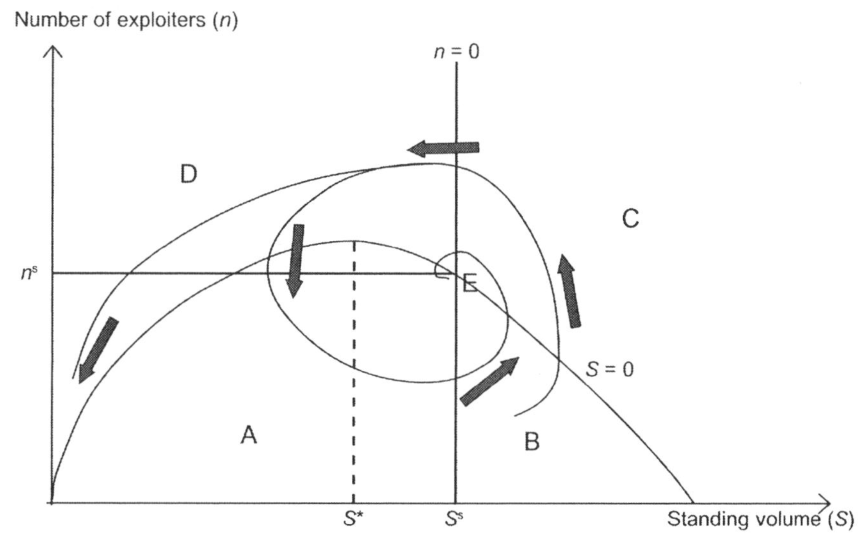

Fig. 5.8. Unregulated exploitation of woodlands

5.4 Dynamic spatial pattern

In some locations, one may easily observe how deforested and degraded areas spread outwards from urban centres over time. To economists who have studied von Thünen's (1875) theory of land allocation, this is not surprising. For simplicity, let forest resource exploitation be modelled in economic terms as if forests are non-renewable. This approach is particularly reasonable when natural regrowth is slow, or when the secondary forest is significantly different or poorer (in some sense) than the original. Furthermore, assume that the demand function is stationary and depends solely on price, and that price elasticity is unitary. Then

$$Q = \alpha / P \qquad (1)$$

where Q = quantity demanded, P = unit price of charcoal delivered at the market, and α = constant. For a practical example, consider Dar es Salaam in 1994. Q was observed to be 2.1 million m^3 of round wood, and P was 1800 Tsh/bag (Hofstad, 1997b); then α = 3780.

If woodland density is homogenous, the area cut, A, determines supply. Observations in 1994 indicate that the average volume harvested per hectare was 65 m^3. If Q = 2.1 million m^3, A = 32,300 hectares. It seems justified to assume that the area harvested is determined by transport costs. Given an initial frontier, charcoal operators will expand that frontier along equal cost transients. If transport costs are the same per km in all directions, these transients represent circles. The radius of these circles is C_t/a, where a is the cost per km, and C_t is transport cost in period t. Price will equal costs:

$$P_t = \beta + a \cdot r = \beta + C_t \tag{2}$$

where r is distance in km. The area harvested in period t is given by the difference in area between the last period's and this period's two circles:

$$A_t = \pi (C_t/a)^2 - \pi (C_{t-1}/a)^2 \tag{3}$$

The solution in each period is for demand to equal supply:

$$3780 / (\beta + C_t) = 65\, \pi\, [(C_t/a)^2 - (C_{t-1}/a)^2] \tag{4}$$

This is a third-order differential equation that is difficult to solve explicitly. It is a special case of a non-renewable resource with a rising extraction cost. If we let population grow while family income remains constant, one may numerically find a path, characterized by price and quantity, which equals supply and demand over time. This is not too difficult, and may, if based on historical figures, be seen as an explanation of previous deforestation or degradation. If the analysis is based on expected future population growth, the path may be seen as a prediction of future deforestation or degradation.

Of course, the spatial pattern found in reality is normally much more complex than predicted by von Thünen's model of homogeneous land use. When land suitability for different uses varies discontinuously from one location to another, the land-use pattern becomes more like a complicated mosaic than concentric circles – as modelled above. Lambin (1997) developed methods to detect various deforestation patterns from satellite images.

5.4.1 Regulating deforestation

Let us assume that it is possible to determine the optimal path by maximizing the net present social value of forests and woodland over a long period of time. We can then compare this path to the one resulting from forest exploitation, clearing and cultivation practices that are undertaken by a multitude of individual agents under various economic externalities. We can then consider whether it might be worthwhile for us to impose taxes on forest clearing, timber harvesting or wood consumption to bring the non-regulated situation in line with the social optimum. This situation implies a consideration of both the non-optimality loss, represented by the difference between the two paths, and the cost of imposing the taxes. We should also try to estimate the amount of tax required to achieve the objective of optimal resource use by simulating the effect of various tax regimes (Baumol and Oates, 1988).

Having said this, we should also keep in mind that property regimes, or sets of user rights, determine the economic externalities (Bromley, 1991) that make it more profitable for the immediate agents to deforest than it is for the society as a whole. For example, if my farm included both the upper and lower parts of a watershed, then it may be less profitable for me to clear the forest on the upper slopes than if my property were situated solely in the upper part (for a more elaborate example, see Herfindahl and Kneese, 1974, pp. 327–329). Therefore, redefining land tenure or the regime of user rights may be a more effective regulatory measure than imposing taxes on agents with the present set of rights. From a practical point of view, however, land reform and other alterations of management regimes may be very hard to implement. Economically efficient solutions may not be politically feasible.

The task of designing an optimal mix of land tenure, user rights, taxes and subsidies is obviously not a simple one. However, many governments, both in the tropics and in the temperate zones, try their best to implement policies designed to regulate forest and woodland exploitation and clearing. Many of these regulating agents (principals) consider deforestation to be one of the most serious processes that they need to address. It is generally recognized that deforestation is more common in poorer countries than in the richer ones. In poor countries the tax base is generally small. Consequently, the budgets available to forest regulating agencies are also small. Financial considerations may, therefore, make socially and economically profitable regulation impossible. Put simply, there might not be funds available to cover the costs of imposing effective tax regimes on deforestation, although the society would be better off if deforestation were halted and the costs of regulation covered.

During 1993 forest royalties and fees equal to about US$11 million were collected in Ghana. The same year, the running of the Forestry Department that collects these taxes cost US$17 million. In 1996 the tax regime had been modified so that the equivalent of US$31 million was collected, while the cost of management was US$18 million (Treue, 2001). In general, the relevant authorities are often very reluctant to transfer funds from other sectors to finance regulatory measures within the forest sector. Such compartmentalization of public services may lead to suboptimal solutions. On the other hand, limiting the expenses of a forestry department to the amount of tax collected within the same sector is probably a better expression of the small political importance of forests rather than relying on embellished forest policy statements.

All such tax regimes would need to make forest exploitation or clearing less profitable than in the non-regulated situation. Tax evasion would – if it were free of cost – make timber harvesting, agriculture or animal husbandry more profitable than legal operations. The immediate profit of tax-evading agents is a function of timber volume removed or land area cleared, prices of products, avoidance costs and the expected value of fine payments. The costs associated with avoidance are incurred through operating at night, operating in generally inaccessible places and the payment of bribes. The probability of detection is partly a function of costs spent on avoidance activities by the agent and partly a function of patrolling activities on the part of the principal. It is not surprising in such situations that many of the companies that are engaged in logging (Sizer and Plouvier, 2000) and agricultural development find that it is profitable to pay bribes in order to reduce the risk of detection. Similarly, it is quite natural that public servants or politicians engaged in the control of tax evasion are sometimes willing to accept bribes or other forms of side payment. Members of the public service are often poorly paid, and salaries are sometimes delayed for several months, so there are sufficient incentives for such officers to accept side payments. Their profits are also dependent on the probability of detection and the expected fines. The costs of illegal behaviour are relatively low in states with inefficient or corrupt judicial systems (Dudley *et al.*, 1995).

The actual royalties collected in the Tanga region of Tanzania during the financial year of 1997/8 were US$45,000 (excluding fines), while the potential royalties of commercially traded forest products, according to the existing rules and regulations, were estimated between US$166,000 and US$567,000, with a 'best guess' figure of US$361,000 (Kobb, 1999). Therefore, the actual tax collection was only 13% of the potential amount that could have been collected. Since both legal and tax-evading operations produce a socially valued output, they should be counted equally in the social calculus. In a social context, both legal and illegal incomes must count as benefits, while avoidance and patrolling costs must count on the negative side. Payments of tax and fines, however, are transfers of wealth within the society, from immediate agents to the regulating principal. Such transfers should not count when searching for the most efficient use of forest resources. On the other hand, externalities generated by overexploitation and deforestation resulting from individual actors' profit maximization, without due payment of tax, must count as a cost.

So far, the implicit assumption has been that the superior regulating agent (i.e. the state) is a law-abiding agent that is interested in economic efficiency and the well-being of all citizens. Criminal loggers, agriculturalists, traders and a few dishonest civil servants are seen to take illegal actions. This description may well fit some countries, but it is not well suited as a general model (Westoby, 1985). One could distinguish between the rational and efficient state on the one hand and the inefficient kleptomaniac state on the other. We may agree that most states are found along a gradient between these two extremes. In some countries, the state apparatus is largely used as an instrument for the appropriation of wealth to a small group of 'elite'. It is quite unrealistic to expect such a state to be interested in the overall economic efficiency of forest resource utilization, or the policies required to minimize illegal felling, or control illegal squatting in forest reserves. In such states, top politicians and civil servants are likely to be more interested in maximizing their own profits from bribes or their direct involvement in the timber trade, rather than in designing efficient resource use policies. Under such conditions one cannot expect social rationality on the part of the central forest service or regional forest authorities.

Some actions should, however, be possible both at the national and international level. FAO (2001b) gives a list of actions that may reduce illegal exploitation of forests. These included increasing the reward for integrity, increasing the probability of detection, reducing the discretionary power of government officials, increasing the use of market mechanisms and increasing the involvement of the media, non-government organizations (NGOs) and the public.

5.5 Conclusion

Deforestation and forest degradation are serious problems in some tropical regions. These same processes have contributed significantly to human development and well-being over time. In some cases, deforestation or forest degradation is still a sustainable development strategy that may contribute to increasing the productivity of land. However, in many other cases the result is often unsustainable, resulting in much higher external costs than the profits earned by the immediate actors. Some deforestation and forest degradation activities result from illegal exploitation. While it is not easy to control illegal forest exploitation, a number of actions can be taken to contain the problem. Controlling deforestation and forest degradation is, however, not primarily a question of reducing illegal logging and forest squatting. Deforestation is, first of all, a result of the expanding cultivation of food crops. If this is to be reduced significantly, the framework conditions that make such expansion profitable and necessary must be addressed. While it may be possible to feed growing populations from a smaller area of cropland than is presently the case, the poorer people of many tropical countries have few alternatives to subsistence agriculture for their survival. To reduce the rate of tropical deforestation significantly, these people must find employment in other sectors (e.g. industry and urban services). However, these issues are central to the concepts of economic growth and development being faced internationally.

References

Angelsen, A. and Kaimowitz, D. (1999) Rethinking the causes of deforestation: lessons from economic models. *The World Bank Observer* 14(1), 73–98.

Angelsen, A. and Kaimowitz, D. (eds) (2001) *Agricultural Technologies and Tropical Deforestation*. CAB International, Wallingford, UK.

Bateman, I.J. and Willis, K.G. (eds) (1999) *Valuing Environmental Preferences: Theory and Practice of the Contingent Valuation Method in the US, EU, and Developing Countries*. Oxford University Press, Oxford.

Baumol, W. and Oates, W. (1988) *The Theory of Environmental Policy*, 2nd edn. Cambridge University Press, Cambridge.

Bromley, D.W. (1991) *Environment and Economy: Property Rights and Public Policy*. Basil Blackwell, Oxford.

Campbell, B., du Toit, R. and Attwell, C. (eds) (1989) *The Save Study: Relationships Between the Environment and Basic Needs Satisfaction in the Save Catchment, Zimbabwe*. University of Zimbabwe, Harare.

Department of Agriculture and Co-operation (DAC) (2000) Statistics at a glance. Ministry of Agriculture, India. [http://www.nic.in/agricoop/statistics/krfood].

Dudley, N., Jeanrenaud, J.-P. and Sullivan, F. (1995) *Bad Harvest? The Timber Trade and the Degradation of the World's Forests*. Earthscan, London.

Ehui, S.K. and Hertel, T.W. (1989) Deforestation and agricultural productivity in the Cote d'Ivoire. *American Journal of Agricultural Economics* 71(3), 703–711.

FAO (1981) *Tropical Forest Resources Assessment Project. Forest Resources of Tropical Africa, Part II: Country briefs*. FAO, Rome.

FAO (1993) Forest resources assessement 1990. Tropical countries. *FAO Forestry Paper* 112, 1–61.

FAO (1997) *State of the World's Forests 1997*. FAO, Rome.
FAO (2001a) [http://www.fao.org/forestry/fo/fra/index.jsp]
FAO (2001b) *State of the World's Forests 2001*. FAO, Rome. [http://www.fao.org/forestry/FO/SOFO/SOFO2001/publ-e.stm]
Fernández-Juricic, E. (1999) An environmental education approach to conservation of the blue-fronted Amazon in Córdoba, Argentina. [http://www.umich.edu/~esupdate/library/99.7-8/fernandez.html]
Grainger, A. (1993) *Controlling Tropical Deforestation*. Earthscan, London.
Grøn, A.H. (1931) *Den Almindelige Skovøkonomis Teori*. Levin Munksgaard, Copenhagen.
Hanemann, W.M. (1994) Valuing the environment through contingent valuation. *Journal of Economic Perspectives* 8(4), 19–43.
Herfindahl, O.C. and Kneese, A.V. (1974) *Economic Theory of Natural Resources*. Merrill, Columbus.
Hofstad, O. (1997a) Balancing public and private in forest policy. *Communications of Skogforsk* 48(8), 145–160.
Hofstad, O. (1997b) Woodland deforestation by charcoal supply to Dar es Salaam. *Journal of Environmental Economics and Management* 33, 17–32.
Horne, P. (1999) Economics of tropical forest land use and global warming. In: Palo, M. (ed.) *Forest Transitions and Carbon Fluxes*. World Development Studies 15, United Nations University, WIDER, Helsinki, pp. 101–116.
Kobb, D. (1999) Forestry royalties in Tanga region: paper versus reality. Unpublished report to East Usambara Catchment Forest Project.
Lambin, E.F. (1997) Modelling and monitoring land-cover change processes in tropical regions. *Progress in Physical Geography* 21, 375–393.
Loomis, J.B. and White, D.S. (1996) Economic benefits of rare and endangered species: a summary and meta-analysis. *Ecological Economics* 18(3), 197–206.
Mitchell, R.C. and Carson, R.T. (1989) *Using Surveys to Value Public Goods: The Contingent Valuation Method*. Resources for the Future, Washington, DC.
Palo, M. (1987) Deforestation perspectives for the tropics: A provisional theory with pilot applications. In: Dykstra, D., Kallio, M. and Binkley, C. (eds) *The Global Forest Sector: An Analytical Perspective*. John Wiley, London, pp. 57–89.
Palo, M. and Lehto, E. (1996) Modeling underlying causes of pantropical deforestation. In: Palo, M. and Mery, G. (eds) *Sustainable Forestry Challenges for Developing Countries*. Kluwer, Dordrecht, pp. 27–61.
Parks, P.J., Barbier, E.B. and Burgess, J.C. (1998) The economics of forest land use in temperate and tropical areas. *Environmental and Resource Economics* 11, 473–487.
Pearse, D., Barbier, E. and Markandya, A. (1990) *Sustainable Development. Economics and Environment in the Third World*. Earthscan, London.
Peters, C.M., Gentry, A.H. and Mendelsohn, R.O. (1989) Valuation of an Amazonian rainforest. *Nature* 33, 655–656.
Raja, R.G. (1996) [http://www.metla.fi/archive/forest/1996/06/msg00145.html]
Rietbergen, S. (1989) Africa. In: Poore, D.J, Burgess, P., Palmer, J., Rietbergen, S. and Synnott, T. (eds) *No Timber Without Trees. Sustainability in the Tropical Forest*. Earthscan, London, pp. 40–73.
Sandler, T. (1993) Tropical deforestation: markets and market failure. *Land Economics* 69, 225–233.
Simpson, R.D., Sedjo, R.A. and Reid, J.W. (1996) Valuing biodiversity for use in pharmaceutical research. *Journal of Political Economy* 104(1), 163–185.
Singh, K.D. (1993) The 1990 tropical forest resources assessment. *Unasylva* 44(174), 10–19.
Sizer, N. and Plouvier, D. (2000) *Increased Investment and Trade by Transnational Logging Companies in Africa, the Caribbean and the Pacific*. A joint report by WWF-Belgium, WRI and WWF-International.

Swaminathan, M.S. (2000) *A Global Vision of Forestry and Society*. Keynote address, 21st IUFRO World Congress, Kuala Lumpur, 7–12 August 2000.

Treue, T. (2001) *Politics and Economics of High Forest Management: A Case Study of Ghana*. Kluwer, Dordrecht.

UNSO (2000) Debt-for-environment swaps for national desertification funds. [http://www.undp.org/seed/unso/pub-htm/swap-eng1.htm#debt]

von Thünen, J.H. (1875) *Der Isolierte Staat in Beziehung auf Landwirtschaft und Nationalökonomie*. Schumacher Zarchlin, Berlin.

Westoby, J.C. (1985) Foresters and politics. *Commonwealth Forestry Review* 64, 105–116.

Whitlow, R. (1988) Potential versus actual erosion in Zimbabwe. *Applied Geography* 8, 87–100.

Wirl, F. (1999) De- and reforestation: stability, instability and limit cycles. *Environmental and Resource Economics* 14, 463–479.

The Application of Forest Zoning as an Alternative to Multiple-use Forestry

6

C.R. Nitschke and J.L. Innes

Faculty of Forestry, University of British Columbia, Vancouver, Canada

Sustainable forest management can take many forms. Early applications of the multiple-use concept suggested that a stand of trees might satisfy many or even all the requirements expected of a forest. While this idea has been largely abandoned, it is possible that well-planned forested landscapes may satisfy the multiple needs of society. Several studies have suggested that zoning may be the optimal way to achieve this. As forest zoning is seen by many managers as the best method to balance social, economic and environmental values, many models for allocating forest land have arisen. With the existence of so many working and proposed models, forest managers must decide which method will work best in the landscape they are managing. The history of zoning models provides help in deciding what type of zoning model will work best in a particular area, and provides a manager with the ability to make a more informed decision about how to allocate forest land in the attempt to achieve sustainable forest management. In this chapter, the historical and current use of zoning in British Columbia, Canada, is discussed along with a review of zoning models used around the world. Zoning models applied in seven different countries on four continents are reviewed and related back to the status and requirements of forest zoning in British Columbia. This chapter also discusses a zoning model, supported by many as the model that will allow for social, economic and environmental values to be maintained at a sustainable level, known as the TRIAD approach. Two case studies of the application of the TRIAD approach are also discussed to provide a realistic assessment of how this model can be applied within the context of economically sustainable forestry.

© CAB International 2005.
Forestry and Environmental Change: Socioeconomic and Political Dimensions
(eds J.L. Innes, G.M. Hickey and H.F. Hoen)

6.1 Introduction

Land allocation through zoning has been a growing practice within forest industries regionally, nationally and internationally over the last decade. Traditionally, many countries have had a policy to divide their forest estates into classes of forests to be managed for different dominant uses such as: timber production, watershed protection, wildlife habitat, recreation and so on (Vincent and Binkley, 1993). However, exploitation of natural forests through commercial logging has allowed timber production to dominate the landscape and non-timber resources to be considered as secondary values that were readily overlooked, leading to long-standing conflicts over resource allocation and land use (e.g. Clayoquot Sound, British Columbia, in the early 1990s).

Within British Columbia, forests have traditionally been managed extensively under the implicit assumption that the majority of the forested land base would be available for timber production (Binkley, 1999). Non-timber values are frequently considered on a stand-by-stand basis and this has led to many economic and ecological challenges at the landscape level by creating a scarcity of forested land. Management on a stand-by-stand basis has placed multiple demands on each hectare of forested land and this has led to an increase in public conflict over forest management policies. Binkley (1997) has stated that all parties involved in forest resource use proceed under the assumption that forest land is abundant, not scarce; therefore, the only way to resolve the resulting conflicts is to allocate land among uses. Binkley goes on to argue that, in the absence of a price system for land and the presence of non-market environmental values, land-use zoning is the logical approach. This approach is supported by research conducted by Vincent and Binkley (1993) and Helfand and Whitney (1994), which found that multiple benefits of forests are best provided by zoning the forest into a series of special use areas corresponding to the range of forest values demanded by society.

In response to public pressures, many jurisdictions have attempted to resolve conflicts over resource allocation and land use through innovative planning techniques (Jackson and Curry, 2002), such as forest zoning. There are many forms of forest zoning being applied, ranging from the consensus-based Land and Resource Management Plans (LRMPs) used in British Columbia to the production–conservation system used in New Zealand. In addition, new systems based on scientific principles are emerging, with one such system being known as the TRIAD approach. This chapter examines the current state and use of forest zoning in British Columbia and elsewhere and examines in greater depth the TRIAD approach to forest zoning, which in recent years has grown into a likely candidate for increasing sustainability and reducing conflicts in resource management in a number of different jurisdictions.

Fig. 6.1. Regional land-use plans in British Columbia (as of May 2005).

6.2 Forest zoning: status and use

The division of forested land to accommodate multiple values is being applied around the world, though at different scales and through different methodologies. The following sections discuss the current use of zoning within British Columbia and in other jurisdictions.

6.2.1 Forest zoning in British Columbia

During the early and mid-1990s the government of British Columbia took its first steps toward land-use zoning by establishing the Commission on Resources and the Environment (CORE). The CORE process established regional land-use plans for five areas of the province (see Fig. 6.1). The purpose of these plans was to divide each region into broad land-use zones and to establish objectives to guide resource management within each zone (Ministry of Sustainable Resource Management (MSRM), 2003a). Fig. 6.2 provides an example of the type of zoning that was applied to the landscape by the government's first attempts at forest zoning. The land-use plan essentially divided the region into four zones: Enhanced Resource Development, Special Resource Development, Integrated Resource Management and Protected Areas. The intensity of timber production within each was varied to account for non-timber values. For example, in the Cariboo–Chilcotin land-use plan, timber production was determined to occur on 83% of the Enhanced Resource Development Zone, 81% of the Integrated Resource Management Zone and on 70% of the Special Resource Development Zone (MSRM, 2003b).

Regional land-use plans, however, have had a short lifespan as they have been augmented by subregional plans known as Land and Resource Management Plans (LRMPs) throughout the remaining regions of the province. Eighteen LRMPs were initiated because of their smaller geographical scale, which provided a better fit for public involvement and shared decision-making (MSRM, 1994). Fig. 6.3 identifies LRMPs that are completed or are in the process of completion as of July 2004 and the five-land use plans that have been developed.

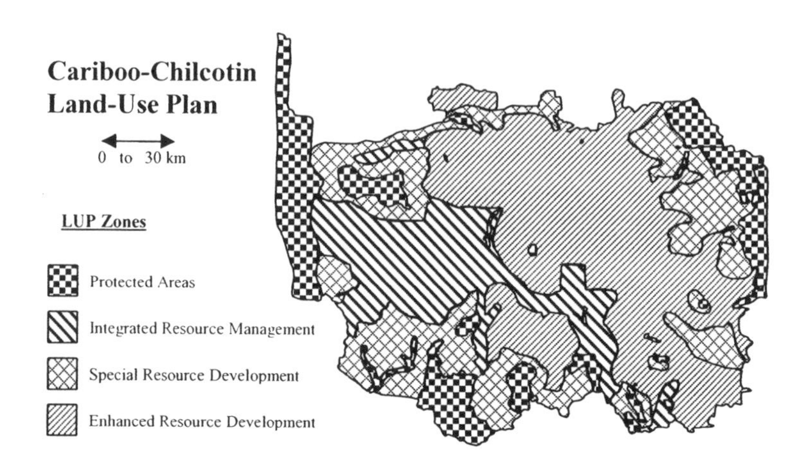

Fig 6.2. Cariboo–Chilcotin Land Use Plan Resource Management Zones.

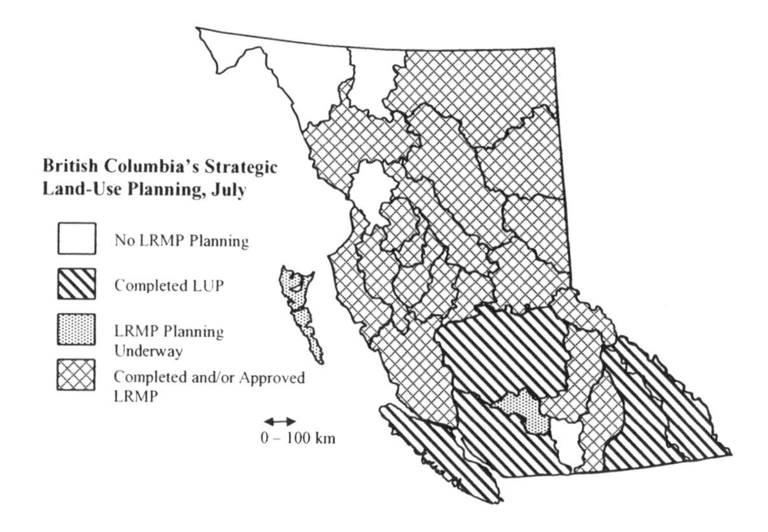

Fig. 6.3. Land-use plans of British Columbia as of July 2004 (adapted from MSRM, 1994).

The LRMP process works at the subregional level and, because of this, land allocation occurs at much finer levels than in the regional land use plans. As in the land-use plans, the goal of the LRMP process was to divide a subregion up into broad land-use zones and to establish objectives to guide resource management within each zone, in order to provide for the needs of the region's communities, economy and environment. Although some subregions have not completed their LRMPs and each LRMP will differ in the allocation of its land, there appears to be an underlying similarity between the zoning processes used in the regions. For example, the Kamloops LRMP has divided the landscape into six zones: General Resource Management, Settlement, Protection, Special Resource Management for Wildlife/Habitat, Special Resource Management for Community Watersheds and Special Resource Management for Recreation/Tourism (see Fig. 6.4 for how these zones are incorporated into the landscape). In comparison, the Okanagan-Shuswap LRMP has divided the region into four distinct zones: General Resource Management, Special Resource Management, Protected Areas and Settlements (MSRM, 2003c). The zoning policy used within each subregion is the same zoning system, a system very closely related to that used in the original land-use plans and, as summarized by Bellamy (1992), quite widely used in British Columbia. The percentage of land allocated to different land-use zones varies from region to region as values differ. Fig. 6.5 provides an example of the percentage of land allocated to each zone within the Kamloops LRMP (MSRM, 2003c).

The differences in the percentages of land allocated to different zones occur because, as the common zoning system becomes more diverse and complex with the presence of more and more non-timber values within a region, sub-zoning must take place within the context of the common zoning system and these subzones can be quite different in order to meet the region's goals of sustaining economic, community and environmental needs. An example of sub-zoning in the context of the LRMP process is the zoning of ungulate habitat into high, medium and low areas within the Special Resource Management Zone, and the zoning of High, Intermediate and Low Biodiversity Zones in the General Management Areas. The spatial presence of these subzones impacts the intensity and type of resource use that may occur within the General Management Zone.

Although forest zoning has increased dramatically over the last 10 years, it has done so only at the subregional level, with millions of hectares of land being considered. For example, the Okanagan-Shuswap subregion covers approximately 2,500,000 hectares. The lands in these regions are not all forested and even forested lands are not all productive forests that could be considered for timber production. In addition, forest companies do not operate throughout an entire region. Forest companies work within Tree Farm Licences or Forest Licence Areas within subregions. The Tree Farm Licence and Forest Licence Areas operate at a much smaller resolution. For example, TFL 49 covers an approximate area of 145,000 hectares and TFL 39 Block 1 has a productive forest area of 89,400 hectares (out of 182,600 hectare total). As a result of this smaller resolution, forest companies that operate under the LRMPs tend to attempt to satisfy the plan's objectives on a stand-by-stand basis.

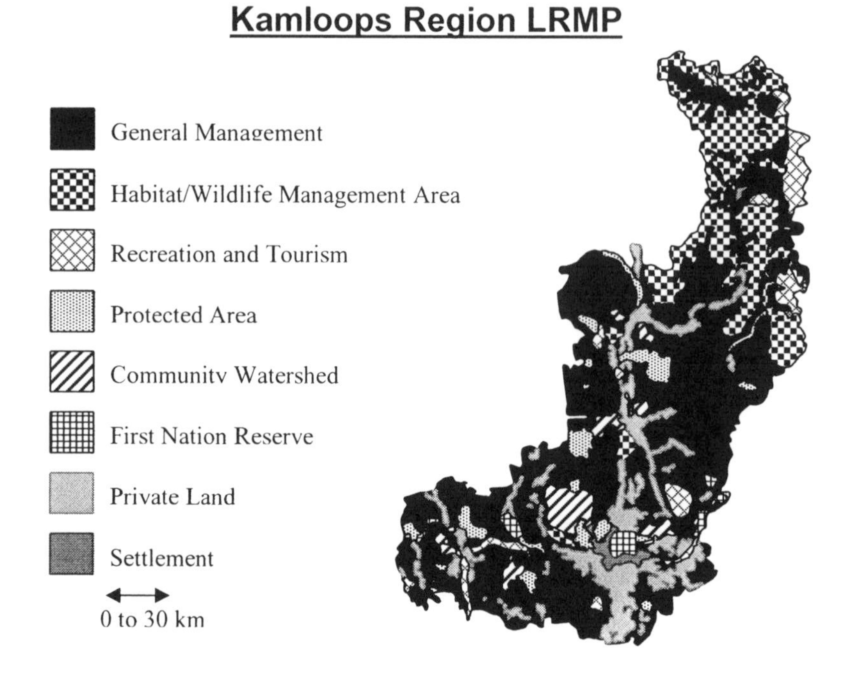

Fig. 6.4. An example of LRMP zoning.

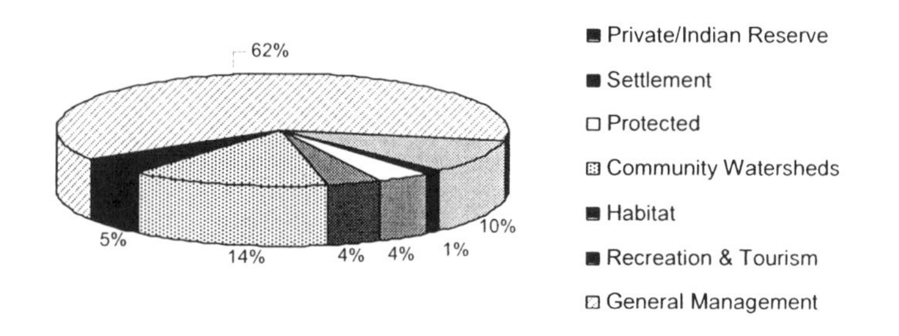

Fig. 6.5. Example of percent land allocation to zones within an LRMP.

Oliver (1992) has stated that to address environmental concerns management must move to the landscape level for an appropriate dynamic balance of stands in diverse structures and patterns that can maintain habitats for a diversity of flora and fauna species. Franklin (1993) supports this, stating that the landscape approach to management is the only way to conserve the overwhelming mass of existing biodiversity. To address these concerns, forest companies in British Columbia are starting to move towards managing forests at the landscape level through the use of Forest Stewardship Plans. One company, Weyerhaeuser Canada

Ltd., has been applying zoning on its timberlands on Vancouver Island since the late 1990s, using a three-zone system. The zones are: Timber, Habitat and Old-Growth (MacMillan Blodel Limited, 1999; Bunnell *et al.*, 2003). Different levels of silvicultural intensity are applied within each zone. Other companies/divisions are approaching the issue of zoning in different ways. For instance, Riverside Forest Products Limited (now Tolko) is looking at the TRIAD approach to zoning for the land area within Tree Farm Licence 49, while Weyerhaeuser's Stillwater Timberlands division has applied a slightly different zoning system through their Forest Stewardship Plan for Block 1 of Tree Farm Licence 39. It is important to note that the Stillwater Forest Stewardship Plan has been developed as a pilot project and is the first of its kind in British Columbia, while the Riverside TRIAD project has not yet progressed to the point where zoning has occurred. Both required special government legislation to allow them to depart from normal practices.

Weyerhaeuser's Stillwater Forest Stewardship Plan utilizes zoning to achieve desired future conditions that will attempt to provide significant conservation of old growth and ensure the broad range of ecological conditions at the landscape level are maintained (Weyerhaeuser Company, 2002a). The zoning system comprises four zones: Timber, Habitat, Old Growth, and Recreation and Tourism (see Fig. 6.6). The Timber Zone focuses on commercial timber production, the Habitat Zone focuses on conserving biological diversity and wildlife habitat in existing old-growth stands through low-intensity harvesting systems (i.e. an extensive approach) and the Old-Growth Zone is the combination of reserve allocation (66% to 100% depending on location in landscape) and restoration of old-growth habitats in previously harvested areas (any harvesting that is allowed will be by ecologically based systems). The final zone is the Recreation and Tourism Zone. This is managed extensively in a way that attempts to be compatible with recreational use of the zone.

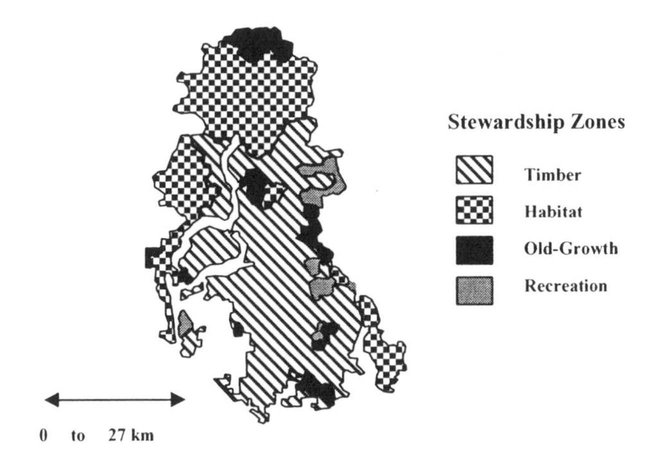

Fig. 6.6. Weyerhaeuser zoning in Tree Farm Licence 39, Block 1 (adapted from Weyerhaeuser Company, 2002b).

The use of landscape zoning in British Columbia is still in its infancy. Regional and subregional planning is well established; however, this level of zoning, though important, does not address the smaller scales that forest companies operate at. For example, LRMPs have been used in conjunction with Innovative Forest Practice Agreements, which promoted increasing the annual allowable cut through 'innovative' stand-level practices. At the same time, landscape issues were addressed through methods set aside in the former British Columbia Forest Practices Code – such as biological diversity targets (International Forest Products Limited and J.S. Thrower and Associates, 2000).

Binkley (1999) stated that the Code has many serious environmental problems as it promotes the dispersal of harvesting across the landscape and this will fragment the landscape and thereby reduce biodiversity. Lindenmayer and Franklin (2002) support Binkley's statement, stating that dispersed-patch-clear-cutting causes rapid fragmentation and makes the landscape vulnerable to many types of disturbances. What this means is that subregional planning is insufficient to provide a sustainable forest industry that meets the requirements of society and the environment. The need is for zoning at a mid-spatial resolution, in particular, the landscape level. In British Columbia, Weyerhaeuser has taken the first steps towards attempting to attain sustainable management at the landscape level through the use of their zoning system on the coast, while other companies, such as Tolko Industries Ltd., are starting to follow suit through the proposed use of alternative zoning techniques.

6.2.2 Forest zoning in Victoria, Australia

Forest management zoning has been implemented in the state forests of Victoria, Australia. In 1999, Victoria's Department of Sustainability and Environment (DSE) created a land classification system, called Forest Management Zoning, in conjunction with the National Parks and Wildlife Service, Resource and Conservation Division of the Department of Urban Affairs and Planning, Department of Mineral Resources and community groups (State Forests, 1999). The primary objective of the zoning strategy is to separate clearly the areas within a landscape to be managed as conservation reserves from the areas to be managed for timber production. The Forest Management Zoning system is a two-tiered system; the first tier divides the landscape into one of eight zones while the second tier provides for special values that may be applied either to the entire zone or to specific sites within the zone. The eight zones used in the Victorian system are provided below along with a description of each zone's main objective (State Forests, 1999):

Zone 1: Special Protection

Management objectives are to maximize protection of very high natural and cultural values.

Zone 2: Special Management

Specific management and protection of natural and cultural conservation values where it is not possible or practicable to include these lands in Zone 1.

Zone 3: Harvesting Exclusions and Special Prescription

Management for conservation of identified values and/or forest ecosystems and their natural processes, while also facilitating other management and production activities.

Zone 4: General Management

Management of native forests for timber production through the use of a full range of silviculture options as appropriate, and for conservation of broad area habitat and environmental values that are not dependent on the structure of the forest.

Zone 5: Hardwood Plantations

Management of hardwood plantations to maximize sustainable timber production on a continuing and cyclic basis.

Zone 6: Softwood Plantations

Management of softwood plantations to maximize sustainable timber production on a continuing and cyclic basis.

Zone 7: Non-Forestry Use

Management of non-forested areas, such as cleared areas and special developments.

Zone 8: Areas of Further Assessment

Interim zoning area, where field investigation will determine the final management zone classification.

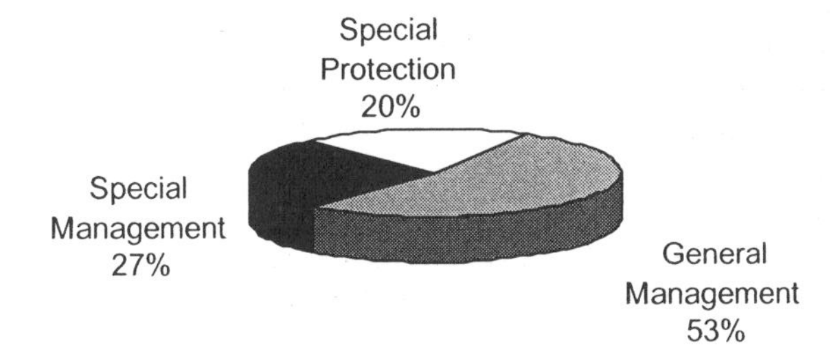

Fig. 6.7. Forest zoning allocation, Midland FMA, Victoria, Australia (adapted from DSE, 2003a).

The Forest Management Zoning system implemented by the DSE is a subregional level system since it works under a hierarchy of higher-level plans known as Regional Forest Agreements, and itself is a hierarchal plan for Forest Management Areas. Forest Management Areas cover between 100,000 ha and 400,000 ha. The zoning system for the Forest Management Areas utilizes three of the eight zones: Special Protection Zone, Special Management Zone and General Management Zone (DSE, 2003a). However, some Forest Management Areas use the following terminology: Conservation Zone, Production Zone and Fire Protection Zone (DSE, 2003b). Fig. 6.7 provides an example of the allocation of the forested area in a typical FMA for the Midlands Forest Management Area (FMA).

The use of forest zoning in Victoria has many similarities to forest zoning in British Columbia. The development of subregional zoning (Forest Management Zoning) relates closely to the LRMP process used in British Columbia; however, the landscape-level planning in Victoria appears to be more advanced than in British Columbia. Even with this, there are similarities between the two. The Forest Management Area zones are developed under Forest Management Plans while in British Columbia Forest Stewardship Plans will require management areas to be zoned to some level, as shown by Weyerhaeuser's Stillwater Timberlands Forest Stewardship Plan. So it appears that even though British Columbia is lagging behind Victoria, the two have closely parallel forest zoning systems.

6.2.3 Forest zoning in New Zealand

The New Zealand government, in late 2000, declared the remaining 180,000 ha of native beech and podocarp forests, located on the West Coast of the South Island as protected areas (Fig. 6.8), even though they were being managed under sustainable management plans. This action was taken because these state-managed native forests have a key conservation role (Reid, 2001). The conservation decision firmly identified the importance of timber grown in exotic plantations to provide for the forest industry (Reid, 2001).

New Zealand's landscape has essentially been divided into two forest zones: Conservation and Production. The Conservation Zone is comprised of state-owned natural forests while the Production Zone is comprised of exotic plantations dominated by *Pinus radiata*. Fig. 6.8 shows that zoning strategies were being applied to the landscape before the government's intensive zoning. The decision to remove native forests placed 98.7% of the forested landscape into protected/conservation areas.

6.2.4 Forest zoning in Sabah, Malaysia

The government of Sabah is in the process of implementing a forest management system in an attempt to achieve sustainable forest management (Government of Sabah, 2003) (Fig. 6.9). A substantial component of the management system is a strategy for forest zoning. The forest management system is applied to Forest Management Units, which are 100,000 ha in area. The forest zoning system applied in Sabah is thus applied at the landscape level. The main objective of the management system is to conserve the functions of the forests. The Sabah government believes these functions can be managed sustainably through forest zoning. The zoning process used divides the forest into four zones: Production, Conservation, Recreation and Community Forestry Zones.

Fig. 6.8. Zoning on the West Coast of New Zealand on Timberland West Coast Ltd.'s operating area, before and after the New Zealand Government's new zoning allocation (adapted from Timberlands, 1999).

Fig. 6.9. Map of Malaysia and the province of Sabah on the island of Borneo.

The Production Zone is used for timber and non-timber production, particularly industrial tree plantations that are comprised either of native or exotic species. These are confined to flat areas, areas with moderately steep slopes (<

46%) and areas with severely degraded forest lands. The Conservation Zone is comprised of areas with slopes > 46%, critical water catchments and areas with < 46% slopes that contain endemic/endangered wildlife and tree species, such as the orang-utan (*Pongo pygmaeus*). The Recreation Zone comprises areas with natural aesthetic and special tourist values, such as scenic views. The final zone is the Community Forestry Zone: these are areas of human settlements adjacent to forests. The Community Forest Zone is set aside to accommodate management for multiple uses that are in line with traditional community land uses such as rattan plantations (*Rotan* sp.) and resin and honey collection by villagers.

There are some similarities between Sabah and British Columbia in the processes that landscapes are managed through. For instance, private companies operating on state land must submit a Forest Management Plan that sets criteria for sustainable forest management; these plans have durations of 10 (minimum) to 20 (maximum) years. In British Columbia, the new Forest Stewardship Plans must set criteria to attain sustainable forest management, but have durations of 5 years, though they are renewable for a period of 20 years. The zoning process applied to Sabah thus provides a good example of how a company can allocate forest land, since many of the management criteria that are used to delineate zones in the forests of Sabah parallel some of the management concerns faced by the forest industry in British Columbia.

6.2.5 Far East Russia and Northeast China

In the mid-1990s, a project was initiated in the Ussuri River Watershed in Far East Russia and Northeast China, with the objective being to bring the Russian Federation and China into close cooperation in environmental and land-use planning (Marcot *et al.*, 1997). The Ussuri River Watershed (URW) covers an area of 26.2 million ha (68% in the Russian Federation, 32% in China). The URW has high biodiversity since it is an area where the boreal taiga conifer forest ecosystems of Siberia intermingle with the temperate hardwood forest ecosystems of Northeast China and Southeast Russia (Marcot *et al.*, 1997). The resulting ecosystem contains 49 tree species (9 coniferous and 40 hardwood species) and many rare and endemic species. For instance, the URW is the only known place where the brown bear (*Ursus arctos*) and the Siberian tiger (*Panthera tigris altaica*) occur together (Dinerstein *et al.*, 1995). The URW is also home to many endangered species including the Siberian tiger and the nearly extinct far east leopard (*Panthera pardus orientalis*) (an estimated 25–40 exist in the wild) (Medetsky, 2001). In addition, the URW is the last remaining watershed in China with spawning populations of salmon and sturgeon. Due to the presence of endangered and endemic species and the demand for resource use in the URW by both indigenous peoples (non-timber values) and industries (timber and non-timber resources), cooperative forest zoning involving both countries was initiated as the only way to manage the URW for social, economic and environmental needs.

Prior to the initiation of the zoning strategy, undisturbed native forests, wetlands and grassland ecosystems were disappearing rapidly, particularly in China, due to the conversion of these ecosystems to intensive agriculture and commercial use (Marcot *et al.*, 1997). To sustainably manage the URW for all values, the watershed was divided into ten resource use zones: Nature Conservation, Recreation, Culture, Watershed Protection, Limited Production Forestry (i.e. extensive forestry), Commercially Managed Forests (i.e. intensive forestry), Wetlands, Settlements, Industry and Mining. However, it is apparent that these ten zones actually form five zonal groupings: Protection Zones, Extensive Use Zones, Intensive Use Zones, Industrial/Settlement/Recreation Zones and Agricultural Zones. As a result of the zoning, 39% of the URW in the Russian Federation was placed into protection zones (nature conservation, parks, wetlands, habitat, cultural), 11% into extensive zones (extensive forestry and watershed protection) and 37% was allocated to intensive forestry zones (commercial forests), 12% to agricultural zones and 1% to industry/recreation/settlement zones. In the Chinese part of the URW the land was allocated differently, with 20% placed in protected areas, 28% in intensive forest use zones (commercial forests), 2% in industry/recreation/settlement zones and 50% in agricultural zones (no land was allocated to the extensive use zone within the Chinese part of the URW).

The zoning of the URW, particularly the forest zoning, provides some important concepts for regional zoning that can be applied at a landscape level. For instance, the protected areas that were set aside include four trans-border reserves to protect and connect important forest and wetland habitats between the two countries. Thus, the zoning process has shown the importance of working with existing landowners or managers to plan a landscape that can sustain multiple values. This is an important concept as many values, such as fish and wildlife, are not static and do not pay attention to boundaries. It is therefore important to ensure that zoning in one landscape is congruent with adjacent landscapes. In addition, the process provides insights into balancing cultural values (particularly of indigenous peoples' use of non-timber resources) with timber production by creating cultural zones where non-timber values can be utilized (for example, mushroom, fern and pine-nut harvesting) by indigenous peoples and by traditional hunters, trappers and gatherers (Marcot *et al.*, 1997).

6.2.6 Forest zoning in Guyana, South America

Eighty-four per cent of Guyana's land base is classified as forest or woodland. Currently, 6.69 million ha of forest are in pristine condition with the remainder allocated to parks or for resource management. Fig. 6.10 shows the current state of the forested land base that was created in 1997 with the release of Guyana's National Forest Policy Statement. Following the 1997 National Forest Policy Statement, the Guyana Forestry Commission began to draft a National Forest Plan and in 2001 a draft of the plan was completed. It proposed that state forests be classified into zones in order to meet the need to balance the dual goals of economic efficiency and conservation with the goal of sustainability (Guyana Forestry Commission, 2002b). Since forests occupy a large proportion of the country (18,477,480 ha (84%)) and indigenous peoples who live and work in the forest occupy many forest areas, there are many conflicts between existing land uses. To accommodate these land uses, the Forestry Commission is in the process of identifying zones of the forest estate that are best suited for each land use (Guyana Forestry Commission, 2002a).

The Guyana Forestry Commission has prescribed that forests be classified into the following: Permanent Production, Permanent Protection and Biodiversity Reserves, Reserve Forests, Extractive Forests, Multiple Forests, Multiple-Use Forests, Permanent Research Forests and Conversion Forests. The Guyana Forestry Commission is in the process of delineating zones and formalizing the zoning process through consultations with both forest sector stakeholders and indigenous people (Guyana Forestry Commission, 2002a).

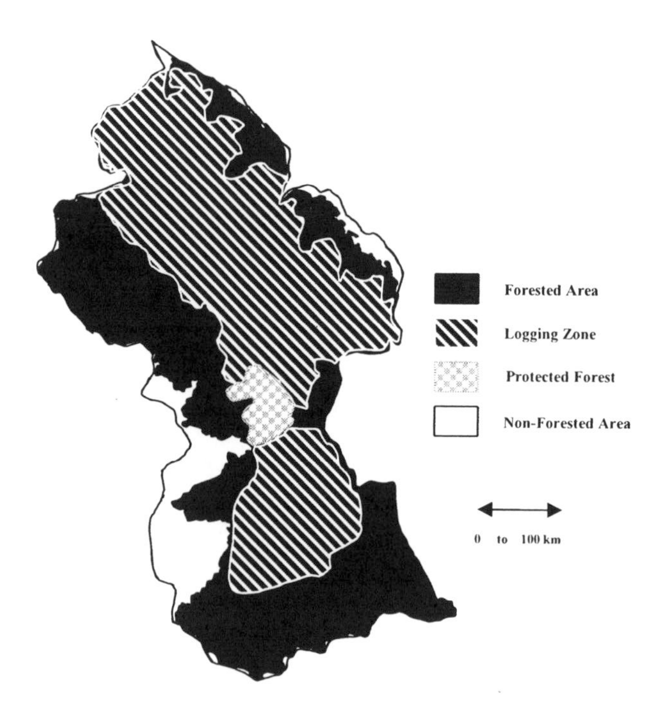

Fig. 6.10. Current state of Guyana's forests.

Guyana is attempting to achieve sustainable management through forest zoning, in much the same way as British Columbia, although there are some important differences. Zoning in Guyana is occurring at a very large spatial scale, approximately 18.5 million ha. However, coupled with the zoning objectives proposed by the Forestry Commission (see Guyana Forestry Commission, 2002a, 2002b), it seems that the large spatial scale will not hinder their objectives. Even though there are eight zones, they really only represent three functions: Production, Multiple Use and Protection. In addition, the government in Guyana has created certainty for forest companies by providing more secure tenures (concessions) for companies that meet Forestry Commission requirements. This defines the differences between British Columbia and Guyana. The secure concessions will promote investment in the production zone while minimizing the impacts in multiple-use areas and will enable the proposed 36% of Guyana's remaining pristine forests to be protected or subjected to harvesting by low-intensity silviculture systems only (single and group tree systems). This frees up the remaining land to be placed into multiple-use and production forests.

In British Columbia, the tenure system does not promote investment; Zhang (1994) found that investment on private land is 81% greater per ha per year than the weakest and most dominant tenure form, the Forest Licence. Zhang also found that Tree Farm Licence holders invested 27.4% more than Forest Licence holders as a result of the greater tenure security; however, this is still 53.6% less of an investment than on private, secure land. Binkley (1999) implicitly states that the key to conservation and sustainable management is to promote production forestry by securing long-term tenure rights.

6.2.7 Forest zoning in Sweden

Existing silviculture systems, from clear-cutting to selection systems, do not incorporate the landscape and stand-level complexity characteristics of natural disturbance regimes (Franklin *et al.*, 2002). In addition, there are temporal contrasts between traditional silviculture systems and natural disturbances (Lindenmayer and Franklin, 2002). The lack of congruency between current harvesting practices and natural disturbance patterns poses many problems when the goal is to maintain and/or conserve landscape biodiversity. Current silvicultural approaches result in a lack of structural complexity at both the stand and landscape levels. As a result of this and the desire to maintain biodiversity at the landscape and stand levels, forest managers in Sweden have applied a new silviculture system that offers greater congruence between harvesting and natural disturbance patterns in Sweden's boreal forest. The system is called ASIO[1] and it applies multiple silviculture applications across the landscape and within single stands to achieve desired levels of structural complexity (Rülcker *et al.*, 1994).

ASIO is a spatially varying zoning system designed to emulate natural disturbances at the landscape and stand levels. It allocates the landscape and stands into zones based on the frequency that fire occurs in that part of the landscape and stand. In 1997, the Swedish boreal forests were classified according to the ASIO system (Angelstam, 1997).

The ASIO system places the following management objectives for the portion of the landscape and stand that falls within a zone with the following attributes (Rülcker *et al.*, 1994):

'Almost Never' Zone: Wetland forests, forests in ravines, small islands in lakes and northeast-facing slopes subjected to fire approximately every 300 years (Angelstam, 1998). These areas are non-forestry operating areas.

[1] ASIO is an acronym that means Almost Never, Seldom, Infrequently and Often. The model was developed by Rülcker *et al.* (1994) and Angelstam *et al.* (1993).

'Seldom' Zone: Riparian forests (lakes and watercourses) and flat, moist areas. These areas are subjected to fire on average every 200 years. Selective and shelterwood harvesting are utilized in this zone.

'Infrequently' Zone: Mesic areas (except those found in Almost Never and Seldom Zones). Fire frequency occurs on average every 100 years in this zone. The silviculture applied in this zone is controlled burning of cutover utilizing a seed tree system.

'Often' Zone: Pine forests on flat terrain with sedimentary soils. Fire frequency occurs approximately every 50 years in this zone. This zone contains species that can survive recurrent low-intensity burns so controlled burning is used. The natural fire regime creates multi-aged stands, so seed tree silviculture systems are applied.

Lindenmayer and Franklin (2002) state that in order to conserve biodiversity in a landscape that is subjected to forest harvesting, a management system that varies silviculture systems across the landscape and varies harvest scheduling both spatially and temporally is necessary. This type of system is essential because it will maintain two of the key tenets required for biodiversity conservation: landscape complexity and heterogeneity, as well as stand structure complexity. Because of these key elements, the zoning system applied in Sweden, which emulates natural disturbance with greater congruence then traditional systems (Rülcker *et al.*, 1994), should be examined in any zoning system applied in British Columbia. The ASIO system provides the opportunity for forest companies to zone at a finer spatial scale. For example, managers could continue managing on a stand level and yet maintain landscape heterogeneity by emulating natural disturbances with greater congruence. This type of zoning is no doubt a mixed method of landscape/stand zoning and because of this it offers flexibility and restraint. However, applying this type of zoning within a two-tiered system such as that used in Australia can allow biodiversity to be maintained at high levels throughout the entire landscape by creating a habitat-rich matrix, and this in turn could allow for the area required for reserves to be reduced (Lindenmayer and Franklin, 2002).

6.3 TRIAD approach to forest zoning

The examination of the status and use of zoning in British Columbia and elsewhere shows that British Columbia is lagging behind most countries in certain areas of forest zoning and yet is on a par with other areas. It is notable that British Columbia is starting to move its zoning to the landscape level and this will bring the province to an equal level in relation to the theory of zoning. However, the way that zoning is applied by various forest companies under varying tenures remains to be seen. One zoning system that has begun to gather praise in recent years for its potential to conserve biodiversity while increasing or maintaining timber yields is the TRIAD approach.

6.3.1 What is the TRIAD approach?

The TRIAD concept was first proposed by Seymour and Hunter (1992), after they realized that replacing current extensive forest practices with ecological forestry would be a positive step towards maintaining biodiversity, but would not solve the larger problem of conserving biodiversity across the landscape (Seymour and Hunter, 1999). Seymour and Hunter (1992) identified the need to set aside adequate, representative reserves while maintaining timber yields through a compensatory increase in timber achieved through intensive management. The proposed system allocates the landscape into three zones with fundamentally different management objectives: Production (Intensive), Extensive (Multiple Use) and Ecological (Reserves), as shown in Fig. 6.11. The objective of the TRIAD approach is to produce a landscape design and management system that will provide for all societal demands (Seymour and Hunter, 1999).

6.3.2 Implementing a landscape TRIAD

The first step in the implementation of the TRIAD system is the acceptance by managers that some of the landscape must be set aside as reserves to ensure that biodiversity is conserved (Seymour and Hunter, 1999). The second step is to assess the potential of the unreserved landscape for commodity production. Where timber demands are low, extensive/ecological forestry should be practised, and, where timber demands are high, portions of the land should managed under production/intensive silviculture to offset reserved land (Seymour and Hunter, 1999). An important caveat is that the TRIAD approach suggests equal allocation of the landscape between zones (Seymour and Hunter, 1999), with the allocation of land to each zone being accomplished on a landscape-to-landscape basis.

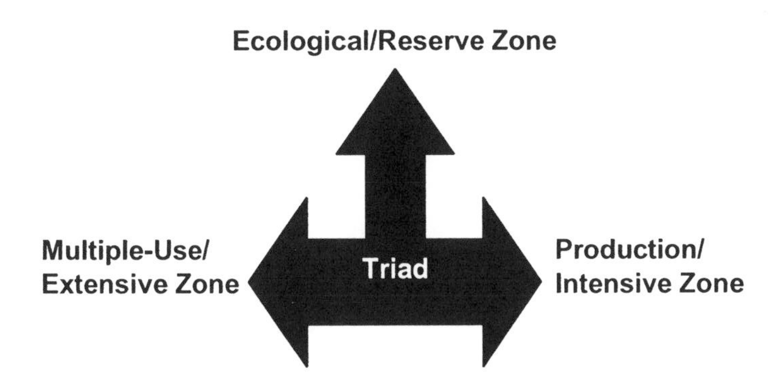

Fig. 6.11. TRIAD approach to forest land allocation.

6.3.3 Ecological/reserve zone

Large tracts of unmanipulated forests are needed to fully protect all the ecological values associated with forest ecosystems (Hunter and Calhoun, 1996). This is supported by Franklin (1993) and Lindenmayer and Franklin (2002), who conclude that habitat reserves are essential elements in any management system that conserves biodiversity for the future. If, as Norton (1999) has argued, the primary function of reserves in the landscape is to ensure that representative examples of all areas of indigenous biodiversity are protected, then any allocation must ensure that all ecosystems that occur on the landscape are incorporated into large-scale or mid-scale reserves. Some managers propose that this zone should comprise large-scale reserves (parks and protected areas) that already exist in the landscape. Although this approach is supported by some (e.g. Binkley, 1997, 1999) it has been adopted less enthusiastically by others. Lindenmayer and Franklin (2002) conclude that a mixture of large reserves, such as those represented by parks, must be supported by a network of mid-spatial-scale reserves that provide support to larger reserves by increasing biodiversity within the surrounding matrix.

In order to reduce edge effects on reserves and to provide connectivity between reserves and habitats within the matrix, it is important to buffer reserves by managing the surrounding matrix (Franklin, 1993; Norton, 1999; Lindenmayer and Franklin, 2002). For example, land-use practices that have managed the matrix around reserves have created buffer zones that enhanced the carrying capacity of existing reserves for the North Island brown kiwi (*Apteryx australis mantelli*) (Potter, 1990). There is the potential to use smaller reserves within the matrix: reserves surrounded by a representative buffer could be up to ten times smaller and still provide the same functions as a large reserve that is not buffered by the matrix (Harris, 1984). The proportion of land allocated to reserves will depend on the landscape, although some percentages have been proposed: 10% for Maine (Hunter, 1994), 12%–15% for Canada's boreal forests (Messier and Brigué, 2002), 13% for Alberta (Proche, 2002), 25% (Burton, 1994) and 73% for British Columbia (Binkley, 1997). Reserves can also function as benchmarks where natural processes can be monitored and thus act as controls to compare management within other zones that attempt to emulate natural processes (Lindenmayer and Franklin, 2002).

6.3.4 Production/intensive-use zone

Many have advocated that incorporating intensively managed areas into the landscape will allow for an increase in protected areas, since the areas subjected to intensive production will raise per hectare yields across a landscape (Gladstone and Ledig, 1990; Vincent and Binkley, 1993; Burton, 1994; Hunter and Calhoun, 1996; Binkley 1997, 1999; Sedjo and Botkin, 1997; Messier and Bigué, 2002; Weber, 2002). In British Columbia, Sahajananthan (1994) has shown that, through moderately increasing management intensity, about 40% of the land base could produce the same yield as 100% of the land base under integrated management of resources. These findings are supported by studies conducted in New Zealand in the mid-1980s (Birchfield and Grant, 1993). Intensive management through plantation forestry has been suggested for more than 10 years because of the gains in yield that occur when production silviculture is applied to a stand. Production silviculture that incorporates site preparation, bedding, fertilization, genetically improved seedlings, mycorrhizal inoculated stock, optimally spaced trees at planting then afterwards by thinning and finally by repeat thinning and fertilization can economically increase yields by two to five times (Gladstone and Ledig, 1990; Binkley, 1997). Intensive forestry in the southern USA has projected increased yields from 4 m^3 ha^{-1} $year^{-1}$ to 10–14 m^3 ha^{-1} $year^{-1}$; in tropical countries, plantations of Caribbean pine (*Pinus caribea*) have increased yields four to ten times to 18 to 34 m^3 ha^{-1} $year^{-1}$ depending on the site, and yields from plantations of white albizia (*Albizia falcataria*) and flooded gum (*Eucalyptus grandis*) have increased four to ten times to yield up to 60 m^3 ha^{-1} $year^{-1}$ (Gladstone and Ledig, 1990). On Weyerhaeuser's Tree Farm License 35 near Kamloops, British Columbia, Tolnai (1991) has shown that stands subjected to more intensive management had a 70.1% increase in yields; some stands on good sites had yield increases from 2.3 m^3 ha^{-1} $year^{-1}$ to 8.3 m^3 ha^{-1} $year^{-1}$ (3.6 times greater yield).

The move to plantation forestry will result in land areas with reduced biodiversity, but the positive side is that, by obtaining maximum production on minimal areas, more land is available for other purposes that are incompatible with timber production (Seymour and Hunter, 1992). Such uses include conserving endangered species requiring old-growth habitat. Seymour and Hunter (1999) argue that increasing yields by three to five fold will allow, in theory, 3–5 hectares extra to be protected for every hectare subjected to intensive management. This is supported by Binkley (1997), who found that intensive production areas could potentially increase protected areas fourfold in British Columbia, and by Messier and Brigué (2002), who found that intensive production can provide 45% of the needed wood supply on 15% of the land, allowing for a large proportion of the land to be set aside for protection (> 15%). In the tropics, intensive forest production could increase the forest area available for protection and other uses by 2.5 times (Gladstone and Ledig, 1990).

The selection of sites for intensive production could cause conflict if not done within the context of ecological processes within the landscape. Many managers will want to select the richest sites for intensive management, as they are the most productive. However, these sites tend to also be the richest in biodiversity (Lindenmayer and Franklin, 2002). Therefore, when selecting sites it is essential not to monopolize the productive sites (Messier and Brigué, 2002) but to incorporate these sites into meso-scale reserves and extensive-use zones. Additionally, it is important to retain biodiversity through the intensive-use zone by maintaining critical structures essential to conserving biodiversity. Connectivity should be maintained through these areas by creating wildlife corridors at appropriate widths and densities, and reserves along streams and rivers at appropriate widths that preserve natural processes (Lindenmayer and Franklin, 2002). Also, wetlands, lakes and ponds should be managed as key areas for biodiversity and subjected to protection. Essentially, the maintenance of biodiversity within production forests involves a landscape approach where the spatial arrangement of plantation areas of varying ages is considered (Norton,

1998). The approach is to mix plantation areas with reserve areas and to place buffers between the two zones. This produces a series of plantations at different rotations surrounding a reserve so that a proportion of mature managed forest always surrounds the buffer and reserve. The proposed system essentially embeds the areas subjected to production forestry into the matrix zone of extensive use and reserves (see Fig. 6.12). Biodiversity in production zones can also be enhanced through retaining standing deadwood (snags), encouraging broadleaf species, increasing the diversity of tree species, extending rotation ages where necessary, and using harvest areas that match the local topography. Increasing the width and structural diversity of permanent edges will also help (Hodge *et al.*, 1998).

Many studies of TRIAD allocation (Burton, 1994; Hunter, 1994; Messier and Brigué, 2002) propose allocating a similar land area for intensive use to that for reserves (10–25%). The final decision regarding the intensive-use zone is the actual degree of intensity the plantations will be managed at. Messier and Kneeshaw (1999) have identified two forms of intensive management: intensive plantation management and super-intensive management. Intensive management is of the type discussed above, while super-intensive management involves growing genetically modified species, either native or exotic. Areas managed under this system are essentially 'fibre farms', where intensive site preparation and vegetation control are used (e.g. herbicide), along with hexagonal spacing of the genetically modified trees (Messier and Brigué, 2002). Managers currently using this system operate at rotation ages of 20 years or less and remove each crop by clear-cutting.

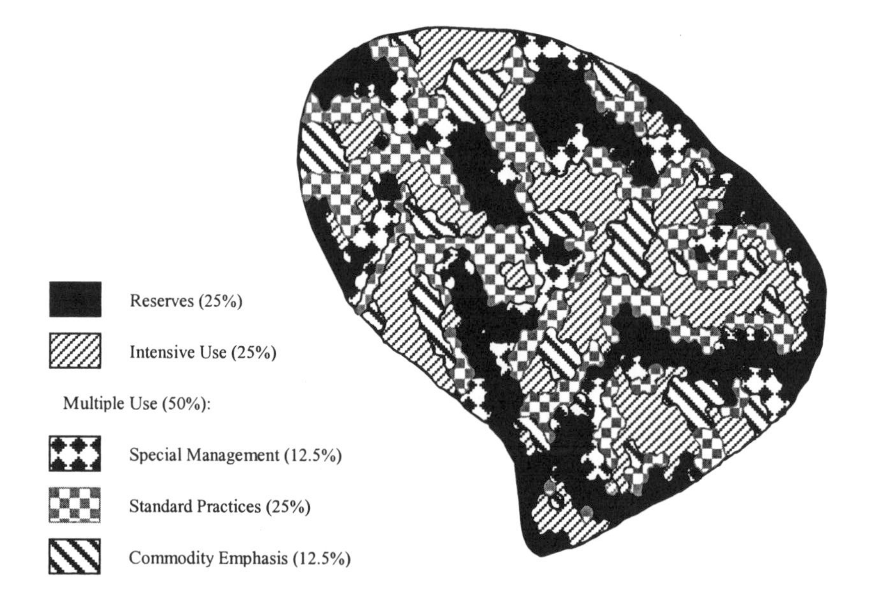

Fig. 6.12. Embedding Intensive-Use Zones within a matrix of extensive use (adapted from Burton, 1994).

Based on the use of these two forms of production forestry, Messier and Kneeshaw (1999) have developed a four-zone spin-off of the TRIAD approach, which they have called the QUAD concept (Fig. 6.13). Due to the intensity of use in fibre farms, Messier and Kneeshaw have allocated only a small portion of the landscape to this zone (3%). However, this 3% can provide 24% of the wood supply from a comparative area (Messier and Brigué, 2002).

Fig. 6.13. QUAD concept to forest land allocation (adapted from Messier and Kneeshaw, 1999).

6.3.5 Extensive/Multiple-Use Zone

The final section of the TRIAD approach is the forest area left over after the landscape has been delineated into reserves and intensive use zones. The remaining forest area comprises the Extensive-Use Zone, also referred to as the Multiple-Use Zone. This zone operates as the matrix, within which the intensive-use areas and reserves are embedded (see Fig. 6.12). This zone comprises the portion of the forest commonly referred to as the matrix, and it has been identified as critical for conserving biological diversity within the landscape (Franklin, 1993; Lindenmayer and Franklin, 2002). Franklin (1993) identified three roles that the landscape matrix plays in conserving biodiversity: (1) by providing habitat at smaller spatial scales; (2) by increasing the effectiveness of reserves; and (3) by controlling connectivity in the landscape, including movements between reserves. Although the matrix is a critical zone for conserving biodiversity, it is also the zone set aside for multiple uses and thus will be exposed to resource extraction. However, the intensity of resource use is reduced in this zone with ecosystem-based management being practised to sustain multiple values. One of the main concepts of ecosystem-based management is the attempt to emulate natural disturbances through appropriate silvicultural systems. Therefore, the type of management within the zone will vary from landscape to landscape according to the dominant disturbance processes. A variety of forestry practices may occur, from clear-cutting to partial cutting, any of which may rely on natural regeneration (Burton, 1994). In addition, management within matrix forests may involve restoration and improvement silvicultural practices to improve productivity and biodiversity.

The management practices within this zone must ensure that ecosystem integrity and biodiversity are sustained. Lindenmayer and Franklin (2002) outline 11 factors needed to accomplish this: seven factors require protection via mid-spatial-scale reserves while the remaining four are management actions. The seven factors requiring protection are: (1) aquatic ecosystems and riparian buffers; (2) wildlife corridors; (3) special habitats (cliffs, meadows, rockslides, caves, vernal features and thermal features); (4) remnant patches of late-successional forest; (5) biological hotspots (calving areas, source areas for coarse woody debris and populations of rare species); (6) culturally sensitive areas; and (7) disturbance refugia (fire, wind and other disturbances). The four management factors are: (1) the road network should be minimized; (2) landscape-level goals should be set for specific structural features (e.g. large snags) or vegetative conditions (e.g. seral stages); (3) spatial and temporal harvesting patterns (dispersed versus aggregated harvest patterns, harvest unit size and rotation lengths); and (4) restoration and recreation of late-successional forests or habitat features. When dealing with landscapes subjected to episodic disturbances it is important that ecosystem-based management techniques applied within the matrix provide the same attributes that result naturally from the disturbance event. Lindenmayer and Franklin (2002) identify five attributes that should be incorporated into matrix management: (1) large (sometimes very large) patch sizes; (2) high levels of within-patch heterogeneity, reflecting spatial variation in disturbance intensity; (3) high levels of biological legacies that vary in quantity and type throughout the patch; (4) edges that are complex in shape (feathered or irregular and diffuse); and (5) variable and often long return intervals between entries. Modification of the matrix through ecosystem management should incorporate these factors not only to conserve biodiversity but also to provide greater management flexibility. Multiple-use zones can and should retain an underlying variability and diversity, even if the zone is apparently dominated by a single use such as timber harvesting.

The underlying diversity associated with managing this zone correctly can provide flexibility and potential for managers (Burton, 1994). Management of the matrix through ecosystem-based techniques provides a system that has the potential to be selective and adaptive in more than one direction. For example, if more lands require protection, the forests of the extensive zone will still have the biological and ecological integrity to support high-quality reserves. Likewise, the same areas could be incorporated into production zones if the opportunity becomes available (Burton, 1994).

Managing the extensive zone for biodiversity will increase the overall amount of land being managed for conservation, and will spread the management over a more representative set of ecosystems. As a result, fewer species will require reserves for protection and what reserves there are will face reduced risks from episodic disturbances (Lindenmayer and Franklin, 2002). Coupling these concepts with those described above by Burton (1994), it becomes clear that management within this zone offers the ability for managers to apply a flexible management system that can continually increase the ability for biodiversity to be conserved within this zone and the intensive-use zone.

6.3.6 Adaptive management and the TRIAD approach

Adaptive management requires a more cautious approach to the use of forest resources because of the need to maintain future options and the difficulty in re-creating those resources quickly once they are gone (Lindenmayer and Franklin, 2002). With ecosystem-based management that attempts to emulate natural disturbances and reserves to act as controls to evaluate natural processes, there is an opportunity to adapt management in the matrix as knowledge increases. This essentially avoids placing the landscape into a state of constancy, considered by Holling (1978) as the cause of landscape fragility. The other strength of the TRIAD approach is that it operates at the landscape level and involves collaboration with stakeholders. In addition, it operates at a scale where all ecological processes are not yet completely understood. These are the key elements required for a management system to operate under adaptive management (Lee, 1993, 1999). The TRIAD approach also provides, within its normal function, a key step for adaptive management to meet its objectives, namely the delineation of experimental units (Rutledge and Lepczyk, 2002). The TRIAD approach provides managers with the ability to implement adaptive management in the landscape from the very beginning of the zoning process, as uncertainty will exist over the ability of the original zoning to conserve biodiversity and maintain timber yields.

The development of zones requires the development of criteria and strategies to reach the goals identified for each zone; if management is to be sustainable, these strategies will require measurable results. The need to have measurable results and indicators of failure or success are the key components of an effective adaptive management system and must be attained through monitoring. The ability to use reserves as controls allows monitoring within the TRIAD approach to be conducted accurately and expediently, particularly if reserves contain representative ecosystems that are also being managed under ecosystem-based systems and intensive systems.

6.3.7 Application of the TRIAD approach

Currently there is much interest in the TRIAD approach, resulting in research being initiated to explore the feasibility of applying the concept to sustainable forest management. Three areas where significant research is being conducted are Maine (USA), New Brunswick (Canada) and Alberta (Canada). The latter two study areas are in by far the most advanced stages of implementation.

J.D. Irving Limited has recently adopted the TRIAD approach for a case study on 190,000 ha of its privately owned forest in northwestern New Brunswick. The goal of the case study is to quantify trade-offs in timber and non-timber values through the combination of intensive, extensive and reserve areas (Montigny and MacLean, 2002). The landscape, known as the Black Brook District, is comprised of over 60,000 ha of plantations, 49,000 ha of tolerant hardwood stands managed for high-quality veneer and saw logs through selection and patch-cut systems, and 7000 ha of scientific reserves. The reserve area comprises two types: core reserves and adaptive management areas. The core reserves are intended to provide ecological benchmark forest conditions to which the managed forest can be compared (controls), while the adaptive management areas will provide areas in which to conduct experimental harvesting based on natural disturbance processes, particularly the spruce budworm (*Choristoneura fumiferana*) and gap dynamics (Dick *et al.*, 2002). This case study is seen as a well-suited candidate to evaluate the management strategies of the TRIAD approach because it has a significant proportion of plantations and extensive forestry areas combined with unmanaged forest.

Alberta-Pacific Forest Industries Inc. (ALPAC), located in north central Alberta, has adopted a TRIAD approach for its 58,000 km^2 Forest Management Area (FMA). In 2001, ALPAC began managing this area utilizing the TRIAD approach. The company has 600,000 ha of existing and proposed reserves within or adjacent to its FMA and is practising ecosystem-based management on the remainder of the FMA, with managers attempting to emulate natural disturbances, predominantly fire, while utilizing an adaptive management framework (ALPAC, 2001). The interesting part of ALPAC's TRIAD approach is the Intensive-Use Zone. The Intensive Use Zone is actually a Super-Intensive Use Zone utilizing poplar species (*Populus* spp.), as proposed by Messier and Kneeshaw (1999). The zone is located on ALPAC's private and leased private land – ALPAC is also currently attempting to lease 25,000 ha of private land within 200 km of its mill. These leased areas will become poplar fibre farms and will be managed on 18–25-year rotations (ALPAC, 2002). ALPAC plans to attain 15–20% of its wood supply from these poplar farms. The company is increasing the amount of land for poplar farms through the Poplar Farm Program, in which local farmers and landowners can grow poplar plantations as a cash crop to be purchased by ALPAC. Early monitoring has shown that their use of ecosystem-based management within an adaptive framework is contributing positively towards sustainable management (ALPAC, 2001).

6.4. Conclusion

Aspects of forest management in British Columbia appear to be similar to other jurisdictions that have adopted forest zoning as a tool to attain sustainable forest management. However, there are many differences in the ways that forests are zoned around the world, even though they all hold the same objectives of balancing social, economic and environmental values. These are important ideals when considering the different zoning strategies that have been applied and deciding which strategy will work best in a particular area. The spatial differences in zoning have shown, with the exception of Guyana, that zoning strategies are moving towards operating at the mid-spatial scale, commonly referred to as the landscape level. Even within British Columbia this has begun, with Weyerhaeuser's Tree Farm Licence 39 being a particularly good example. Although the basic concepts are understood, zoning systems such as the TRIAD approach are just beginning to be researched. This has resulted in a state of constancy and fragility in British Columbia at a time when the drive for sustainable management is at the forefront of government and public demands upon the industry. Management strategies, in particular the TRIAD approach, provide forest companies with tools that allow for flexibility through active adaptive management and will allow companies to go further, faster, along the path to sustainable forest management.

References

Alberta-Pacific Forest Industries Inc. (ALPAC) (2001) *Forest Management.* Available at: http://www.alpac.ca/Forest_Management.htm (accessed 16 March, 2003).

Angelstam, P. (1997) Landscape analysis as a tool for the scientific management of biodiversity. *Ecological Bulletins* 46: 140–170.

Angelstam, P.K. (1998) Maintaining and restoring biodiversity in European boreal forests by developing natural disturbance regimes. *Journal of Vegetation Science* 9: 593–602.

Angelstam, P., Rosenberg, P. and Rülcker, C. (1993) Aldrig, sällan, ibland, ofta. *Skog och forsking* 93 (1): 34–41.

Bellamy, S. (1992). *Land Use Zoning Techniques Used in Forest Planning in British Columbia.* Resource Planning Section, Integrated Resources Branch, Ministry of Forests, Victoria.

Binkley, C.S. (1997) Preserving nature through intensive plantation forestry: the case for forestland allocation with illustrations from British Columbia. *The Forestry Chronicle* 73 (5): 553–559.

Binkley, C.S. (1999) Ecosystem management and plantation forestry: new directions in British Columbia. *New Forests* 18: 75–88.

Birchfield, R.J. and Grant, I.F. (1993) *Out of the Woods: The Restructuring and Sale of New Zealand's State Forests.* GP Publications, Wellington.

Bunnell, F.L., Dunsworth, B.G., Huggard, D.J. and Kremsater. L.L. (2003) *Learning to Sustain Biological Diversity on Weyerhaeuser's Coastal Tenure.* The Forest Project, Weyerhaeuser, Nanaimo, BC.

Burton, P.J. (1994) The Mendelian compromise: a vision for equitable land use allocation. *Land Use Policy* 12(1): 63–68.

Department of Sustainability and Environment (2003a) *Striking a Balance.* Midlands Forest Management Area. Available at: http://www.nre.vic.gov.au/4A25676D00235B51/BCView/2F9C598E89A504494A256777002369BC?OpenDocument (accessed on 13 March, 2003).

Department of Sustainability and Environment (2003b) *Otway Forest Management Area.* Available at: http://www.nre.vic.gov.au/4A25676D00235B51/BCView/2F9C598E89A504494A256777002369BC?OpenDocument (accessed on 13 March, 2003).

Dick, A.R., Maclean, D.A. and Hennigar, C.R. (2002) *New Brunswick TRIAD Case Study: Implement Harvesting Inspired by Natural Disturbance.* Conference Proceedings: Eastern CANUSA Forest Science Conference, 19–20 October, University of Maine, Orono, Maine.

Dinerstein, E., Krever, V., Olson, D.M. and Williams, L. (1995) An emergency strategy to rescue Russia's biological diversity. *Conservation Biology* 8(4): 934–939.

Franklin, J.F. (1993) Preserving biodiversity: species, ecosystems, or landscapes? *Ecological Applications* 3(2): 202–205.

Franklin, J.F., Spies, T.A., Van Pelt, R., Carey, A.B., Thornburgh, D.A., Berg, D.R., Lindenmayer, D.B., Harmon, M.E., Keeton, W.S., Shaw, D.C., Bible, K. and Chen, J. (2002) Disturbances and structural development of natural forest ecosystems with silviculture implications, using Douglas-fir forests as an example. *Forest Ecology and Management* 155: 399–423.

Gladstone, W.T. and Ledig, F.T. (1990) Reducing pressure on natural forests through high-yield forestry. *Forest Ecology and Management* 35: 69–78.

Government of Sabah, Malayasia. Sustainable Forest Management (2003) Available at: http://www.sabah.gov.my/htan/english/sfm/SFM1.HTM (accessed 14 March, 2003).

Guyana Forestry Commission (2002a) *Forest Zonation Paper: Summary and Recommendations*. Available at: http://www.forestry.gov.gy/forestzonation.htm (accessed, 17 August, 2004).

Guyana Forestry Commission (2002b) Projects. Available at: http://www.forestry.gov.gy/projects.htm (accessed on 17 August, 2004).

Harris, L.D. (1984) *The Fragmented Forest*. University of Chicago Press, Chicago.

Helfand, G.E. and Whitney, M.D. (1994) Efficient multiple-use may require land use specialisation: comment. *Land Economics* 70(3): 391–395.

Hodge, S.J., Patterson, G. and McIntosh, R. (1998) The approach of the British Forestry Commission to the conservation of forest biodiversity. *Scottish Forestry* 52: 30–36.

Holling, C.S. (1978) *Adaptive Environmental Assessment and Adaptive Management*. John Wiley and Sons, Oxford

Hunter, M.L., Jr (1994) *Fundamentals of Conservation Biology*. Blackwell Science, Cambridge, Mass.

Hunter, M.L. and Calhoun, A. (1996) A triad approach to land use allocation. In Szaro, R. and Johnston, D.W. (eds) *Biodiversity in Managed Landscapes: Theory and Practice*. Oxford University Press, Oxford, pp. 477–491.

International Forest Products Ltd. And J.S. Thrower and Associates (2000) Forestry Plan #1 for the Adams Lake Innovative Forest practices Agreement. Available at http://www.adamslake-ifpa.ca/docs/forestryplan2001.pdf (accessed 13 March 2003).

Jackson, T., and Curry, J. (2002) Regional development and land use planning in rural British Columbia: peace in the woods? *Regional Studies* 36(4): 439–443.

Lee, K.N. (1993) *Compass and Gyroscope: Integrating Science and Politics for the Environment*. Island Press, Washington, D.C.

Lee, K.N. (1999) Appraising adaptive management. *Conservation Ecology 3(2)*.

Lindenmayer, D.B. and Franklin, J.F. (2002) *Conserving Forest Biodiversity A Comprehensive Multiscaled Approach*. Washington, D.C., Island Press.

MacMillan Blodel Limited. (1999) *Summary of first year critique workshop on the MacMillan Blodel BC Coastal Forest Project, July 14–16 1999*. Prepared by Dovetail Consulting for MacMillan Blodel, Vancouver, British Columbia

Marcot, B.G., Ganzei, S.S., Zhang, T. and Voronov, B.A. (1997) A sustainable plan for conserving forest biodiversity in far East Russia and northeast China. *The Forestry Chronicle* 73 (5): 565–571.

Medetsky, A. (2001) *Zoos may help save leopards in wild*. Vladivostok News Available at: vn.vladnews.ru/Arch/2001/ ISS261/News/News04.HTM (accessed on March 14, 2003).

Messier, C. and Brigué, B. (2002) Using fast-growing plantations to promote forest ecosystem protection in Canada. Presented at IUFRO meeting: *Management of Fast Growing Plantations*. Izmit, Turkey, September 11–13, 2002.

Messier, C. and Kneeshaw, D. (1999) Thinking and acting differently for a sustainable management of the boreal forest. *The Forestry Chronicle* 75: 929–938.

Ministry of Sustainable Resource Management (1994) *Caribou-Chilcotin Land Use Plan*. Available at: http://srmwww.gov.bc.ca/rmd/regional/carichil/3.htm#b (accessed on March 12, 2003).

Ministry of Sustainable Resource Management (2003a) Strategic Land Use Planning. Available at: http://srmwww.gov.bc.ca/rmd/regional/ (accessed on March 12, 2003).

Ministry of Sustainable Resource Management (2003b) Strategic Land Use Planning. http://srmwww.gov.bc.ca/rmd/regional/carichil/3.htm#b (accessed on March 12, 2003).

Ministry of Sustainable Resource Management (2003c) Strategic Land Use Planning. Available at: http://srmwww.gov.bc.ca/rmd/lrmp/images/lrmpmap0102.jpg (accessed on March 12, 2003).

Montigny, M.K. and MacLean, D.A. (2002) *TRIAD Approach to Forest Management: Land Use Zoning in Northern New Brunswick*. Conference Proceedings: Eastern CANUSA Forest Science Conference, October 19–20, University of Maine, Orono, Maine.

Norton, D.A. (1998) Indigenous biodiversity conservation and plantation forestry: Options for the future. *New Zealand Forestry* 43(2): 34–39.

Norton, D.A. (1999) Forest reserves. In Hunter, M.L. Jr. (ed): *Maintaining Biodiversity in Forest Ecosystems*. Cambridge University Press, Cambridge, pp. 525–555.

Oliver, C.D. (1992) A landscape approach: Achieving and maintaining biodiversity and economic productivity. *Journal of Forestry* 90: 20–25.

Potter, M.A. (1990) Movement of North Island brown kiwi (*Apteryx australis mantelli*) between forest remnants. *New Zealand Journal of Ecology* 14: 17–24.

Proche, J. (2002) *The TRIAD Approach to Landscape Management, Achieving Ecological* Objectives. RENR 601 Seminar, November 26, 2002. University of Alberta Available at: http://www.rr.ualberta.ca/courses/RENR601/Jason%20Proche%20TRIAD%20seminar.htm (accessed 15 March, 2003)

Reid, A. (2001) Impacts and effectiveness of logging bands in natural forests: New Zealand. In: Durst, P.B., Waggener, T.R., Enters, T. and Cheng, T.L. (eds): *Forests Out of Bounds: Impacts and Effectiveness of Logging Bans in Natural Forests in Asia-Pacific*. RAP Publication 2001/08, Asia-Pacific Forestry Commission, FAO, Bangkok.

Rülcker, C., Angelstam, P. and Rosenberg, P. (1994) Natural forest fire dynamics can guide conservation and silviculture in boreal forests. *SkogForsk* 2: 1–4.

Rutledge, D.T. and Lepczyk, C.A. (2002) Landscape change: patterns, effects and implications for adaptive management of wildlife resources. In: Liu, J. and Taylor, W.W. (eds) *Integrating Landscape Ecology into Natural Resource Management*. Cambridge University Press, Cambridge, pp. 312–333.

Sahajananthan, S. (1994) *Single and Multiple Use of Forest Lands in British Columbia: The Case of the Revelstoke Forest District*. Report submitted to the BC Ministry of Forests, Revelstoke Forest District

Sedjo, R.A. and Botkin, D. (1997) Using forest plantations to spare natural forests. *Environment* 39 (10): 14–20, 30.

Seymour, R.S. and Hunter, M.L. Jr. (1992) *New Forestry in Eastern Spruce-Fir Forests: Principles and Applications in Maine*. Maine Agriculture and Forestry Experiment Station Miscellaneous Publication 716. Orono, Maine, 36 pp.

Seymour, R.S. and Hunter, M.L., Jr (1999) Principles of ecological forestry. In: Hunter, M.L., Jr (ed.): *Maintaining Biodiversity in Forest Ecosystems*. Cambridge University Press, Cambridge, pp. 22–61.

State Forests (1999) *Forest Management Zoning in State Forests*. Victoria. Available at: http://www.nre.vic.gov.au/web/root/domino/cm_da/nrenfor.nsf/frameset/NRE+Forestry?OpenDocument (accessed on 13 March, 2003).

Timberlands (1999) *Beech Management Overview*. Available at: http://www.timberlands.co.nz/forest/sustforestry/mgtplans/bchoverview/Part_1/Section_2/ (accessed 14 March, 2003).
Tolnai, S. (1991) Addition value to our heritage through silviculture. Paper presented to Western Silvicultural Contractors Association, Vancouver, British Columbia.
Vincent, J.R. and Binkley, C.S. (1993) Efficient multiple-use may require land use specialisation. *Land Economics* 69(4): 370–376.
Weber, M. (2002) The challenge of biodiversity conservation in Canada's boreal forests. *Tomorrow's Forests, The Sustainable Forest Management News Letter* Winter 2002, 1–3.
Weyerhaeuser Company (2002a) *Forest Stewardship Plan. Tree Farm License 39, Block 1.* Stillwater Timberlands Business Unit, Powel River, British Columbia.
Weyerhaeuser Company (2002b) *Stillwater Timberlands Pilot Project.* Available at: http://www.stillwaterpilot.ca/fsp/fsp.htm (accessed on 12 March, 2003).
Zhang, D. (1994) Implications of tenure for forest land value and management in BC. Ph.D. thesis, Faculty of Forestry, University of British Columbia, Vancouver, BC.

Capital Management – the Forests in Countries in Transition – Welfare Impacts

7

P. Csóka

State Forest Service, Hungary

Accelerated destabilization of forest ecosystems was observed over large parts of Europe in the mid-1980s, with especially high amplitudes in Central-Eastern Europe. This new type of forest decline, induced or at least triggered by effects of human activities such as air pollution, helped to draw public and political attention to the need for environmentally sound management of natural resources. The forests in the former socialist countries of Central-Eastern Europe faced new challenges that had to be met under the constraints of the overall economic and social transition – a process by which the centrally planned economy was transformed into a market economy and private ownership was partially re-established in several fields, including forests and forestry.

Forest protection and industry renewal had to be accomplished in an economic environment characterized by a sharp decline in production and consumption, social tension, unemployment and general indebtedness. Due to intensive forest policy and legal framework reform, new forest laws were enacted in almost all of the countries, followed by reforms to institutional frameworks. The state forest companies were also reformed in almost all of the countries and, after some difficult years, are now showing signs of recovery and development. It has become obvious that sustainable forest management (SFM) will only be practised if economic sustainability is ensured. The proper economic environment for SFM has yet to be established in the Central-Eastern European countries.

© CAB International 2005.
Forestry and Environmental Change: Socioeconomic and Political Dimensions
(eds J.L. Innes, G.M. Hickey and H.F. Hoen)

7.1 The forests of countries in transition – a case study

Through its restless history, the 20th century provided many opportunities for conducting case studies on how the largest terrestrial ecosystem is influenced by man and vice versa. The last two decades have brought unique developments in both the environment (i.e. forest decline and climate change) and in the structure of Central-Eastern Europe and large parts of Asia (i.e. economic and political transition).

This chapter concentrates on the current area of Central-Eastern Europe, which, in 2004, included 27 countries (together with the Newly Independent States (NIS) of the former USSR). This area was distributed between only nine countries at the beginning of the structural changes. According to the Temperate and Boreal Forest Resources Assessment (TBFRA) 2000 report (Anon., 2000), the latest information available on forest resources in Central-Eastern Europe, the geographical area of the region is approximately 2566 million ha, 898 million ha of which are covered with forest. This is about one-fourth of the world's total forested land area. Moreover, the large areas of virgin, old-growth, primarily coniferous forests (located mainly in the Russian Federation) make the region even more important at the global level.[1] Additionally, there are about 80 million hectares of other wooded land in the region, less important as a source of wood, but extremely important in environmental terms.

Although often classified as one group, 'countries in transition' (CiTs) are heterogeneous in their respective history, traditions, economy and forest resources. For example, while the forest cover is 54.5% in Slovenia, 48.1% in Estonia or 46.4% in Latvia, this ratio is as low as 9.8% in the Republic of Moldova (Fig. 7.1). The other countries of the former USSR are also very different in this respect; for example, the forest cover is 49.9% in the Russian Federation, but only 2.8% in Tajikistan (Fig. 7.2). The area of forest per capita varies between 1.41 ha (Estonia) and 0.07 ha (Moldova) in the European CiTs, and between 5.54 ha (Russian Federation) and 0.07 ha (Tajikistan) in the former USSR. The problems in the European CiTs reflect their heterogeneity, and include terrestrial eutrophication and desertification, agricultural overproduction and poverty alleviation, accumulation of growing stock and overexploitation of resources.

7.2 Environmental stresses

Forest decline has been known in the region throughout history. Records from earlier centuries described periods when forests were particularly affected by disease and pests, or by anthropogenic stresses, such as exploitation or industrial activities, that led to the destruction of some forest areas. However, symptoms of heavy defoliation and mass dieback of forests observed in several parts of Europe and elsewhere from the late 1970s have been surprising, in both extent and the speed of development. The term 'novel forest decline' was introduced at that time to identify this new phenomenon. The area of the, then socialist,

[1] These areas are very important gene pools and resources of biological diversity.

countries was rather heavily affected by this new type of dieback, particularly in the coniferous stands of Czechoslovakia, the German Democratic Republic (GDR) and Poland. Other countries, such as Hungary, Romania and Ukraine, acknowledged signs of deterioration in their oak forests.

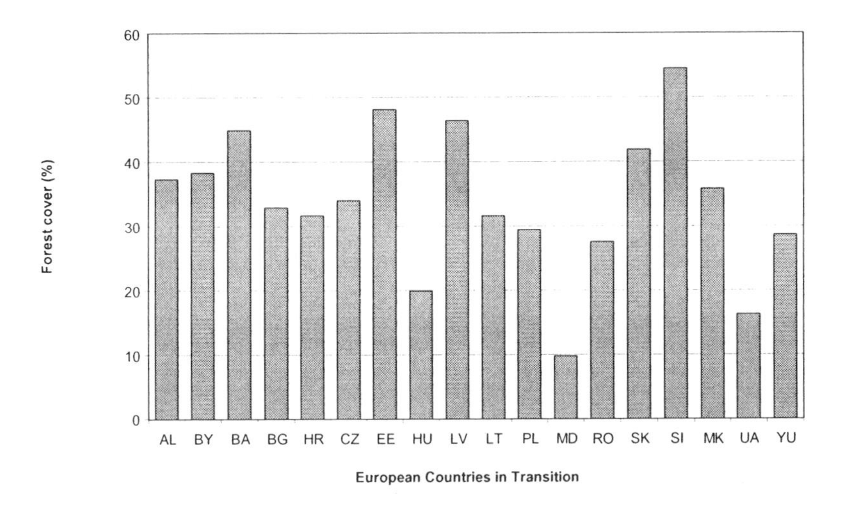

Fig. 7.1. Forest cover in the European countries in transition.

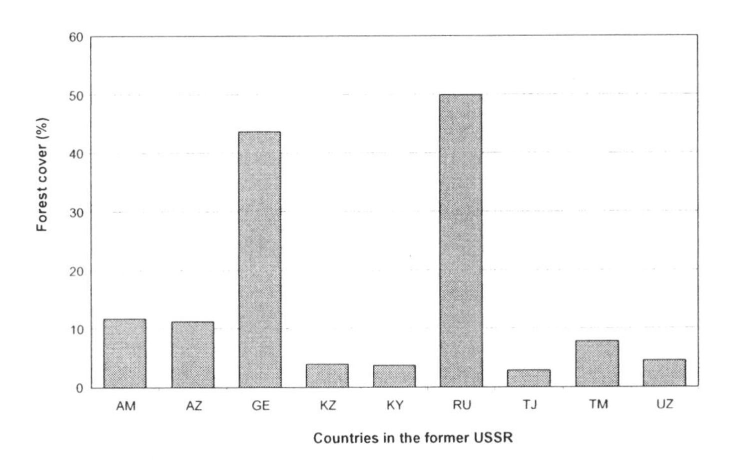

Fig. 7.2. Forest cover in the former USSR. AM: Armeina; AZ: Azerbaijan; GE: Georgia; KZ: Kazakhstan; KY: Kyrgystan; Ru: Russia; Tj: Tajikistan; TM: Turkmenistan; UZ: Uzbekistan.

When the International Cooperative Programme on the Assessment and Monitoring of Air Pollution Effects on Forests programme (ICP Forests) was established under the Geneva Convention on Long-Range Transboundary Air Pollution in 1985, Bulgaria, Czechoslovakia, Hungary, Poland and Yugoslavia joined the programme, followed by the GDR, Lithuania, Estonia, the Kaliningrad Region, Belarus, Latvia, Slovenia, Romania, the Russian Federation and Ukraine (Fig. 7.3). Their membership, partly following and partly preceding the political changes in the region, lasted until 1992. From 1990 onwards, the countries voluntarily decided to apply the regulations defined by the European Union Scheme for the Protection of Forests Against Atmospheric Pollution (EEC 3528/86 and its amendments).

Results from the early observations reflected serious problems and described a very dark picture for the future of montane coniferous forests. Correlations were found between high levels of SO_2, NO_x and ammonia emission (caused by industrial activities, transport and combustion of fossil fuels) and deterioration of forest health in certain locations: the so-called Black Triangle of south-eastern GDR, southern Poland and the northern Czech Republic. The relation between air quality and forest health was less evident elsewhere. Other factors, including adverse weather conditions, were also believed to be important factors weakening forest health. As more data enabled detailed analysis of spatial and temporal change in forest condition, the picture became more complex, but less alarming at the same time. Crown defoliation increased in the whole region in both conifers and broadleaves, but followed different patterns by countries and species groups, often including marked fluctuations. The highest level of defoliation occurred in the period 1993–1996 in most of the countries, with somewhat different patterns in the Czech Republic and Bulgaria, and in Ukraine, where an increase of defoliation was observed at the end of the 1990s (see Fig. 7.3).

Other damage factors, such as high winds, frost, drought, insects and disease, had a considerable impact on the forests of the region. A general observation is that these forms of damage have been increasing in recent decades, and that the damage caused by biotic agents seems to be strongly linked with that caused by abiotic factors, many of which act as predisposing or triggering factors that directly contribute to the outbreak of epidemics.

As environmental stresses continued to influence forests, many observations supported the hypothesis that forest ecosystems with semi-natural structures showed a higher level of resistance, and were less sensitive to these stresses. Many of the heavily affected forest areas went through considerable transformation in recent centuries, and in many places coniferous forests were established on the clear-felled sites of former broadleaved stands. Both the tree-species composition and the age structures were altered, increasing the forests' sensitivity to stresses such as pest attacks or wind-throw.

An important – but, up until now, unique – stress factor should be mentioned separately: radioactive contamination. After the Chernobil disaster in April 1996, a huge amount of radioactive substance was emitted from the reactor that had exploded, contaminating vast areas in Belarus, Ukraine, and the Russian Federation. The direct effects of radioactivity on forest ecosystems are not fully known, but, from forestry's point of view, both forest area and wood were lost to high radiation. Moreover, the whole region, which is highly susceptible to forest fires, represents a continuous threat of releasing deposited radionuclides into the atmosphere.

7.3 The transition process

As human activities initiated ecological changes, many of which were feared to be irreversible, the deteriorating environment induced changes in society, such as increasing the need for environmentally responsible thinking and behaviour at all levels. In Central and Eastern Europe, this global process coincided with the political movements of the late 1980s. The former centrally planned economies chose a new direction for social and political development, and entered a new era identified as 'the transition toward market economy conditions'. The first wave of changes in the East European countries began in 1989, and eventually led to the re-unification of Germany, followed by the disintegration of the USSR in 1991, and the removal of the socialist governments in Czechoslovakia and Yugoslavia. These changes were relatively peaceful. The crisis in the Balkans, however, seriously affected the forests in that region, and their rehabilitation will take a long time to complete.

The process of transition was characterized by very similar symptoms across the whole region. Economic production decreased considerably in almost all of the countries in the early 1990s. Indebtedness, high budget deficits, accelerated inflation and increased social tensions were typical. A slow recovery began in the mid-1990s in some of the CiTs. After one decade of transition, the region has become diversified economically; some countries are producing an impressive growth in gross national product of about 5% while others are still struggling with severe structural problems. During this time, forestry found itself in a very complicated situation. Sitting at the crossroads of different intentions, it had several roles to play: a property to be privatized, a source of income for rural people, a possible response to the challenge of agricultural overproduction, a source of industrial raw materials, a provider of environmental and spiritual values, and an extremely important, endangered, ecosystem. This very real challenge determined the direction of the forest sector in the CiTs at the beginning of the 1990s. Forest management was separated from wood processing, and almost all branches of the forest industries were privatized. Many activities associated with silviculture and forest management, such as logging and transport, were left to private entrepreneurs. Most European CiTs decided to re-establish private forestry. Re-privatization and compensation were both used to achieve this. However, the Russian Federation and other countries of the former USSR have so far chosen to keep their forests completely in the public domain.

Privatization, in the classical sense, only took place to a limited extent. The change of forest ownership, shown in Fig. 7.4, is the result of the restitution process. Some of the most recent information (i.e. country reports prepared for the 29th meeting of the Food and Agriculture Organization (FAO) European Forestry Commission) suggested that considerable changes are expected in Bulgaria and

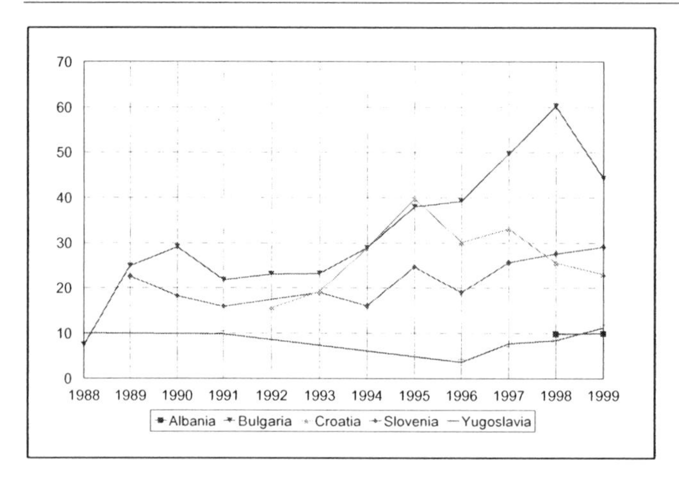

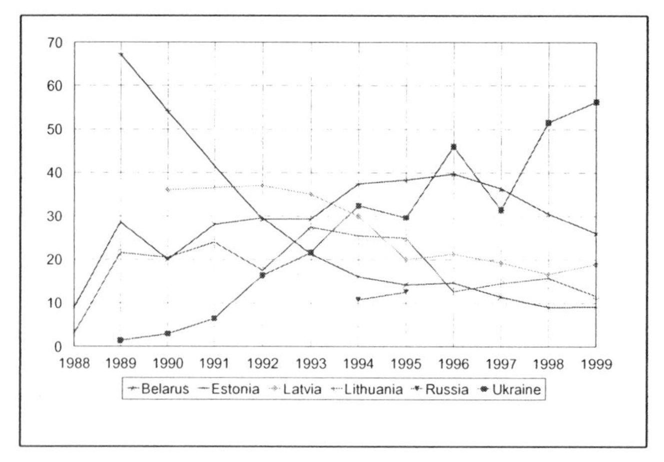

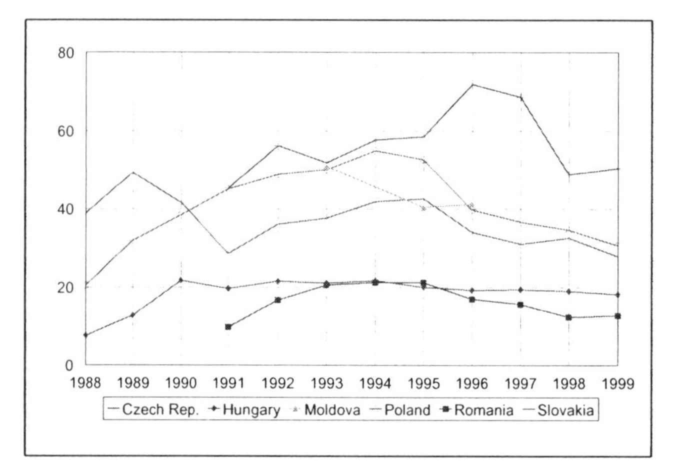

Fig. 7.3. Development of crown defoliation (above 25%) of all species (for Latvia only conifers) in the CiTs, according to ICP Forests.

Romania, where restitution is still in an early phase. The complexity of the issue is indicated by questions about such issues as the property of the church in several countries, which will require a very long time to resolve. While privatization of state forests is expected to play a minor role in the future, these issues, together with the planned afforestation on private lands, may considerably increase the proportion of private forests in many countries of the region.

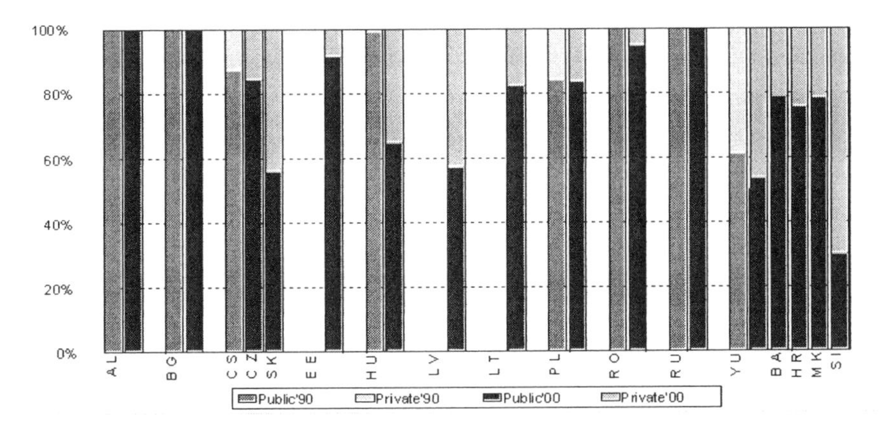

Fig. 7.4. Forest ownership documented in the TBFRA 1990 (Anon., 1992) and TBFRA 2000 (Anon., 2000).

7.3.1 Changes in forest cover during the transition period

The restructuring of the forest sector in CiTs coincided with other societal processes directly or indirectly affecting the production and processing of wood. The sharp decrease in domestic economic output resulted in decreased demand for wood and wood products, primarily in the construction, board, pulp and paper production and furniture manufacturing areas. This decrease in demand, together with other problems, led to a strong reduction of the annual cut and, correspondingly, the annual income derived from forests. In the Russian Federation alone, the annual cut decreased by 124 million m^3 between 1990 and 1993, to 180 million m^3, 59.2% of the 1990 level (Lipman, 1994). Similar processes — but less extensive ones — were observed in other countries. The fifth UN-ECE/FAO European Timber Trends Study (Anon., undated) (ETTS V), while warning of the high level of uncertainty in the forecast, projected a slight recovery towards the end of the 1990s. However, recent statistical data show that only some countries were able to increase their wood production; for others, the forecast was too optimistic, although the potential for growth still exists.

In the mid-1990s, the felling intensity, expressed as a percentage of the increment, varied between 83% (Albania, Czech Republic) and 36% (Slovenia). In most cases, felling intensity was well below the 70% European average. Not only did the intensity of felling change in this period, but its composition changed as well. Unplanned tree felling (salvage felling) seems to be a long-term problem in areas with high air pollution and potential for high winds, snow and frost (e.g. Czech Republic, Slovakia, Poland, Romania, Slovenia). Biotic agents, pests and disease in the Baltic states and fires in the south-eastern countries have been identified as the main reasons for unplanned felling.

While the volume of wood coming from sustainably managed forests decreased considerably, other factors have resulted in an undesirable increase in forest exploitation. The collapse of the traditional domestic and international markets for agricultural products caused serious problems for much of the rural population. This, together with the high energy prices, led to the increasing threat of illegal felling, primarily for local use. In addition to these problems, a number of new forest owners were doubtful about the long-term nature of the changes. As a result of this uncertainty, they chose to materialize their new-found wealth as quickly as possible. This was achieved by harvesting and selling wood, and the needs of sustainable forest management rarely featured in such decisions. While these activities did not result in the volume limits of harvesting as set by the current increment being exceeded, they could have caused serious local problems by overexploiting certain areas, hindering regeneration and paving the way to site deterioration, erosion and unfavourable changes in the water regime.

One study, the Phare programme ‘Conservation and sustainable management of forests in central and eastern European countries’, has considered the various factors affecting the forest sector of the Phare countries. Based on national reports and consultations with national correspondents, the study summarized the current state and future development of felling (Table 7.1). The study drew the following conclusion on the present state and future development of felling in the Phare countries:

> Air pollution, the growth of conifers outside their natural range, and many other interventions make the forests less stable and susceptible to diseases and pests. The synergy of these factors reduces the life expectations of forest tree species especially in Central Europe, where growing stocks are the highest in Europe, and rotation ages of Norway spruce and Scots pine, for example, are found to be longer than in southern Finland. As a consequence, not only air pollution but also the practice of retaining forests which are biologically too old contributes to inferior forest condition in the region. (Phare Environment Consortium, 1999)

Although the region’s forests have been exposed to several different threats, the extent of forested land has grown in the European CiTs. Contrary to the Dobris Assessment (European Environment Agency, 1995), the Phare study concluded that, over 50 years, forest area increased considerably in almost all of the countries, reaching an impressive 125% in Estonia, 110% in Macedonia, 75% in Lithuania and 56% in Hungary. In Poland alone, the forest area increased by 2.3 million hectares. Exceptions included Albania and Bosnia-Herzegovina.

According to the TBFRA 2000 (Anon., 2000), Yugoslavia reported an annual 1450 ha decrease of forest cover between 1979 and 1995. The forest cover is stable or increasing in the countries of the former USSR with the exception of the Russian Federation, where statistics show a considerable loss of forest cover amounting to 1.1 million ha per year. This is partly counterbalanced by an increase of other wooded land, resulting in a net growth of the wooded areas in statistical terms. The loss of forest cover, however, may have far-reaching consequences although, unfortunately, different sources provide different explanations, from overexploitation to alternative land uses to changes in terms and definitions. All in all, the forest area of the European CiTs increased by 85,000 ha annually during the reference period.

With the exception of Yugoslavia and the Russian Federation, the growing stock in the European CiTs has been increasing considerably, despite the forest decline mentioned above. This is explained by the increase in forest cover, increasing growth rates (partly ascribed to the fertilizing effect of the deposition of air pollutants such as nitrogen compounds), decreased harvesting and the successful application of sustainable management practices in many of the countries.

Table 7.1. Facts and provisions about felling in the CiTs. AL: Albania; BA: Belarus; BG: Bulgaria; CZ: Czech Republic; EE: Estonia; HU: Hungary; LV: Latvia; LT: Lithuania; MK: FYR of Macedonia; PL: Poland; RO: Romania; SK: Slovakia; SI: Slovenia.

	Date/ trend	AL	BA	BG	CZ	EE	HU	LV	LT	MK	PL	RO	SK	SI
Felling Intensity %	Mid-1990s	83	52	40	83	52	60	51	54	61	74	39	53	36
Salvage felling and illegal felling	Period	'94	'90	'90	'90-'96	'95	'96	'96	'91-'96	'95	'96	'90-'96	'90-'96	'96
	% of felling	40	16	16	60	8	12	21	30	13	36	30	53	47
Total felling	Short term	+	0	-	0	+	-	+	+	0	+	0	-	-
	Long term	+	+	+	+	+	0	0	+	+	+	0	0	+
Salvage felling	Short term	+-	+		0	+	0	0	+	+-	0	+	+	+
	Long term	-	+		+	+	+	+	+-	+-	+	0	+	

7.4 Setting up new policies and infrastructure

Increasing concern about the condition of forests and the recognition that forests are the source of a multitude of environmental values resulted in concerted actions in Europe that led to the launch of the Ministerial Conferences on the Protection of Forests. The strong commitment of the Central and Eastern European countries to fighting unfavourable environmental change was illustrated by their participation at the 1st Ministerial Conference in Strasbourg in 1990. The Conference coincided with major political changes in some of the European CiTs, but this did not prevent them from signing the six resolutions aimed at different aspects of protecting the continent's forests (see Table 7.2).

By the time of the 2nd Ministerial Conference in 1993, the signatory countries, or their successors (i.e. Czech Republic, Slovak Republic, Russian Federation), were able to report considerable development in the monitoring of forest ecosystems, the conservation of genetic resources and the establishment of a forest fires database. Most reports on the management of mountainous forests, tree physiology and forest ecosystem research stressed the critical level of forest deterioration and the need for financial resources, which were difficult to find during the early phases of transition. The reports expressed a strong desire for international cooperation.

Table 7.2. Signatory CiTs to the Resolutions of the Ministerial Conferences on the Protection of Forest in Europe.

	S1	S2	S3	S4	S5	S6	H1	H2	H3	H4	L1	L2
Albania	+	+	+	+	+	+	+	+	+	+	+	+
Belarus							+	+	+	+	+	+
Bulgaria	>	+	+	+	>	+	+	+	+	+	+	+
Croatia											+	+
Czechoslovakia	+	+		+	+	+						
Czech Republic							+	+	+	+	+	+
Estonia							+	+	+	+	+	+
Hungary	+	+	>	+	+	+	+	+	+	+	+	+
Latvia							+	+	+	+	+	+
Lithuania							+	+	+	+	+	+
Poland	+	+	+	+	+	+	+	+	+	+	+	+
FYR of Moldova											+	+
Romania	+	+	+	+	+	+	+	+	+	+	+	+
Russian Fed.							+	+	+	+	+	+
Slovakia							+	+	+	+	+	+
Slovenia							+	+	+	+	+	+
Soviet Union	+	+	+	+	+	+						
Ukraine							+	+	+	+	+	+
Yugoslavia	+	+	+	+	+	+						

> signed later

Existed as federal partner of another country
Cease to exist
Same name refers to a changed territory

The 2nd Conference in Helsinki devoted its attention to the sustainable management of forests (SFM), including protection of their genetic resources and cooperation with the CiTs in achieving these. The 3rd Conference in Lisbon in 1998 was focused on the socioeconomic aspects of forestry and the implementation of SFM. The topical structure of the conferences clearly shows the development of the international forestry dialogue and increasing public awareness of forestry and forest-related environmental issues.

The interim and follow-up reports after the Helsinki Conference reflected very considerable progress in the European CiTs, listing the completed and ongoing activities and future plans. They also demonstrated a significant change in tone and language from the first reports of the CiTs. This shift expressed a return in self-confidence and the belief in the manageability of the problems that obviously continue to exist in several fields. This is in very sharp contrast to the first report[2] published in 1993, and is undoubtedly the result of the work done in restructuring the forest sector.

Along with their international commitments, the European CiTs continued efforts to reform their legal and institutional frameworks. Not only were traditional forest laws included in this process, but so were the policies and legal instruments ensuring the protection of nature and the environment in general, the utilization of forest products and services other than wood, the promotion of the economic sustainability of forestry, the contribution to rural development, the promotion of the use of wood and, where it became an issue, preparation for accession to the European Union. Table 7.3 lists the predominant forest policy and legislation instruments in the European CiTs. All these measures should be seen in light of the general restructuring within agriculture which, intentionally or otherwise, resulted in decreasing production and the setting aside of cultivated land. Afforestation of such lands is expected to provide environmental benefits on the one hand, and employment and income for rural populations on the other. This, together with the struggle against environmental problems, increased the importance of forestry, but many countries continue to stress that budget allocations do not reflect this tendency satisfactorily.

The development of the policy and legal frameworks, outlined in Table 7.3, was followed by the transformation of the institutional background, which was substantial in the countries where private ownership was re-established. It proved difficult as the strengths of forestry in the centrally planned system were often overestimated while its weaknesses were underestimated. Many feared that the introduction of market conditions would threaten SFM, even in the state sector.

From 1992 to 1999, the countries with some private forestry created new entities to manage the state forests, or transformed the existing ones. In most cases, the new state forestry organization is a large, hierarchically structured company with different levels of dependence on the state budget, focused mainly on silviculture and the other activities associated with multiple-use forestry. Work is performed either by their own employees (if they have a considerable labour force) or subcontractors. Exceptions to this exist in Hungary, where 21 independent state holdings are responsible for the management of the state forests, and in Slovenia, where almost all activities are privatized and the state forestry organization is restricted to contracting and controlling the implementation of the contracts. These new state agencies are expected to be good examples of proper forest management, and could have several special tasks, such as managing protected and social forests, maintaining promotional forests, providing forestry extension and education, and managing private forests on request. In some cases, they even have limited normative or authoritative roles and are able to issue regulations (e.g. Albania and Poland) or professionally control private forest management (e.g. Croatia). In most countries, the state has retained jurisdiction over forest inventory and the planning and control of policy implementation. However, in Albania, the state's responsibilities have been merged with management activities, while elsewhere, such as in the Czech Republic, everything except control is privatized and performed on a contract basis.

[2] *Report on the Follow-up of the Strasbourg Resolutions* (Liaison Unit, 1993).

Table 7.3. Forest-related policies and legal instruments in selected CiTs

ALBANIA	
The Green Strategy	Law on Forestry and the Forest Service Police, 1992
Law on Environmental Protection, 1993	Law on Pastures and Meadows, 1995
Wildlife Law, 1995	Law on Leasing the State Owned Agricultural Land, Meadows, Pastures and Forests, 1998
Law on Forest Revenue, 1998	
BELARUS	
Forest Code	Concept of Sustainable Development of Forestry up to 2015
BULGARIA	
Forest Law, 1997	Law for the Restitution of Forests, 1997
Hunting Law, 1997	
CROATIA	
Forest Law, 1990, 1993	Law on Fire Protection, 1993
Law on Environmental Protection, 1994	Law on Plant Protection, 1994
Law on Hunting, 1994	Law on Water, 1995
Law on Forest Seed and Planting Material, 1995	National Forest Policy, 1991
CZECH REPUBLIC	
Act on Inspection of the Environment and its Competence in Forest Protection	Act on the Protection of Nature and the Landscape, 1992
Act on the Environment	Principles of the State Forest Policy 1994
Forest Act, 1995	
ESTONIA	
Forest Act, 1993, 1997	Forest Development Programme, 1995
Estonian Forest Policy, 1997	Development Plan for the Forest Sector, 1997–2001
HUNGARY	
Afforestation Program, 1985–2000	Law on Forests and the Protection of Forests, 1996
Law on Nature Protection, 1996	Law on Joint Forest Tenure, 1996
Law on Hunting, 1996	National Programme for Agriculture, 1
LATVIA	
Law on Forest Management and Utilization, 1994	Law on Natural Resources Tax, 1995
National Biodiversity Action Plan, 1995	
LITHUANIA	
Forest Law, 1994, updated in 1996	Regulations on Forest Enterprises 1995
Reg. on General Forest Enterprise, 1996	Reg. on State Forest Service, 1995
Reg. on State Forest Inspection, 1995	Reg. on Forming and Using the Forest Fund, 1995
Forestry and Forest Industry Development Programme, 1994, updated in 1996	

Table 7.3. (Cont.) Forest-related policies and legal instruments in selected CiTs

FYR of MACEDONIA	
Strategy for the Development of Agriculture, Forestry and Water Economy, 1995	Law on Hunting, 1996
Law on Forest, 1997	
POLAND	
Act on Forests, 1991, 1997	Act on Protection of Nature, 1991
Act on Protection of Environment, 1980, 1997	Act on Protection of Arable and Forest Land, 1995
Polish Policy for Comprehensive Forest Resource Protection, 1995	National Forest Density Advancement Programme
Land Development Law, 1994	National Policy on Forests, 1997
ROMANIA	
Law on Environmental Protection, 1995, 1999	Law on Hunting Fund and Protection of Game, 1996
Forestry Code, 1996	
RUSSIAN FEDERATION	
Forest Code, 1997	
SLOVAK REPUBLIC	
Act on the State Fund of Improvement of Forests of the Slovak Republic,1991	Act on Forest Management and State Forestry Administration
Act on Forests, 1977	Act on the Slovak Chamber of Forestry, 1993
Conception of Forest Policy by the year 2005	
SLOVENIA	
Act on Forests, 1977	Act on the Slovenian Chamber of Forestry, 1993
Conception of Forest Policy by the year 2005	
UKRAINE	
Law on Environmental Protection, 1991	Forest Code, 1994

The state forest administrations have also been restructured in countries without private forestry, but the restructuring in these countries has mainly been aimed at dividing the vertical production lines and privatizing certain activities. One of the most interesting processes has been the ongoing transformation of the Federal Forest Service in the Russian Federation, the largest forest service in the region, which had been led by a Minister until May 2001. The agency was then merged with the Ministry of Natural Resources with the aim of reforming its entire structure and function. While changes to the institutional structure of a forest service responsible for the largest forest property in the world are of high interest, no results of this transformation have yet been made public.

Structural change has had marked effects on the labour force employed in forestry. A decrease in production resulted in decreased income, and fewer people were employed. Most forestry workers were released from the state service and became entrepreneurs, and the number of professionals also decreased. Privatized services, such as transport, logging, inventory and some types of planning, helped to absorb this surplus of mental capital, but financial instruments facilitating the start of these new enterprises were very limited. The state companies sold or leased the capital equipment to the new entrepreneurs, but concerns over whether these new activities would be profitable enough to cover the cost of modernization or replacement of the capital equipment were justified by subsequent developments.

The changes in employment levels were quite similar across the CiTs and helped to establish market economy conditions in the forestry and forest industries sector. However, income levels, in general, remained below the national average in most of the European CiTs.

Setting up the appropriate infrastructure for private forestry seems to have been more difficult than expected at the beginning of the transition. The restitution process is not complete, even in those countries where it started in the early 1990s, partly due to the gaps in the legal system and partly due to the technical difficulties of identifying the new owners of the former collectively owned properties. Also, the regulations were not oriented towards the creation of private forest holdings of a suitable size for commercial forestry operations. Consequently, the fragmentation of forested land has become one of the main limiting factors of the economic viability of SFM (see Table 7.4). Legal and financial instruments were too weak to counter the strong negative feelings associated with some types of common forest management, which were feared simply as the reincarnation of the previous political system.

Table 7.4. Average size of private holdings on forests and other wooded land (FOWL) in selected countries (source: Anon., 2000).

Selected EU countries		Selected CiTs	
Country	Average size (ha)	Country	Average size (ha)
Belgium	2.5	Croatia	0.9
Denmark	17.9	Czech Republic	3.0
Germany	9.5	Estonia	10.3
Greece	934.4	Latvia	10.5
Ireland	13.1	Lithuania	2.5
Italy	8.8	Slovakia	31.0
Luxembourg	3.4	Slovenia	2.7
Netherlands	5.8		
Spain	30.1		
Sweden	92.6		
United Kingdom	13.4		

Following the transition, the number of forest owners in the European CiTs has increased considerably compared to the pre-transition period. This has exposed a substantial weakness in the new forestry administrations; they were unprepared for the new tasks or to deal with new clients, who, in turn, were in need of various kinds of assistance, such as extension, education, financial support to cooperation, and special management objectives. To the extent that general economic development allowed, several instruments were introduced in the following years to bridge these gaps, but, as concluded by several recent analyses, a firm economic basis for sustainable private forestry has yet to be created in the European CiTs. Surprisingly, the economic problems faced by both state and private forestry enterprises are very similar in the region.

7.5 Economic aspects

Although the economic transition has provided solutions for some of the environmental problems (e.g. a decrease in air pollution as a result of the closure of many power plants and factories), most of the problems have required considerable financial resources. Rehabilitation of degraded forest ecosystems, changes to the tree-species composition in favour of native species, limitation of clear-cuts, erosion control, special management for protection and conservation and production of non-wood goods and services are expected to be partly financed by the forest sector itself, while activities aimed at rural development (e.g. afforestation of abandoned land) would require external resources.

In the European CiTs, wood production still generates the highest proportion of income in the forest sector; therefore, its efficiency is crucial. A paradoxical situation is that, while forest resources allow for increased production, and the proximity of timber major markets and relatively low labour costs provide a suitable environment for efficient timber production, outdated or inefficient capital equipment and underinvestment in the sector make it very difficult to attain efficiency. While forestry's contribution to the GDP in the European CiTs remained stable or increased slightly in the 1990s, the investment in the sector amounted to about 50% of the GDP contribution, and was as low as 25% in Slovenia and less than 20% in Slovakia. In the early stages of its establishment, private forestry was extremely sensitive to shortages in capital. In addition, wood production could not increase without increasing the domestic demand for wood, which drastically decreased at the beginning of the transition period. The promotion of wood use and its competition with less environmentally friendly substitutes, such as plastics, require concerted efforts and additional resources.

Improving the marketability of forest products and services, such as protection, recreation or aesthetic values, is an even more challenging task in the European CiTs. These traditionally 'free goods' need to be properly valued, and the cost of their continued supply should be covered by the users (e.g. local communities or the whole society) since the goods may be the main, sometimes only, products of several forest holdings. Like the vast areas of virgin forest in the Russian Federation, the forests of the European CiTs have enormous value, as the relatively low level of exploitation and the existence of a permanent forest cover in recent centuries have helped to maintain the forests in a semi-natural state. These forests are quite rich in species and are habitats for some rare predator species that have been extirpated across much of their former ranges in Europe. Forest policies have recognized this and, according to the TBFRA 2000, the area of forests not available for wood supply in the European CiTs has increased considerably between the last two reference periods. Conservation was identified by most of the countries as the main reason for protection. However, forest owners are often not compensated for the limitations that are imposed on their forests. This lack of coordination between goals and methods decreases the efficiency, or even the viability, of the implementation of the policies related to public services.

The market for forest services is underdeveloped in almost all European CiTs, although the most recent rural development plans have placed emphasis on activities such as ecotourism. In the absence of such markets, these services, traditionally available free of charge, are expected to be financed by the state budget, which in turn increases the role of the large central redistribution systems and narrows the space for market mechanisms.

Also, the role of forest products other than wood is not clearly defined in achieving SFM. The European CiTs are important providers of non-timber products, but most of them have fragmented information on quantity and quality. For example, the Czech Republic, which is the country with the most complete forest inventory, has found it hard to determine the real volume and value since self-consumption, an important element, is not monitored. Hunting and related services were traditionally the most developed areas of non-wood goods and services, and many of the CiTs have become preferred destinations for hunting tourists. Although hunting has a high level of economic efficiency, it generates problems by encouraging and maintaining high populations of game. Future policies concerning non-wood products need to consider their impact on the environment and be based on a more complete analysis of their varying advantages and disavantages.

Due to improvements in their economic situation, many European CiTs have been able to allocate some resources to emergency preparedness (e.g. response to forest damage). Compensation includes covering the costs associated with abatement, tax relief on the income from sanitation felling, contributions to the extra costs associated with using certain tree species or management techniques and the financing of projects designed to improve forests or production capacity. However, the aims of financial policies are often not clearly defined, and these policies tend to be formulated for the short term. Unfortunately, without economically sound and complex forest valuation techniques, the target set for financial policies in the European CiTs will remain biased and driven by wishful thinking.

References

Anon. (1992) *The Forest Resources of the Temperate Zones. The UN-ECE/FAO 1990 Forest resources Assessment. Vol. 1, General Forest Resource Information.* ECE/TIM/62. United Nations Economic Commission for Europe, Food and Agriculture Organization of the United Nations, Rome, Italy.

Anon. (2000) *Main Report. Forest Resources of Europe, CIS, North America, Australia, Japan and New Zealand (Industrialized Temperate/Boreal Countries).* Geneva Timber and Forest Study Papers 17. United Nations Economic Commission for Europe, Geneva.

Anon. (undated). *European Timber Trends and Prospects into the 21st Century*. ECE/TIM/SP/11. United Nations Economic Commission for Europe, Food and Agriculture Organization of the United Nations, Rome, Italy.

European Environment Agency (1995) *Europe's Environment*. European Environment Agency, Copenhagen.

Liaison Unit (1993) *Report on the Follow-up of the Strasbourg Resolutions*. Ministerial Conference on the Protection of Forests Liaison Unit, Helsinki, Finland.

Lipman, D. (1994) The Russian forest industry during the transitional period. *Unasylva* 45, 179.

Phare Environment Consortium (1999) *Conservation and Sustainable Management of Forests in Central and Eastern European Countries*. EC Phare Programme. Available at: http://www.fris.sk/Phare/Projects/ManagementOfForestInEurope/index.html (accessed 22 June, 2005).

UN/ECE/FAO (1996) *The European Timber Trend and Prospects: into the 21st Century*. ECE/TIM/SP11. New York and Geneva.

Certification of Forest Management and Wood Products

8

J.L. Innes and G.M. Hickey

Faculty of Forestry, University of British Columbia, Vancouver, Canada

During the 1990s, a remarkable change occurred in forest management, namely the emergence of independent certification of forest management and forest products. The certification process covers the management of specified forest areas and, if chain of custody is demonstrated, the products from these forests. Certification is closely linked to sustainable forest management, and several certification schemes are based on the criteria and indicators developed through international negotiation. The area of third-party certified forests has increased extremely rapidly, and by the end of 2002 had reached almost 125 million hectares worldwide. However, the standards to which these forests are certified vary considerably. While the mid-1990s saw a proliferation of certification schemes internationally, by 2004 there was evidence of consolidation, with a few dominant schemes or frameworks emerging. As certification becomes more widespread, it is being increasingly recognized as a tool for promoting sustainable forest management. This represents a major change in the socioeconomic and political environment of forestry in more industrialized countries. Market-based mechanisms to promote particular forest management practices have not been used widely in the past. As a result the active support by some governments for the certification process represents a major shift in the way in which forest policy is being developed and applied. However, many problems remain, and, while the forest sector is moving rapidly towards certification, the environmental groups that initially promoted the concept are having considerable difficulty in ensuring that their original goals are met. Increased dialogue between the different stakeholders of forest certification will help to resolve some of these problems with time.

© CAB International 2005.
Forestry and Environmental Change: Socioeconomic and Political Dimensions
(eds J.L. Innes, G.M. Hickey and H.F. Hoen)

8.1 Introduction

Over the past decade, third-party certification has become a new force affecting forest management internationally. Forest certification involves management audits, undertaken by independent groups, designed to determine the extent to which a forest is managed 'sustainably'. These audits can extend from the forest to the entire supply chain, enabling the labelling of specific products as having come from 'sustainably managed' sources (Viana *et al.*, 1996; Wallis *et al.*, 1997). A unique feature of forest certification is that the management standards are often defined by groups independent of the government in whose jurisdiction the audit takes place (e.g. Weber *et al.*, 2000; Hickey, 2004). Government representatives are sometimes present when the certification standards are being set, and in some countries, such as Sweden, Finland and the Netherlands, third-party certification may be seen as complementing general environmental legislation (Terstad, 1999). This represents a major change in government approaches to improving forest management standards. According to Murray and Abt (2001), market-based incentives for sustainable forestry have traditionally been seen as an instrument used by government to control use in the form of tradable permits and/or externality taxes.

The idea of certifying that forest management practices adhere to a particular standard can be credited to environmental non-governmental organizations (ENGOs). Frustrated at the lack of progress made at the 1992 United Nations Conference on Environment and Development (UNCED), and increasingly concerned about continued deforestation and forest degradation, ENGOs set up the Forest Stewardship Council (FSC) in 1993. Since then, certification has grown rapidly. For example, in a survey conducted in Ontario, Canada, in 1997, Spinazze and Kant (1999) found that awareness of certification by primary wood producers and re-manufacturers in Canada was very low. By 2001, this situation had changed dramatically, such that Ontario was seeking province-wide certification from FSC International. Much the same situation existed in the United Kingdom and Finland towards the end of the 1990s, with private landowners being initially suspicious and sceptical of the certification process (Lindstrom *et al.*, 1999). However, today certification is generally seen by forest managers as a means of increasing public confidence in forest management, ensuring market access for forest products and responding to pressures from ENGOs. As such, it is sometimes seen as a means of ensuring that resource managers maintain their social licence to practise forestry.

A major hindrance to the development of forest certification has been disagreement over objectives and the subsequent standards that should be set (e.g. Weber *et al.*, 2000). The primary objective of third-party certification is to let a consumer know that a particular wood product is derived from a sustainably managed forest (e.g. McIntyre and Beaumont, 1999; Teisl and Roe, 2000). The problem, however, lies in defining what constitutes a sustainably managed forest. As a result, the standards adopted by the various certification schemes differ, with those being promoted by industry generally having different environmental standards than those promoted by ENGOs. For example, in 2002, some members of the forest industry certified by the Programme for the Endorsement of Forest

Certification (PEFC) in Sweden and Finland experienced negative attention from environmentalists due to their treatment of indigenous peoples and old-growth harvesting issues respectively (Raunetsalo *et al.,* 2002). In each case the PEFC was said to be weak, non-credible, industry-biased and containing weak social and ecological criteria (Vallejo and Hauselmann, 2001; World Wide Fund for Nature (WWF), 2003).

Even within schemes, there are differences. For example, one reason that the province of British Columbia, Canada, took 5 years and three drafts to develop a regional standard for the FSC scheme was the insistence of exceptionally high environmental standards by the environmental representatives in the steering group. The creation of a very high starting standard is implicit in the assumption that certification can only occur once sustainable forest management has been achieved (Lawes *et al.*, 1999). However, it fails to acknowledge that most certification schemes will allow the certification of a forest area if there is a commitment on the part of the manager to continually improve towards sustainable forest management. There is no requirement to have actually achieved this complex and often nebulous goal. There is a major practical difficulty here. If the standards are set too high, and there is little demand for certified wood products, certification will not be seen as economically viable for forestry companies, leading to the failure of the scheme. As a result, the desired improvement in forest management standards internationally will not be achieved through certification. Similarly, if standards are set too low, there will be no consensus between environmental groups and industry on the credibility of a particular standard of forest management, resulting in the failure of certification. Clearly a balance needs to be achieved.

This chapter examines the nature of the certification available to the forest industry internationally and describes a range of different schemes. It also examines some of the issues facing certification as a mechanism to improve forest management standards internationally.

8.2 Different forms of certification

There are two basic forms of certification. The first, often referred to as process-based certification, focuses on the development of a systematic approach to management (e.g. the International Organization for Standardization's (ISO) 14001 standard). The other, referred to as performance-based certification, specifies performance standards that a management operation must meet (e.g. the FSC). Many of the performance-based schemes also contain process-based elements, and some process-based schemes contain a few performance measures. Consequently, the classification can be fairly indistinct. Process-based schemes give much greater flexibility to the manager over issues such as the choice of technology to be used in the forest, the silvicultural practices adopted, the volume of timber to be removed and the extent to which other values (e.g. water) are protected (Gale and Burda, 1998). As a result, environmental groups are often sceptical of process-based approaches as they do not necessarily specify the standards of environmental protection that are to be achieved.

Certification schemes also vary in the rigour of the standards and the degree of enforcement. First-party certification is undertaken by the forest company or management organization itself, which then determines whether or not its management is achieving the desired outcome. Such certification systems generally have very little credibility outside the industry itself (e.g. McIntyre and Beaumont, 1999). Second-party certification is undertaken by an organization to which the manager belongs. Standards are generally more rigorous than with first-party certification, but there may still be substantial credibility issues. Third-party certification occurs when an organization without any connection to the manager undertakes the audit. Increasingly, third-party certification is the only type that is considered to be credible.

The proportions of forest certified by different schemes vary among jurisdictions, depending on availability and whether or not appropriate regional standards have been developed. During the 1990s, there was a proliferation of certification schemes, each with its own set of standards. These are gradually converging, although the distance between some schemes remains substantial. Various attempts have been made to provide comparisons between schemes [e.g. Confederation of European Paper Industries (CEPI), 2000], but these comparisons have quickly become dated. The large number of certification schemes has the potential to cause confusion, although this may be resolved in part through mutual recognition or similar approaches. For example, the FSC recognizes a number of local schemes, including the CFV (Consejo Boliviano para la Certificación Forestal Voluntaria) and the UKWAS (UK Woodland Assurance Scheme Steering Group), amongst others. The PEFC recognizes the Finnish Forest Certification Scheme (FFCS), Living Forests (Norway), and LRF Skogsagarna (Sweden) amongst others. These schemes can then use the logo of the recognizing institution, facilitating identification in the marketplace.

The following sections describe a number of different certification schemes and certification frameworks that have emerged to play an important role in forestry. It is not designed to be exhaustive. It should be noted that some organizations develop a specific set of standards that auditors then use when certifying a particular forest (e.g. ISO 14001 and the Sustainable Forestry Initiative (SFI)). These are referred to throughout this chapter as certification 'schemes'. Other organizations approve particular standards, rather than developing their own, although they may have generic standards that can be used as guidelines. These are referred to here as certification 'frameworks', and include the PEFC and the FSC. To distinguish among the initiatives presently available internationally, issues related to the geographical coverage (i.e. global, regional or national), the existence of on-product labels, customer demand for certified products, the use of third-party auditors and support from the forest industry and/or environmental groups need to be considered (Vogt *et al.*, 2000).

8.2.1 International Organization for Standardization

The International Organization for Standardization (ISO) has developed voluntary standards for environmental management through its ISO 14000 series. Forest management is certified through the ISO 14001 standard (formally endorsed in 1996), which defines the criteria for an Environmental Management System (EMS). The objective of the EMS is to provide a framework that will ensure that managers identify and address the important environmental aspects of their management and the environmental impacts that may result. The system does not set or define specific performance criteria. Rather, a manager seeking ISO 14001 certification can establish unique standards for environmental performance. As with most other schemes, a commitment to continual improvement is required. Forestry is addressed specifically through ISO/TR 14061, which is a document that links ISO 14001 to woodland operations through criteria and indicators for sustainable forest management. Since 1996, a significant number of forestry companies have taken up the scheme (e.g. as of April 2005, over 137 million ha of Canada's forest had been certified under ISO 14001). The scheme has also been fairly successful in Australia, New Zealand and Japan.

One of the biggest problems associated with ISO 14001 certification is that the environmental performance criteria are set by the manager rather than by an independent body. This means that the initial standards can be set quite low, making it relatively easy for a manager to achieve the performance criteria. Furthermore, there is no requirement to seek an external audit of compliance. This lack of any external standard is the reason that environmental groups often discredit the scheme. However, if the general goal of certification is to improve the standard of forest management, then ISO 14001 plays an important role. The documentation required by the EMS is compatible with other certification schemes, so, once ISO 14001 has been achieved, other certifications can often be obtained at a lower cost.

8.2.2 Forest Stewardship Council

The Forest Stewardship Council (FSC) was established in 1993, and is currently (2004) based in Bonn, Germany. The FSC originated from a broad base of interests, with environmental groups (e.g. the World Wide Fund for Nature (WWF), the Rainforest Alliance and Greenpeace) and timber companies (e.g. AssiDomän AB, Sweden) involved in its conception. As of June 2004, over 43 million hectares of forest in more than 60 countries had received FSC certification.

The organization is governed by a General Assembly that meets regularly. The voting powers were initially divided between two chambers: one dealing with economic issues (25% of votes) and the other dealing with social and environmental issues (75% of votes). Since 1996, there have been three chambers, with the voting powers divided equally between the economic, environmental and social stakeholders. Within each of the chambers, voting is also divided equally between northern and southern sub-chambers. This structure ensures that no one group has dominance, yet there is uncertainty where some groups fit (e.g. community-run forests). The organization is run by a Board and Secretariat, with the Board comprising nine members that represent the environmental and social (seven members) and economic (two members) interests. Thus, while voting in the

General Assembly is divided equally amongst interests, the Board remains weighted towards environmental and social stakeholders.

FSC certification is undertaken by third-party auditors that have been accredited by the FSC. In each case, the forest manager must comply with a set of principles, criteria and indicators. National and regional standards must be finally approved by FSC International in Bonn, Germany. This provides some level of standardization; however, it remains clear that major differences in regional standards exist (Holvoet and Muys, 2004; Hickey *et al.*, 2005). If a manager wishes to obtain FSC certification before a national or regional standard has been developed, it is possible to do so using the generic standard. However, once a regional/national standard had been put in place, the manager must ensure that they comply within a specific time frame.

Table 8.1. Countries with over 1 million hectares of FSC-certified forest (June 2004) (source: FSC, 2004)

Country	Area of forest certified (million hectares)
Sweden	10.3
Poland	6.2
USA	4.8
Canada	4.3
Croatia	2.0
Latvia	1.7
Brazil	1.6
South Africa	1.5
Russian Federation	1.4
United Kingdom	1.2
Bolivia	1.1
Estonia	1.1

Sweden provides an excellent example of the role that ENGOs can play in forest certification. There, WWF initially promoted the concept of FSC certification to forest owners, discussions that later involved the Swedish Society for Nature Conservation (SNF). However, by 1995, there was an apparent impasse between forest owners and the environmental organizations. As a result, the ENGOs presented their own 'Preliminary criteria for environmental certification in Swedish forestry' in May 1995 (Van Kooten *et al.*, 1999). Discussions with forest industry representatives were later resumed, although there was a strong industry push to move towards a Nordic Forest Certification Programme. Here, the divergence between the operations of large and small companies became apparent, with larger companies supporting FSC certification and small owners (represented by the Forest Owners Association) pushing for the Nordic system. An FSC national standard for Sweden was approved in 1998, and Sweden now has the greatest area of FSC-certified forest in the world (10.34 million ha in June 2004). The other countries with over 1 million hectares of FSC-certified forest are presented in Table 8.1.

In the United Kingdom, 84% of the certified area is managed by Forest Enterprise, the state-owned agency for forestry. State-owned forest enterprises are also responsible for the majority of FSC-certified forest in Poland, Croatia and Estonia.

8.2.3 Programme for the Endorsement of Forest Certification

The Programme for the Endorsement of Forest Certification (PEFC) (formerly the Pan-European Forest Certification framework) represents an alternative approach to several of the other systems described in this chapter. It was launched by national forest owner and forest sector interest groups (from Austria, Finland, France, Germany, Norway and Sweden) in 1999, and initially emphasized the needs of relatively small private owners. The PEFC framework works by recognizing that nationally developed standards adhere to certain performance criteria and indicators. It also provides an on-product 'eco-label' for certified forest products in an effort to assure stakeholders that certain forest management standards have been met.

Since the inception of the PEFC, a number of national certification schemes in Europe have been developed (some voluntarily and others as a result of external pressures). A good example of a country that experienced external pressures to develop a standard is Sweden (Terstad, 1999; Elliott and Schlaepfer, 2001). The PEFC framework has received rapid acceptance from the forest industry and government sectors in Europe, although some environmental groups (e.g. the WWF, Greenpeace, the Rainforest Action Network) remain heavily critical of it. As of April 2005, the PEFC framework covered the greatest forest area internationally (122 million ha). The forest management performance criteria are based on the criteria and indicators developed through the 'Helsinki Process' in 1994, referred to as the Lisbon Pan-European Operational Guidelines (1998).

Table 8.2. National schemes endorsed by the PEFC (May 2005) (source: PEFC, 2005)

Country	Area of forest certified (million hectares)
Australia	1.8
Austria	3.9
Belgium	0.2
Canada	63.7
Czech Republic	1.9
Denmark	0.01
Finland	22.3
France	3.6
Germany	7.0
Italy	0.3
Japan	–
Latvia	0.03
Netherlands	–
Norway	9.2
Spain	0.4
Sweden	6.4
Switzerland	0.3
United Kingdom	0.01

One of the impediments to the use of the PEFC label on wood products has been a lack of chain-of-custody certificates. However, since 2000, many wood products manufacturers have attained these certificates to enable the labelling of their wood products in the marketplace. As of May 2005, the number of PEFC logo users was 14,419, the majority of which were in Germany (7156) and France (6297). Finland has by far the largest PEFC-certified forest area in Europe, with 22.3 million hectares, and Canada has the greatest area of PEFC-certified forests (63.7 million ha). Table 8.2 shows the countries with national schemes certified to the PEFC standards by July 2004.

The scheme has been poorly received by some environmental groups, who see it as a direct competitor to the FSC. For example, Greenpeace Italy, objecting to the possibility of Italy joining the PEFC, stated in a press release dated 26.7.2001:

> A number of weak and ineffective certification systems have been developed in response to the growing success of the FSC. One of these, the PEFC, was created on the impetus of forest owners and has resulted in highly questionable outcomes, certifying primary forest operations of low quality or involving practices that are actually destructive.

Although this statement is highly debatable, such campaigns are likely to continue as long as environmental groups see the PEFC as a threat to the success of the FSC.

Although it was originally developed in Europe, the PEFC has established itself as the international standard for certification schemes, and a number of non-European certification schemes have been endorsed (such as the Australian Forest Certification Scheme and CertforChile).

8.2.4 Canadian Standards Association

The Canadian Standards Association (CSA) provides an example of a certification scheme developed by a single jurisdiction (Canada). In September 1996, the Standards Council of Canada approved a set of standards for sustainable forest management (also known as CSA-Z809). In 2001/2002 these standards were revised to include chain of custody certification (also known as the Forest Products Marketing Program). The CSA standards require third-party auditing by an organization accredited by the Standards Council of Canada. The CSA approach requires an EMS that is consistent with the ISO 14001 standard, a significant public participation component and field performance measures (Lapointe, 1998). The field performance measures are developed during the public consultation process and are specific to each management unit (termed a 'defined forest area').

The CSA-Z809 standard follows the criteria and indicators for sustainable forest management developed by the Canadian Council of Forest Ministers (CCFM) (based on the seven criteria developed through the Montreal Process (1995)). These are accompanied by 21 'critical elements' that a manager is obliged to address through performance monitoring. There is also a requirement for continual improvement. As of June 2004, over 32 million hectares of Canadian forest had been certified under the CSA-Z809 standard. As of 2004, the CSA scheme had not received widespread support from the environmental community, partly because of the strong involvement of the Canadian forest products industry. The scheme does, however, require a major effort to be made by industry in a public consultation process. The CSA scheme has been recognized under the PEFC framework, but this framework also lacks the widespread acceptance of many environmental groups.

8.2.5 Finnish Forest Certification Scheme

The Finnish Forest Certification Scheme (FFCS) is similar to the CSA-Z809 in that it was set up as a national-level scheme and it involves a combination of process and performance measures. The FFCS standard was initially developed with the support of Finland's ENGOs (WWF-Finland and the Finnish Association for Nature Conservation) and forestry producers (the Finnish Central Union of Agricultural Producers and Forest Owners (MTK) and the Finnish Forest Industries Federation). However, since its initiation, ENGOs have had a difficult relationship with the scheme. This has resulted in several, including WWF-Finland, leaving the process. With the recognition of the FFCS by the PEFC framework in 2000, the FFCS became a direct competitor to the FSC. This has resulted in an acrimonious dialogue between supporters of the national scheme and supporters of the FSC standard for Finland.

The FFCS standards are based on the European Union's Eco-Management and Audit Scheme, ISO 14001, the Forest Stewardship principles and criteria, and the pan-European criteria and indicators. The scheme has achieved some independent recognition outside Finland, with a major British hardware retail chain (B&Q) recognizing it in 1999. In August 2001, a Working Group on Forest Certification Standards was established to revise the forest management requirements in the FFCS and to better incorporate the idea of continual improvement (FFCS, 2002).

8.2.6 Sustainable Forestry Initiative

The Sustainable Forestry Initiative (SFI) was established by the American Forest and Paper Association (AF&PA) in 1994. All members of the AF&PA must participate in the SFI programme, with any members failing to comply being expelled. Originally designed to improve industrial practices and reporting, the SFI programme is based on 9 key principles, namely sustainable forestry, responsible practices, reforestation and productive capacity, forest health and productivity, long-term forest and soil productivity, protection of water resources, protection of special sites and biological diversity, legal compliance and continuous improvement. It does this through a series of objectives and performance measures. In an attempt to improve the SFI's credibility with ENGOs, the AFPA created the Sustainable Forestry Board in 2000. This 15-member, multi-stakeholder board (66% of its members are from non-forest industry interests) manages the SFI Standard, the SFI Verification Procedures and the SFI program compliance. In 2001, the Board became a separate, not-for-profit, entity and the revised 2002–2004 SFI Standard was accepted. A significantly revised version of the Standard was issued in January 2005. Adherence to the SFI principles can be assessed at various levels: by the company itself, by a timber buyer, by another forestry company or by an independent third-party audit. In 2002 a major step towards international recognition was made with the establishment of an SFI Canadian Review and Interpretations Task Force (CTF) to ensure that the differing operating environments were considered when the SFI was applied in the Canadian context (Raunetsalo *et al.*, 2002). As of July 2004, 21.4 million hectares of forest in Canada had been certified under the SFI scheme. In the USA, more than 17 million hectares had been third-party-certified by April 2004.

The SFI scheme has been strongly opposed by some environmental non-governmental organizations, who see it as having lower environmental standards than the FSC and who also see it as a competitor. For example, in December 2001, plans by the SFI scheme to issue a label for timber originating from certified forests were postponed following threats of a negative advertising campaign by the Rainforest Action Network. Nevertheless, in 2002 the SFI legally registered an on-product labelling mark titled 'A Good Sign that Somebody Cares' in a number of countries (Raunetsalo *et al.*, 2002). In 2001, the US-based Meridian Institute published a comparison of the SFI and FSC, the result of a study commissioned jointly by Home Depot, the AF&PA and the FSC. The two certification schemes were found to be very similar in most of the scientific aspects related to forest management (Meridian Institute, 2001).

8.2.7 American Tree Farm System

The American Tree Farm System (ATFS) was established in 1941 and is the longest running certification scheme available today. It is a performance-based scheme that involves predominantly first-party auditing of forest management practices, although in 2002 third-party audits were introduced. It has been especially designed for small landowners in the USA. At the end of 2002 there were 65,549 certified 'Tree Farms' covering 10.5 million hectares, (approximately 60 percent of forestland), in 48 states. The ATFS has established standards and guidelines that property owners must meet to become a certified Tree Farm. In order to achieve certification, private forest owners must develop a management plan based on environmental standards and pass an inspection by an ATFS volunteer forester every five years.

8.2.8 Green Tag forestry

Green Tag is an example of one of the small-scale certification systems that emerged during the 1990s. It is a performance-based certification scheme developed by the US National Forestry Association, in consultation with the

Association of Consulting Foresters and the National Woodlands Association. The first certificate was awarded in August 1998. It is based on second-party certification and third-party verification (by a consulting forester).

The costs of this scheme have been kept deliberately low. Certification requires a forest management plan to be in place; however, this is not a major issue, since a pre-approved state or federal stewardship plan may be used, or the manager may utilize state 'best management practices' as the plan. The system has had difficulty with chain of custody. Because of this, and because the standards are generally at the minimum required within a state, the scheme is unlikely to have much support from ENGOs in the near future.

8.2.9 Australian Forestry Standard

The Australian Forestry Standard (AFS) was developed in 2002 to provide a basis for third-party certification against a performance-based standard of criteria and indicators. The AFS has been designed to evaluate management practices in both native and plantation forests. In 2004, the AFS was endorsed by the PEFC.

Importantly, Australia's National Association of Forest Industries (NAFI) has indicated a full commitment to the AFS. Based on the historically negative stance of the NAFI with regard to third-party certification, this commitment has made many non-industry stakeholders wary of the credibility and transparency of the standard.

8.2.10 Lembaga Ekolabel Indonesia

Lembaga Ekolabel Indonesia (LEI) has received considerable support from government, industry and environmental groups with its attempts to develop a national certification standard for Indonesia. Criteria and indicators were developed and field-tested in 1994 and revised and tested again in 1996. By 1997, a certification programme had been developed, termed the 'Eco-labelling Certification Programme for the Sustainable Management of Production Forest at the Forest Management Unit Level'. This was a full certification scheme, including standards of management, chain of custody procedures and a set of principles for certifying institutions. The scheme is based on the principles for sustainable forest management developed by the International Tropical Timber Organization (ITTO), adapted for Indonesian conditions. The LEI also has the backing of the World Bank. Certification in Indonesia faces a number of problems. Not least is the lack of demand for certified forest products. The major markets for Indonesian timber do not currently demand certified timber so price premiums are rare. Another major problem is the lack of skilled personnel capable of undertaking the audits. Costs of certification to the LEI standard have compared favourably with the price paid for the FSC certification in the region.

8.3 Costs and benefits associated with forest certification

Without exception, third-party certification adds a direct cost to forest management. This is because an audit team must be employed to make the certification. It is often argued that the increased efficiency associated with certified management offsets the costs, or that the costs can be recovered through a price premium for certified products. There is, however, very little evidence to support the former, and the latter is only true in a few cases. For example, the town of Solothurn in Switzerland has more than recouped its certification costs through a premium obtained for its certified timber. This section will provide an overview of the costs and benefits associated with forest certification internationally.

8.3.1 Direct and indirect cost of certification

The direct costs of certification are extremely difficult to determine. Forest managers are reluctant to indicate their expenditures in this area as it is commercially sensitive, while auditors decline to provide such information as it varies depending on the nature of the management unit to be certified. Murray and Abt (2001) claim that costs are 'quite small' for large ownerships, at less than US$1 per hectare, but may be proportionally larger for smaller ownerships. This is the case in Canada, where at least one ISO certification (conducted in 2000) has involved a cost of ca. US$0.63 ha^{-1} for an area of approximately 100,000 ha. Interestingly, the FSC certifications for Collins Pine Co. cost US$0.63 ha^{-1} for their Californian operation (38,041 ha, certified in 1993) and US$0.62 ha^{-1} for their Pennsylvanian operation (49,373 ha, certified in 1994) (Jenkins and Smith, 1999). Collins Pine also incurred additional costs. Annual fees of US$7,200 are incurred for each location, forest management costs at Almanor Forest, California, are estimated to have doubled after certification, and chain-of-custody costs are estimated to have added a further US$150,000 annually. Elsewhere, direct costs have varied widely, from US$0.02 ha^{-1} for FSC certification of the 1.27 million ha Mazuma Crafts Ltd. forest in Zambia, to US$0.90 ha^{-1} for the 52,000 ha Chiquitano forest enterprise at Lomerio, Bolivia (Thornber and Markopoulos, 2000). In Chile, the direct cost of obtaining ISO 14001 certification in 1996 for an area of 82,564 ha of plantation forest was US$1.77 ha^{-1}.These figures suggest that costs vary quite widely, although this might be expected given the variety of factors that will influence the cost of a particular certification.

Economically viable certification for small landowners is possible through schemes that allow 'group certification', a process that has been successfully implemented in many European countries. In 2002 the FSC awarded the world's first group chain-of-custody certification in the UK to a group of small-scale sawmills and timber merchants known as Independent Forestry Ltd. (Raunetsalo *et al.*, 2002). Group certification has also been successful in some Latin American countries, where timber is often produced by a large number of small-scale, informal operations (e.g. Sierra and Stallings, 1998). However, certification in several Latin American countries (e.g. Bolivia and Costa Rica) has been accompanied by major reforms in forest policy that have institutionalized the costs of certification and have linked certification to the auditing of compliance with concession requirements (Nittler and Nash, 1999). This is an interesting development that could be used in a number of situations where forest policy tends towards results-based management.

The indirect costs of certification can be related to the management actions that are required to meet the documented standards. These costs vary according to the nature of the management operation (i.e. it will be more costly to modify intensive forestry operations than to change less-intensive practices) (Murray and Abt, 2001). An additional factor that needs to be considered is that certification may be a means by which a manager improves practices that might have been improved anyway. For example, ISO certification requires the development of an EMS, something that is seen by most managers as useful in day-to-day business. Consequently, many managers that have actually undertaken a certification exercise have found it useful (Hayward and Vertinsky, 1999).

One way in which costs may be reduced in the future is through joint auditing. For example, in 2001, Diamond Raya Timber in Indonesia obtained a joint FSC–LEI certificate (Vilhunen *et al.*, 2001). In 2002, the Estonian State Forest Management Centre was certified jointly to FSC and ISO 14001 standards. Such joint certifications offer considerable potential for cost saving; however, mutual recognition would most likely provide an even better approach to resolving this problem.

8.3.2 Price premiums

To date, most of the studies related to price premiums have been based on surveys of consumers' willingness to pay (e.g. Winterhalter and Cassens, 1993; Ozanne and Vlosky, 1997; Vlosky and Ozanne, 1997; Stevens *et al.*, 1998; Spinazze and Kant, 1999; Vlosky *et al.*, 1999). These have tended to reveal widely varying estimates, reflecting the inherent difficulties associated with willingness-to-pay surveys, especially in the environmental field. From the research, a figure of ca. 10% seems to be the generally agreed amount that consumers say that they would be willing to pay; however, this willingness to pay is not common to all consumers. For example, only about 3% of the customer base of US companies not selling certified wood products was found to be willing to pay a premium for certified timber (Stevens *et al.*, 1998). In a more recent survey conducted for *National Home Centre News* (USA) in 2001, 62% of respondents were willing to pay up to 5% more for certified forest products, but 30% were not willing to pay a premium. Stevens *et al.* (1998) indicate that in the USA, the average premium charged to certified wood buyers is 4.7% and the average premium paid to manufacturers is 6.6% above the price of comparable non-certified products.

The failure of consumers to pay price premiums for certified timber products may reflect a lack of product education on the part of the consumers. It may also highlight a failure of the forest industry to understand the characteristics of 'green' consumers (see Coddington (1993) and Shrum *et al.* (1995)). For example, Teisl and Roe (2000) noted that, for a consumer to buy a certified product, consumers must understand, believe and care about the information being presented. Unfortunately, at the present time, there is a high level of uncertainty, even amongst specialists, as to what constitutes a sustainably managed forest. Some of the reluctance may also be attributable to consumers' scepticism over claims about sustainability (Zinkhan and Carlson, 1995; Kiekens, 2000), although the efforts of the WWF to promote the FSC may allay some of these fears. It should also be noted that the marketing opportunities offered by certification are not always fully exploited. For example, Karna *et al.* (2000) have found that Finnish forestry companies are much more likely to take advantage of the marketing opportunities presented by certification than Swedish companies.

Where price premiums exist, it is most likely because demand for a particular wood product exceeds supply. This appears to be the case for British Columbian red cedar being sold in the Netherlands (5% premium in 2001). However, as more forestry companies achieve certification, any price premium is likely to disappear (Côté, 1999). This has led some (e.g. Rice *et al.*, 1997) to question whether certification is of any value as a driver of change in forest management. However, market premiums are not the only motivation for certification – market access is also an issue.

8.3.3 Market access

Vilhunen *et al.* (2001) argue that it is not the possibility of promoting sustainable forest management that is influencing forest companies to adopt certification. More important is an expected loss of market access and loss of income. For many companies, certification has been encouraged because of decisions by timber purchasers to insist on forest products from certified sources only.

Internationally, Western European do-it-yourself (DIY) retailers have been the major catalysts in moving markets towards certified products (e.g. IKEA, OBI). More recently this has also occurred for some of the larger American purchasers, such as HomeBase, Lowes and The Home Depot. ENGOs are currently advocating the use of only a few certification schemes, with preference being given to the FSC. For example, the WWF have encouraged this policy through the formation of buyers groups that will preferentially purchase FSC-certified timber. However, forest product buyers in countries such as the USA are increasingly supporting a range of certification schemes. For example, Hallmark Cards announced in 2001 that they would recognize CSA, FSC, SFI, the ATFS and the PEFC standards. Timber producers have noted these demands, and therefore continued market access is a major factor influencing decisions to start the certification process (Stevens *et al.*, 1998).

The Global Forest and Trade Network (GFTN) has been at the forefront of encouraging partnerships between companies that produce, use, buy, sell or specify timber and paper products. Specifically, it encourages companies to produce or purchase increasing quantities of wood from well-managed forests that are independently certified. The network consists of a loose association of national groups such as Club Pro Forêts in France, the WWF 1995+ Group in the UK and the Certified Forest Products Council in the USA. As of December 2002, groups had been or were being set up in 18 countries (GFTN, 2002). The GFTN claims to have no formal, legal or institutional linkages with FSC, and has played an important role in developing mutual recognition amongst certification schemes. For example, it recognizes as partners not only the FSC, but also other 'emerging credible schemes', although, up to June 2004, none of the existing major schemes had been included in this category. This is important because the success of certification will depend on obtaining a 'critical mass' of certified wood in the marketplace (see Simula *et al.*, 2001).

8.4 Problems associated with forest certification

There are many problems associated with the implementation of certification in forestry. A large proportion of these stem from the difficulties associated with some of the criteria. For example, it is very difficult to measure the extent to which a manager has shown respect to indigenous populations, because what constitutes respect to one culture may be different to what another culture expects. In this section, some of the major issues relating to forest certification internationally are examined.

8.4.1 Standard setting

Some of the problems associated with certification can be related to the lack of scientific knowledge applied during the setting of standards. Links between scientists and those setting the standards have sometimes been weak, although, in some countries, strong links have been forged (e.g. Sweden, see Axelsson and Ostlund (2001)). Related to this is the way in which certification standards are being developed and applied. Certification was initially designed to be an audit, not a set of instructions on how to practise forestry. It should not be confused with the development of criteria and indicators for sustainable forest management (Lawes *et al.*, 1999), although this is precisely what has happened during the development of some regional standards.

Some of the value-based indicators that have been selected may be inappropriate as indicators of sustainability. Vogt *et al.* (1999) argue that the selection of value-based indicators may result in the incorrect identification of the important constraints that control ecosystem function. They may also be poor predictors of ecosystem sustainability. They argue that it is important to select non-human value-based constraining variables and that the same suite of indicators cannot be used automatically at every site. This creates problems for performance-based standard-setting groups, as the participants invariably use human-based values and indicator suites that are applied universally within a particular region. An example is provided by the emphasis that many standards place on stand-level characteristics, and the relative lack of attention paid to landscape-level characteristics. However, the latter may be much more important for the conservation of biodiversity (Brown *et al.*, 2001).

In areas where the debate over forest management practices has become polarized, the setting of standards may become a difficult and drawn-out process. This was the case for the FSC regional standard for British Columbia, Canada. The process was initiated in 1996 by a group of environmentalists, with no involvement of the forest industry (Hoberg, 2001). Draft standards were issued in June 1999, but the lack of industry involvement meant that they were largely ignored. A revised process, involving industry, resulted in a second draft standard being produced in 2001 and ultimately a third draft in 2002. Debate over certification standards in Europe has also been significant. For example, in north-eastern Finland, logging in an area of PEFC certified, state-owned forest has drawn considerable criticism from environmentalists due to differing views of what constitutes 'old-growth' forest. According to both the WWF and Greenpeace, clear-cutting in these forests provided a clear example of the ecological and social weakness of the PEFC (Raunetsalo *et al.*, 2002).

8.4.2 Standards set by different schemes

Given the range of different certification schemes available to the forest industry, it is important for consumers to know how schemes rate against each other. This is achieved through a process known as mutual recognition, whereby one scheme recognizes that another scheme is equivalent to its own standards (Elliott and Hackman, 1996; CEPI, 2000). So far, only limited progress has been made in this area (see Raunetsalo, *et al.*, 2002). However, as of February 2005 both the CSA and SFI schemes (amongst others) were being considered for recognition by the PEFC. In North America the FSC and SFI have been shown to have very similar standards (Meridian Institute, 2001). Environmental groups (especially as represented by the FERN coalition) have been generally unhappy with such comparisons. In April 2001 the FERN claimed that 'the Forest Stewardship Council is currently the only independent and credible certification system in the market' (Ozinga, 2001).

Systems for comparing different certification schemes have been established by the International Forest Industry Roundtable (IFIR) and the CEPI in 2001 (Griffiths, 2001). The IFIR system aims to differentiate credible and non-credible certification schemes and is intended to have very high entry standards. Similarly the CEPI framework for mutual recognition has tried to identify the strengths and weaknesses of each scheme available with a view to establishing comparability. In 2002, the CEPI released a third edition of its comparative matrix of certification schemes.

8.4.3 Percentage-based claims

Chain-of-custody certification allows a wood product to be marketed with the label of a particular certification scheme. However, for many products, it is extremely difficult to ensure that 100% of the wood fibre comes from certified forests. For example, many paper products are made up of a mix of new and recycled wood fibre. The PEFC has a minimum percentage of 70% by volume or weight for all forest-based products. The FSC has also developed a complex set of guidelines for describing the percentage of certified wood in a product. For solid wood products, at least 70% of the wood used in manufacturing the product must come from FSC-certified forests. For chip and fibre products, at least 17.5% by weight of the total chip or fibre and at least 30% by weight of the new virgin wood chip of fibre used in manufacturing the product must come from FSC-certified forests. In all cases, where only a percentage comes from certified forests, this percentage must be stated in the same place as the label is displayed. This may not always be obvious. For example, magazines may have the FSC logo on the front cover, but the details only appear inside the magazine. An example is provided by the magazine *BBC Wildlife*. Its April 2001 issue showed the FSC logo on the front cover. On page 4, there is a small box containing details of the source of the paper (Ortviken Mill, Sweden) and the statement: 'At least 30% of the wood fibre in *BBC Wildlife* magazine comes from well-managed forests, independently certified according to the rules of the Forest Stewardship Council.'

8.4.4 Acceptance by consumers

Forest certification has not achieved universal acceptance. For example, as discussed above, consumers appear to favour the concept but are unwilling in practice to pay a premium for certified wood products. Considerable doubts exist amongst some groups, such as American public foresters and land managers (Vlosky, 2000), about the efficacy of certification and the degree to which it will be accepted by the general public. This, in part, may be an issue of public education (Teisl and Roe, 2000), and is being addressed through advertising campaigns, (e.g. the FSC). Certification is also seen by many policymakers as an ineffective tool for improving forest management at the international level (Haener and Luckert, 1998; Côté, 1999), but, as indicated above, this is not the primary purpose of the process. Analyses of consumer response to certification have revealed very different patterns, with some groups being more in favour of certified forest products than others (Ozanne and Smith, 1998). It is clear that the particular features of consumers vary with the economic, social and environmental climate in each case.

Collins Pine Co. provides an example of the difficulties faced by a producer in gaining acceptance for certified products in the USA. This company was one of the first in the world to pursue certification, having obtained Scientific Certification Systems (SCS) certification for its Almanor Forest in Chester, California, in 1993. This was followed by certification for its Kane Hardwoods forest in Pennsylvania in 1994 and by certification of its sawmill and timberlands at Lakeview, Oregon, in 1998 (Jenkins and Smith, 1999). Despite its early entry into certification, and its association with the FSC, Collins Pine has experienced considerable difficulties in the marketing of its products. It has identified particular markets where there is a demand for certified products, including Santa Fe, New Mexico; Austin, Texas; Vail and Aspen, Colorado; and the United

Kingdom. In other areas, including Oregon, attempts to market certified timber products have been unsuccessful.

This situation contrasts starkly with the situation faced by forestry companies in Europe in the 1990's where a range of buyers groups were quickly established to ensure that a market existed for certified products (especially in Sweden, the UK and the Netherlands).

8.4.5 Acceptance by forest managers

An issue that is of concern to forest certification stakeholders is the level of awareness that forestry companies have regarding the issue. While forest-related stakeholders in some areas (e.g. Sweden) are very much aware of certification, it is less well known in many developing countries. For example, Vlosky *et al.* (1999) suggested that lack of awareness was a major factor limiting the adoption of certification in Honduras. There, 90.5% of consumers interviewed were unaware of certification, as were ca. 66% of primary and secondary manufacturers. While there is an interest in certification amongst some developing countries (Stoian and Carrera, 2001), progress 'on the ground' has been slow, often occurring only with the financial support of outside agencies.

There are also differences between sectors within the forest industry in relation to certification. Large companies, concerned about continued market access, are vulnerable to changes in timber procurement policies by large buyers. Small-scale companies and private landowners may be less concerned about certification, as their timber is more likely to be processed locally (or even on site) by small-scale mills. Surveys of private landowners have revealed scepticism and uncertainty over certification (Lindstrom *et al.*, 1999). However, in the USA and many Western European countries, FSC certification has been sought by a number of forest owners that are committed to improving their forest stewardship (Hayward and Vertinsky, 1999).

There has been evidence that some forestry companies have been having difficulty with third-party certification. For example, in 2000, Rayonier Inc. did not renew its SGS/FSC certificate in New Zealand, and JD Irving Ltd. withdrew from the FSC in the Maritimes of Canada. The latter situation arose because JD Irving had achieved certification before the regional FSC standard was developed. The standard, once developed, included management requirements that JD Irving were unable to comply with. The issue involved the use of chemicals in the forest, and is one of the clearest examples of differences between jurisdictions, since limited use of chemicals is permitted by the FSC standard in the adjoining US forests.

The role of public forests and university research forests in acting as demonstration areas for certification is still uncertain. Certification of public forests has not always been viewed favourably by environmental organizations (Mater *et al.*, 1999), and there are a number of concerns that some of the activities in such forests may be incompatible with the aspirations of certification. For example, research forests are frequently the sites of genetic research. While this research may be limited to traditional selection methods, a general lack of understanding related to genetic research in the community can result in the rejection of an application for certification.

Another reason why certification may not have spread further is the generally conservative nature of most forest managers. It seems likely that the certification of a number of key forests could encourage others to follow suit. For example, at Quabbin Forest, Massachusetts, certification was achieved with the specific goal of providing a lead to other forest managers in the area (Barten *et al.*, 1998). This motive is also likely to have been behind the FSC-certification of the state-owned forests in Great Britain. Once a number of such forests have achieved certification, and both the costs and benefits are clear, other managers may well follow suit.

8.4.6 Acceptance by local people

In a number of cases, international donor agencies are encouraging the certification of forest management through subsidies and tied aid. This approach has been criticized (Thornber and Markopoulos, 2000), and it would probably be more efficient for the agencies to improve forest management practices at a pace that local communities can adapt to. It is worth noting that, in 2002, almost a quarter of all FSC certifications were for community forests, with the majority located in developing countries, where outside aid plays a critical role (Bass and Simula, 1999).

One of the problems with most certification schemes, including the FSC, is that southern and small enterprises have been poorly represented during the standard-setting process (Thornber *et al.*, 1999). This has resulted in concerns that the process has been dominated by northern industrial enterprises and ENGOs. Even within northern countries, there have been concerns amongst some small enterprises about the nature and costs of certification. In Sweden, this led to the smaller enterprises setting up their own certification scheme, rather than following the FSC standard (which had been developed between environmental groups and larger industrial enterprises, such as AssiDomän AB).

8.4.7 Acceptance by environmental organizations

To date, most ENGOs have strongly supported the FSC. For example, Gale and Burda (1998) argue that: 'It is vital ... that FSC win the logo competition, and it is here that the environmental network can help. Environmental organizations around the world can support the FSC process and work to discredit alternative processes.' However, by 2004 there was increasing dialogue between some of the supporters of the FSC framework and those of other frameworks and schemes.

The FSC, originally set up by environmental groups, has gradually moved away from some of its initial principles. This is evident in the evolution of its attitudes towards 'old-growth' forests and 'high conservation value' forests (FSC Principle 9). For certification to work, certified timber must be available to consumers. This requires the cooperation of forest managers. This cooperation will only be forthcoming if the standards set by the FSC are seen as being reasonable. To achieve this, industry has been allowed to participate in the FSC, resulting in concerns that the interests of environmental groups and are being marginalized (Gale and Burda, 1998). As a result, tensions have developed within the FSC that sometimes spill over into the public arena.

The acceptance of a standard by individual chapters within organizations is sometimes dependent more on the policy of that chapter than on the standards. For example, the problems created for the SFI programme by the Rainforest Action Network in 2001 can be directly related to that group's opposition to the awarding of SFI certification to a Canadian company (International Forest Products Ltd.) involved in logging old-growth forests in coastal British Columbia, Canada. This issue also arose in Finland when, after initially supporting and participating in the development of the FFCS, environmental groups pulled out in favour of the FSC process. In a joint press release on 27 October 1998, Finnish environmental organizations made the following comments on the FFCS: 'Finnish environmental organisations do not support the certification initiative promoted by forest industry and private forest owners – the level of standards in the Finnish self-certification initiative is inadequate for environmental NGOs' (Vallejo and Hauselmann, 2001). While the allegations of environmental groups may or may not be justified, the initiation of 'smear' campaigns and the publication of reports during the certification process represents a direct challenge to the validity of third-party auditing and must be seriously questioned.

8.4.8 Acceptance by government

By definition, forest certification involves independent, third-party audits of forest management practices. It is widely seen as a market-based tool. However, it can also be seen as a usurpation of the traditional role of government. The resulting variation in the reactions of differing governments has been interesting. In some cases, governments (e.g. the National Timber Certification Council in Malaysia) have promoted their own schemes as alternatives to third-party certification. In others, such as with the FFCS, the AFS and the CSA initiative, there has been strong government support. Some regional and national governments have actively promoted third-party certification (e.g. Ontario, Canada; the Netherlands;, the United Kingdom), while others have been more hesitant (e.g. British Columbia, Canada; Australia; USA). Another example is Bolivia, where certification has been seen as a method to increase timber exports and evaluate concession compliance by forestry companies (Nittler and Nash, 1999). It appears that many governments are increasingly recognizing certification as a means of encouraging sustainable forest management with more endorsements of the certification process being published by government bodies.

A number of international governmental and non-governmental organizations have also promoted certification. Certification was initially viewed with considerable scepticism, but has been progressively accepted. International organizations are now becoming involved, with the WWF–World Bank Alliance for Forest Conservation and Sustainable Use being an example (this group has the aim of achieving certification for 200 million hectares of sustainably managed forests). The involvement of the World Bank is especially important in the less industrialized countries, and a number of other donor agencies have emulated this approach (e.g. the Deutsche Gesellschaft für Technische Zusammenarbeit (GTZ) in Costa Rica). The European Union (EU) has been particularly supportive of forest management certification, both within member states and elsewhere. Formal recognition of the need to develop the capacity for certification in developing countries was laid out in Protocol 10 to the Fourth Lomé Convention (1995–2000), which regulates the cooperation between the European Commission (EC) and 77 developing states. One of the identified action areas was: 'Supporting the definition and the development of certification systems for timber production from tropical forests bearing in mind sustainable forest management principles as part of envisaged internationally harmonized certification systems for all kinds of timber and timber products.' Certification has also been promoted in EC forest sector policy through the 1995 *Council Regulation on Operations to Promote Tropical Forests* and the 1996 *Guidelines for Forest Sector Development Cooperation.*

The support for certification promoted through the EU is evident in the actions of the member states. For example, the UK Department for International Development (DFID) has supported the Soil Association's international Woodmark certification scheme, particularly its application to tropical forestry. It has also helped to develop an FSC certification scheme in Mexico by assisting with the costs of training in certification and accreditation procedures and the costs of marketing certified forest products. Other EU countries, including Germany and the Netherlands, have taken similar actions.

8.5 Conclusion

Certification has developed rapidly since its inception at the beginning of the 1990s. Within a period of 10 years, approximately 124 million ha of forest have been certified to one or more standards (Raunetsalo *et al.*, 2002). This area is likely to increase as some national and regional certification standards, such as those of the FSC, are finalized. With such a large area already certified, and so many different countries becoming actively involved, it seems that certification has become an important part of forest management. Whether it will remain is uncertain. It is still not clear what will happen if the majority of productive forests become certified, or what will happen to non-certified forests.

In many cases, certification has succeeded in improving forest management, with a move towards a more scientifically valid model of forest management being adopted by certified companies (Thornber and Markopoulos, 2000). This is to be encouraged. However, certification has been hampered by a lack of scientific agreement over what constitutes a sustainably managed forest in practice. Scientists still need to work to resolve a range of important issues and then communicate their findings to other interested parties. This is particularly true of environmental issues. Environmental groups were the initial driving force behind forest certification, yet some of the principles that have been promoted have not been based on sound scientific evidence. There is, therefore, a strong need for better dialogue between scientists and environmental organizations internationally.

The economics of certification are also still unclear. There are discrepancies in the costs of certification in different jurisdictions, as one might expect in a rapidly developing industry. While there are going to be differences in the costs associated with certifying different forests, some degree of standardization is advisable, and this is a role that the parent organizations (i.e. those accrediting the auditors) could play. With many certifications in tropical countries being sponsored by aid agencies, it is important to ensure that opportunistic auditors do not overly inflate the costs.

Certification still faces a number of problems, amongst which is its credibility amongst different stakeholders. The development and implementation of mutual recognition between schemes would go a long way to resolve this problem, helpful for both the industry and for the consumer. However, such recognition must be based on a fair, scientifically rigorous and transparent process. At the present time, rivalries and competition between schemes makes such a recognition process unlikely.

Today forest certification represents a major change in the forestry-operating environment. Some managers have been quick to respond, others are still considering their response. Many governments and interest groups have been taken by surprise by the speed at which certification has developed. Environmental organizations are increasingly making use of new technologies such as the Internet, and this has undoubtedly helped to educate stakeholders. However, the change is also indicative of the growing challenge that native forest managers are facing from non-traditional sources (e.g. plantation timber from New Zealand). Unless they are able to meet these challenges, it seems likely that forest certification will be another step by which forest managers lose their freedom to adaptively manage forest resources.

References

Axelsson, A.L. and Ostlund, L. (2001) Retrospective gap analysis in a Swedish boreal forest landscape using historical data. *Forest Ecology and Management* 147, 109–122.

Barten, P.K., Kykersnowman, T., Lyons, P.J., Mahlstedt, T., Ocoonor, R. and Spencer, B.A. (1998) Massachusetts – managing a watershed protection forest. *Journal of Forestry* 96(8), 10–15.

Bass, S. and Simula, M. (1999) *Independent Certification/Verification of Forest Management*. Background paper for WB/WWF Alliance Workshop, 8–9 November, Washington, DC. [http://www.esd.worldbank.org/wwf/certwkshp.htm].

Brown N.R., Noss R.F., Diamond D.D. and Myers M.N. (2001) Conservation biology and forest certification: working together toward ecological sustainability. *Journal of Forestry* 99, 18–25.

Coddington, W. (1993) *Environmental Marketing*. McGraw-Hill, New York.

Confederation of European Paper Industries (CEPI) (2000) *Comparative Matrix of Forest Certification Schemes*. CEPI, Brussels.

Côté, M.A. (1999) Possible impact of forest product certification on the worldwide forest environment. *Forestry Chronicle* 75, 208–212.

Elliott, C. and Hackman, A. (1996) *Current Issues in Forest Certification in Canada*. WWF Canada, Toronto.

Elliott, C. and Schlaepfer, R. (2001) The advocacy coalition framework: application to the policy process for the development of forest certification in Sweden. *Journal of European Public Policy* 8, 642–661.

Finnish Forest Ceritfication Scheme (FFCS) (2002) Website. Available at: http://www.ffcs-finland.org/eng/index.htm, accessed 17 June 2004.

Forest Stewardship Council (FSC) (2004) Website. Available at: www.fsc.org/, accessed 3 August 2004.

Gale, F. and Burda, C. (1998) The pitfalls and potential of eco-certification as a market incentive for sustainable forest management. In: Tollefson, C. (ed.) *The Wealth of Forests. Markets, Regulation, and Sustainable Forestry*. UBC Press, Vancouver, pp. 278–296.

Global Forest and Trade Network (2002) *About the Network*. Available from http://www.panda.org/forestandtrade/; accessed 21 December 2002.

Griffiths, J. (ed.) (2001) *Proposing an International Mutual Recognition Framework. Report of the Working Group on Mutual Recognition between Credible Sustainable Forest Management Certification Systems and Standards*. International Forest Industry Roundtable.

Haener, M.K. and Luckert, M.K. (1998) Forest certification – economic issues and welfare implications. *Canadian Public Policy – Analyse de Politiques* 24 (Suppl. 2), S83–S94.

Hayward, J. and Vertinsky, I. (1999) High expectations, unexpected benefits – what managers and owners think of certification. *Journal of Forestry* 97, 13–17.

Hickey, G.M. (2004) Regulatory approaches to monitoring sustainable forest management. *International Forestry Review* 6, 89–98.

Hickey, G.M., Innes, J.L, Kozak, R.A., Bull, G.Q. and Vertinsky, I. (2005) Monitoring and information reporting for sustainable forest management: an inter-jurisdictional comparison of soft law standards. *Forest Policy and Economics*, in press.

Hoberg, G. (2001) The 6 percent solution: the Forest Practices Code. In: Cashore, B., Hoberg, G., Howlett, M., Rayner, J. and Wilson, J. (eds) *In Search of Sustainability. British Columbia Forest Policy in the 1990s*. UBC Press, Vancouver, pp. 61–93.

Holvoet, B. and Muys, B. (2004) Sustainable forest management worldwide: a comparative assessment of standards. *International Forestry Review* 6, 99–122.

Jenkins, M.B. and Smith, E.T. (1999) *The Business of Sustainable Forestry. Strategies for an Industry in Transition*. Island Press, Washington, DC.

Karna, J., Juslin, H. and Steineck, F. (2000) Green marketing of paper products in Finland and Sweden. *Paperi Ja Puu – Paper and Timber* 82(3), 182–188.

Kiekens, J.-P. (2000) Global trends in forest certification. *Forestry Chronicle* 76, 399.

Lapointe, G. (1998) Sustainable forest management certification – the Canadian programme. *Forestry Chronicle* 74, 227–230.

Lawes, M.J., Everard, D. and Eeley, H.A.C. (1999) Developing environmental criteria and indicators for sustainable plantation management: the South African perspective. *South African Journal of Science* 95, 461–469.

Lindstrom, T., Hansen, E. and Juslin, H. (1999) Forest certification: the view from Europe's NIPFs. *Journal of Forestry* 97, 25–30.

Mater, C.M., Sample, V.A., Grace, J.R. and Rose, G.A. (1999) Third-party, performance-based certification. *Journal of Forestry* 97, 6–12.

McIntyre, B.I. and Beaumont, R. (1999) Benchmarking sustainable forest management performance through voluntary certification systems. *Pulp & Paper – Canada* 100(11), 38–41.

Meridian Institute (2001) Comparative analysis of the Forest Stewardship Council© and Sustainable Forestry Initiative® certification programs. October 2001. Available from http://www.merid.org/comparison; accessed 12 November 2001.

Murray, B.C. and Abt, R.C. (2001) Estimating price compensation requirements for eco-certified forestry. *Ecological Economics* 36, 149–163.

Nittler, J.B. and Nash, D.W. (1999) The certification model for forestry in Bolivia. *Journal of Forestry* 97(3), 32–36.

Ozanne, L.K. and Smith, P.M. (1998) Segmenting the market for environmentally certified wood products. *Forest Science* 44, 379–389.

Ozanne, L.K. and Vlosky, R.P. (1997) Willingness to pay for environmentally certified wood products: the consumer perspective. *Forest Products Journal* 47, 39–48.

Ozinga, S. (2001) *Behind the Logo: An Environmental and Social Assessment of Forest Certification Schemes.* Fern, United Kingdom. Available from http://www.fern.org/pubs/reports/behind/btlrep.pdf

Programme for the Endorsement of Forest Certification (PEFC) (2004) Website. Available at: http://www.perfc.org, accessed 2 August 2004.

Raunetsalo, J., Juslin, H., Hansen, E. and Forsyth, K. (2002) *Forest Certification Update for the ECE Region, Summer 2002.* Geneva Timber and Forest Discussion Papers, United Nations Economic Commission for Europe and the Food and Agriculture Organization of the United Nations, Geneva and Rome.

Rice, R.E., Gullison, R.E. and Reid, J.W. (1997) Can sustainable management save tropical forests? *Scientific American* April, 34–39.

Shrum, L.J., McCarty, J.A. and Lowrey, T.M. (1995) Buyer characteristics of green consumers and their implications for advertising strategy. *Journal of Advertising* 121, 71–82.

Sierra, R. and Stallings, J. (1998) The dynamics and social organization of tropical deforestation in Northwest Ecuador, 1983–1995. *Human Ecology* 26, 135–161.

Simula, M., Rametsteiner E., Blåsten, A., Green, T. and Pajara, B. (eds) (2001) *Forest Certification: Forging Novel Incentives for the Environment and Sustainable Forest Management.* Proceedings of the International Workshop, Brussels, Belgium, 6–7 September 2001. European Union, Brussels.

Spinazze, M.C. and Kant, S. (1999) Market potential for certified forest (wood) products in Ontario, Canada. *Forestry Chronicle* 75, 39–47.

Stevens, J., Ahmad, M. and Ruddell, S. (1998) Forest products certification: a survey of manufacturers. *Forest Products Journal* 48, 43–49.

Stoian, D. and Carrera, F. (2001) La certificación forestall en la encrucijada. Entre la panacea y un callejón sin salida. *Revista Forestal Centroamericana* 34, April–June, 6–11.

Teisl, M.F. and Roe, B. (2000) Environmental certification – informing consumers about forest products. *Journal of Forestry* 98(2), 36–42.

Terstad, J. (1999) Swedish experiences of incentives for the protection of nature. *Science of the Total Environment* 240, 189–196.

Thornber, K. and Markopoulos, M.D. (2000) *Certification: Its Impact and Prospects for Community Forests, Stakeholders and Markets.* International Institute for Environment and Development, London.

Thornber, K., Plouvier, D. and Bass, S. (1999) *Certification: Barriers to Benefits. A Discussion of Equity Implications.* Discussion Paper 8. European Forest Institute, Joensuu.

Vallejo, N. and Hauselmann, P. (2001) *PEFC: an Analysis.* WWF Discussion Paper. Pi Environmental Consulting, Switzerland. The World Wide Fund for Nature, European Forest Team, Gland, Switzerland.

Van Kooten, G.C., Wilson, B. and Vertinsky, I. (1999) Sweden. In: Wilson, B., van Kooten, G.C., Vertinsky, I. and Arthur, L. (eds) *Forest Policy. International Case Studies.* CABI Publishing, Wallingford, pp. 155–185.

Viana, V.M., Ervin, J., Donovan, R.Z., Elliott, C. and Gholz, H. (1996) *Certification of Forest Products. Issues and Perspectives.* Island Press, Washington, DC.

Vilhunen, L., Hansen, E., Juslin, H. and Forsyth, K. (2001) *Forest Certification Update for the ECE Region, Summer 2001*. Geneva Timber and Forest Discussion Papers ECE/TIM/DP/23. United Nations, Geneva.

Vlosky, R.P. (2000) USDA Forest Service, Bureau of Land Management, and state forester perspectives on certification. *Forest Products Journal* 50, 21–27.

Vlosky, R.P. and Ozanne, L.K. (1997) Forest products certification: the business customer perspective. *Wood Fibre Science* 29, 195–208.

Vlosky, R.P., Aguirre, J.A., Montes, E.C.S., Ozanne, L.K. and Silva, G. (1999) Certification in Honduras: perspectives of wood product manufacturers, consumers, NGOs and government policy makers. *Forestry Chronicle* 75, 646–654.

Vogt, K.A., Vogt, D.J., Boon, P., Fanzeres, A., Wargo, P., Palmiotto P.A., Larson, B., O'Hara, J.L., Patel-Weynand, T., Cuadrado, E. and Berry, J. (1999) A non-value based framework for assessing ecosystem integrity. In: Meurisse, R.T., Ypsilantis, W.G. and Seybold, C. (eds) *Proceedings: Pacific Northwest Forest and Rangeland Soil Organism Symposium*. US Department of Agriculture Forest Service General Technical Report Pacific Northwest 461. USDA Forest Service, Portland, pp. 3–20.

Vogt, K.A., Larson, B.C., Gordon, J.C., Vogt, D.A. and Fanzeres, A. (2000) *Forest Certification: Roots, Issues, Challenges and Benefits*. CRC Press, Boca Raton.

Wallis, A.D., Stokes, D., Westcott, G. and McGee, T. (1997) Certification and labelling as a new tool for sustainable forest management. *Australian Journal of Environmental Management* 4, 224–238.

Weber, N., Hardter, U., Rother, A. and Weisshaupt, M. (2000) Activities of environmental NGOs in German forest policy – a preliminary stocktaking. *Allgemeine Forst- und Jagdzeitung* 171(8), 144–153.

Winterhalter, D. and Cassens, D. (1993) Telling the sustainable forest from the trees. *Furniture Design and Manufacturing*, August 1993, 101–106.

World Wide Fund for Nature (WWF) (2003) *State of Europe's Forest Protection*. WWF European Forest Programme. Available from: http://www.panda.org/downloads/forests/stateeuropeforests.pdf, accessed 10 November 2004.

Zinkhan, G.M. and Carlson, L. (1995) Green advertising and the reluctant consumer. *Journal of Advertising* 121, 27–29.

How will New Large-scale Nature Reserves in Temperate and Boreal Forests Affect the Global Structural Wood Products Sector? 9

J.M. Perez-Garcia

University of Washington, USA

Rising income levels, the growth in population and globalization have all led to greater demands for wood products and nature reserves. These conditions raise the question of what effect new large-scale nature reserves would have on the forest products sector. An analysis of the demand and supply projections for forest products and the demand for nature reserves identifies potential areas of conflicts. Because of the low-cost nature of wood fibre production and their export market shares, new reserves established in Canada and Scandinavia would have both global and local impacts, more so than other regions due to their importance as exporters of wood products. New reserves in other regions would likely lead to impacts when they compete with timber output, but their global effect would be smaller as their importance as low-cost wood suppliers and excess producers is lower.

9.1 Introduction

In the past, growing human population and rising incomes in many countries have led to vast areas of natural forests being used for agriculture or timber production. Today, in many countries that have developed agricultural and forestry sectors, growing human population and rising incomes are leading to the establishment of large-scale nature reserves. The establishment of new, large-scale nature reserves in temperate and boreal forests are likely to have large impacts, locally and globally, on forest sector markets and biological resources. For example, the preservation of spotted owl (*Strix occidentalis*) habitat in western North America led to the displacement of raw material inputs for the forest products industries in both the Pacific North-west and worldwide (Sedjo *et al.*, 1994; Perez-Garcia, 1995). Potential conflicts in balancing the management between timber and non-

© CAB International 2005.
Forestry and Environmental Change: Socioeconomic and Political Dimensions
(eds J.L. Innes, G.M. Hickey and H.F. Hoen)

timber resources from forest lands will continue as human populations increase and economies expand.

This study examines three questions surrounding these potential conflicts. The questions are developed from a forest sector perspective and are necessarily limited in scope. The questions do not address the potential benefits from preservation, but rather measure the potential costs to the forest sector associated with the establishment of large-scale nature reserves in the temperate and boreal forests. These questions are as follows: What are the costs to the forest products sector from establishing new nature reserves? In which timber supply areas and markets are the higher costs likely to be observed? How can society minimize the impact associated with the establishment of large nature reserves while preserving a given amount of biological resources?

To address these three questions, the study examines the demand for nature reserves and wood products, where potential conflicts in land use may occur and how they may affect the supply of wood products and other forest resources. The study provides a brief discussion of the economics of nature reserves and wood products. The chapter then utilizes an economic framework to discuss the impact of large-scale nature reserves. The CINTRAFOR Global Trade Model (CGTM) is an economic model that simulates the major softwood and hardwood markets in structural wood products around the globe. Hence the discussion that follows pertains to impacts on structural timber markets, primarily softwood lumber markets, and includes the supply, demand and trade aspects of nature reserves and forest products. The study's intention, however, is not to conduct a cost-benefit analysis. Rather, the objective is to assess the economic impact of establishing new reserves on the timber-based forest products sector. Furthermore, the discussion is restricted to temperate and boreal forests, ignoring the importance of tropical forests as a source of wood and the need to preserve their biological diversity. Much of the discussion can be extended to include tropical forests when specific data on the forest economies and nature reserves in tropical regions are introduced.

9.2 The economics of nature reserves and wood products

The demands for nature reserves and wood products are mutually related to population and income levels, both of which are growing. While the supply of forested nature reserves and that of wood products rely on the same input (i.e. forest land), they are distinctly different. We can expand wood products supply through technological advances, effectively improving the production processes associated with growing timber and producing wood products. Technological advances in the production process associated with a nature reserve, however, are non-existent and, to a large extent, exclusionary by the nature of natural reserves. So, as populations expand and income levels rise, the demand for nature reserves and wood products increases. For upward-sloping supply curves, this shift in demand would raise the price for wood products and the value of nature reserves. Higher prices, over the long term, attract investments that increase productivity. This process happens only in the forest products market, however, and not in the nature reserves market. The expansion of nature reserves relies on political processes, rules and regulations that allocate new reserve areas since price mechanisms are lacking in nature reserves markets.

One further distinction between the economics of wood products and nature reserves lies in the ability for trade in forest products to dissipate price disparities in different regions. Wood products trade occurs when price differentials are large so as to allow one region to export its products to another region and cover its costs of transportation. Demand differences in nature reserves, those that arise due to income differences among regions, currently have no such mechanism that would allow regions with excess nature reserves to export these services to areas with excess demand for reserves. In principle, nature reserves should be more suitable for global trade and 'global equilibrium', since there is no need for any transportation services to establish and maintain a nature reserve. What is lacking is the establishment of a proper and functioning market.

9.3 Market signals (or the lack thereof) for industrial raw materials and biological reserves

Forests produce various competing resources. On the one hand, these forests are a store of biological resources, whose demand for preservation is growing. On the other hand, competing with biological resources, these forests provide raw material inputs for consumptive uses, such as housing, energy and other wood products, whose demand is also growing. Active trade in wood products connects these temperate and boreal forests around the world. Even though forest lands are not traded, timber prices reflect the changing market conditions in other regions due to trade activity and transparency in product markets. So, prices provide signals to producers to either expand or contract their wood processing activity.

Temperate and boreal forests are also found in many of the developed economies of the world. Values regarding nature reserves may be similar due to the high-income levels associated with developed economies, which link the call for the preservation of large forest areas across countries. As economies in temperate and boreal regions have developed their agricultural and forestry sectors, and raised their income levels, the pressures to preserve the remaining temperate and boreal forests have increased. These signals, however, are not coming from markets. Political processes drive the need for greater areas of nature reserves, some of which are international in nature. Under what conditions do these two activities come into conflict?

9.4 The demand and supply for raw material inputs

This study links potential new nature reserves to areas where forests are being managed for timber products to analyse the potential effects these new reserves might have on the structural forest products sector. In temperate regions, North American forests are undergoing a change in management objectives that reflect change in societal values (Haynes *et al.*, 1998). Studies on sustainable forest management and wood fibre use by the forest industry have suggested that changes in management are needed (Solberg *et al.*, 1996; Nilson *et al.*, 1999). More publicly held forest lands are being managed for preservation purposes. The change in policy, which in effect creates new reserves on existing public and some private lands, will likely have a large impact on the forest products sector. The size of the impact, measured as a cost to the industry, is related to timberland and processing capital investments, which are prematurely taken out of service. Where trade flows in forest products occur, new nature reserves are likely to have an international impact. Key pieces of information to determine the potential impacts of new reserves are global and regional demands on timber.

Future timber consumption projections to 2010 by region were developed to determine where demand and supply imbalances are likely to occur and where investments are likely to take place in the forest products sector. The analysis was implemented with the CGTM, which includes lumber and wood-based panel product markets and their corresponding timber demand and supply projections (Perez-Garcia *et al.*, 1999). Demand projections were based on projected economic growth in specific regions and its consequences for the demand for wood fibre. The growth and development of the pulp and paper sector was also included in the analysis since this sector draws wood fibre from the forestry and solid wood products manufacturing sectors. The resulting projections of timber supply, product supply and product demand illustrate an equilibrium solution for forecasted demand growth in wood end-using sectors and trade in forest products. The results presented in Table 9.1, as annual percentage growth rates, were then used to identify potential areas where conflicts between forest-land uses may occur.

Table 9.1 provides projected annual growth rates in both the supply and demand of lumber, plywood and timber used to produce these products for selected regions of the world (to 2010). In some instances, countries were aggregated into regions to reduce the number of entries in the table and capture general trends for larger regions. Regional breakouts were included where clear distinctions in trends were evident.

China is the first entry in the table. Policy decisions, rather than market forces, have driven Chinese supply and demand for wood products. As such, growth rates for product supplies and demands, based on government plans, are similar. Even so, there is evidence that China will need to import softwood logs to meet its timber demand. Assuming similar production, consumption and annual average growth rates for the next decade, China's production of timber will decline by 1.6% while the demand for timber will grow by 1.7%. The growth in timber consumption feeds into the growth in the supply of lumber and plywood. The forecast assumes that, to a large extent, China is likely to import its raw material input and take advantage of low labour costs to manufacture lumber and plywood, rather than import the lumber and plywood directly.

For Europe, the annual percentage growth in the demand for lumber is projected to increase by 2.2%. Production will increase by 2.5% to meet this demand growth and create an excess in production, mainly in Sweden and Finland, which is then directed to export markets in Asia and North America. The European region as a whole will likely meet its product demand growth with its own domestic timber resources and export its surplus.

Table 9.1. Annual growth rates (1997–2010)

Region	Lumber Supply	Lumber Demand	Plywood Supply	Plywood Demand	Roundwood Supply	Roundwood Demand	Sawlog Supply	Sawlog Demand
China	2.3%	2.3%	1.9%	1.9%	–1.6%	1.7%	–1.7%	2.2%
Europe	2.5%	2.2%	9.9%	2.1%	2.4%	2.4%	2.9%	2.9%
Japan	-2.9%	-2.8%	0.0%	0.0%	-2.0%	-0.6%	-2.2%	-2.9%
North America	0.5%	1.6%	-3.6%	-2.5%	0.9%	0.7%	0.2%	0.1%
Oceania	6.8%	-0.5%	2.8%	2.8%	4.8%	4.3%	8.2%	6.2%
Other	-0.7%	-0.1%	8.0%	8.0%	0.9%	0.0%	0.7%	-0.2%
Russian Federation	-0.4%	0.0%	-3.3%	2.2%	-0.1%	0.3%	0.0%	0.0%
South America	5.1%	1.2%	4.1%	-0.3%	3.6%	3.3%	5.4%	5.0%

Japan has a different projection for demand than other regions. Demand projections based on wooden housing starts indicate an annual decline in the consumption of wood products of 2.8%. Similar declines in production and derived demand for timber are also projected.

North American production of lumber products will continue to increase, but at a slower rate than observed during the decade of the 1990s: 0.5% to –3.6%. Demand will increase by 1.6% in lumber whereas there will be a decline in plywood demand of –2.5%, indicating that there will be an increase in imports of lumber products from outside the continent. These imports will come from Europe, Oceania and South America. Higher wood fibre cost for processing constrains the growth rate of new investments in processing capacity in the USA.

Both the Oceanic region and South America are projected to increase their production rate substantially, primarily due to maturing timber plantations. Demand growth in these regions, less than 1% for Oceania (lumber) and 1.2% in South America (lumber), will not expand at the same rate as production rates (over 5% for both regions); hence, much of the production will be exported. Both the demand for timber and the timber harvests will increase to meet higher production rates and export activity.

Very little change in the Russian Federation is assumed in the above projections. An increase in plywood demand is assumed based on projections for greater construction activity. There is too much uncertainty in the near term, however, to make any prediction for Russia's forest sector.

In summary, projections of demand and supply suggest that the North American market will be a wood deficit market with greater imports from Europe, Oceania and Latin America. Hence, investments in the forest products sectors are likely to take place in Europe (mainly Scandinavia), Oceania (mainly New Zealand and Australia) and Latin America (mainly Chile). Within North America, Canada is a major producer of wood products with low wood fibre costs. Canada's US market share of lumber products, however, is capped by administrative agreements rather than market forces. Northern China is a second area where demand is larger than its own domestic supply of forest products. Temperate forests in China are under strain. New wood sources to meet the Chinese demand would come from the Russian Federation and New Zealand. Investment in forest management is taking place, but only in the southern Chinese provinces.

9.5 Global cost curves for softwood fibre

The costs of producing wood fibre can be described with a curve that ranks regions by the average cost of production. This cost curve is the sum of each region's production of softwood fibre to previous production with each region ranked from lower to higher cost. The process leads to the cumulative cost curves illustrated in Fig. 9.1. Two curves are constructed, one each for 1983 and 1993. We then compare the cost curves to describe the impact of an increase in softwood raw material input costs due to restricted timber harvests in the Pacific Northwest region of the USA. We do so to illustrate what is likely to occur if new large-scale nature preserves are developed in regions where timber is used for forest products.

There are two distinctions to be made with the chart. First, the upward shift in the global cost curve is primarily the result of reducing timber harvests from the Pacific North-west (PNW) region of the USA. The reduction in harvest caused higher log prices for the PNW region, and, because of its trading position, it also raised the prices of timber elsewhere. As a result, we observe a shift upward in the cost curve, a global effect. Locally, log prices in the PNW region were raised substantially and the region became a higher-cost region (an increase in the ranking of cost from lower to higher), but the region continues to supply wood fibre to domestic and international markets. Other wood fibre producers servicing the US and international markets also saw an increase in their log prices, such as Chile and New Zealand, which are not labelled in the chart. The US South also witnessed higher log prices, but not to the extent observed in other regions. As a result the US South is now in a lower cost position on the 1993 global cost curve. The cost curve also illustrates how much costs rise for the last units produced. The sharp upturn in the curve after 600 million cubic metres is the result of meeting the demand for wood products in regions of high-cost production.

9.6 The supply for nature reserves

Having characterized the cost structure of the global forest sector and provided projections of supply and demand volumes across regions, the study asks the question: Where are the reserves likely to take place? Existing large intact ecosystems are likely candidates for new nature reserves.

Recent work by the World Resources Institute (WRI) has identified regions of frontier forests, the remaining areas of original forests (Bryant *et al.*, 1997). These frontier forests are only one area where reserves are likely to be established. According to the WRI study, there are 12 countries with most of the remaining frontier forests. The top two countries, the Russian Federation and

Canada, contain over 50% of the remaining frontier forests. Other boreal and temperate regions with significant areas of frontier forests are the USA and Chile. The WRI study also indicates which regions are at risk of losing the remaining frontier forest. Among them, Finland and Sweden are two boreal and temperate forest regions with 99% and 97% respectively of their original frontier forest lost. The WRI study classifies these two countries, among others, as requiring immediate actions. Argentina, New Zealand and China will require action to contain the continued conversion of frontier forests, but are lower in priority than Finland and Sweden.

While the frontier forest initiative strives to preserve the remaining large intact tracts of undisturbed ecosystems, different approaches to conserving ecosystems are being pursued elsewhere. In Australia, for example, a draft policy position recognizes nature reserves as part of large regional management plans (Anon., 1999). Under the policy, the Australian government has agreed to manage and conserve forests for a full set of values that forests can provide for current and future generations. Certification schemes are also being sought so that forests are managed to provide a full set of values, including those that safeguard biodiversity, old growth, wilderness and other natural and cultural values of forests. Many of these efforts will likely take place in developed economies, where boreal and temperate forests abound. Even in the Russian Federation, there are efforts to establish such management schemes with the help of international assistance. Finland and Sweden are two countries that contend that they manage their resources in a sustainable fashion and can be certified, negating somewhat the call for preserving remaining forests in these countries.

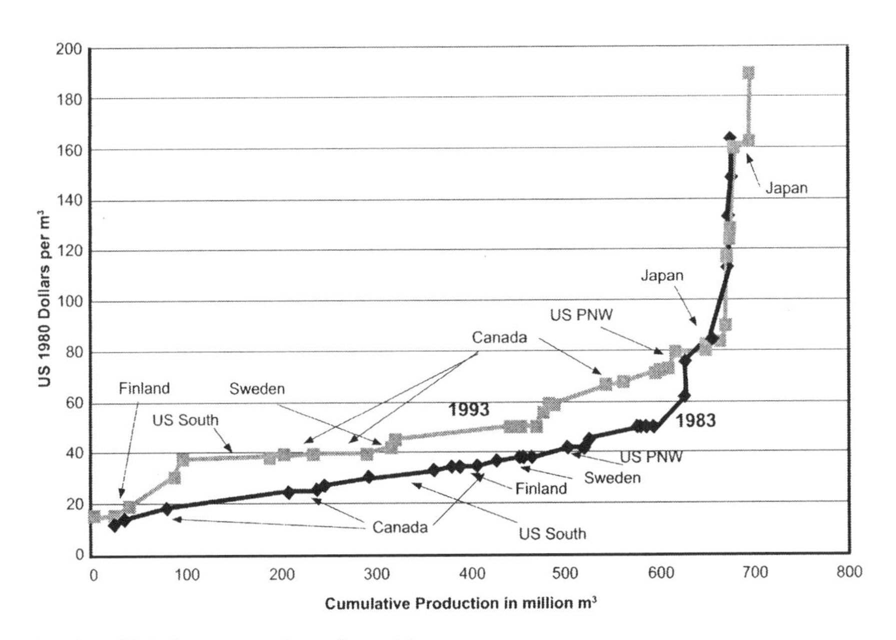

Fig. 9.1. Global cost curve for softwood logs.

9.7 Putting supply and demand together

Armed with both cost and demand information, the study identifies areas where conflicts between new reserves and timber production might occur. Fig. 9.1 illustrates how new nature reserves can affect the cost structure of the timber-based industry depending on which region is impacted. It uses the reduction in timber harvest from the Pacific Northwest region of the USA as the example. We extrapolate from this example to suggest the following hypothesis. If newly initiated reserves increase costs for low-cost producers, then it is likely to raise the floor for timber product prices and have both a large local and significant global impact. This was the case for the spotted owl habitat preservation, where a 5% reduction in timber harvests at the global level imposed large welfare losses in both the USA and internationally (i.e. an estimated one-third of the economic impact from the harvest restriction in the US PNW was felt in international markets (Perez-Garcia and Lippke, 1997)). Trade patterns adjust as a result of new prices leading to the transmission of cost impacts to other wood fibre producing regions. So the creation of new reserves in regions characterized by low wood fibre costs would increase log prices locally, and then globally as wood product flows from other regions adjust to a new set of wood fibre prices.

Fig. 9.1 ranks important producers of raw materials by their costs. They are located in Canada and Scandinavia, as well as in the US South. The first two regions — Canada and Scandinavia — are important producers for international markets. Their excess supplies of wood help maintain the cost of wood-using industries low, primarily in the USA, Europe and Japan. Establishing new reserves in these regions is likely to have substantial impacts locally and globally whereas the establishment of new nature reserves in the US South will likely lead to large local but small global impacts, since the US South region exports little of its production outside the USA. For other regions, especially those that have a higher rank on the cost curve, establishing new reserves is likely to cause large local impacts, but smaller global effects, since their contribution to total global production is small.

9.8 Future changes in wood consumption

The relationship between income growth and consumption of wood has changed in part as the result of previous environmental restrictions on timber harvests. Fig. 9.2 plots the annual consumption of timber raw material input for the production of forest products, including paper and paperboard, versus annual gross domestic product (GDP) from the period 1965 to 1998.

The figure illustrates several points. First, the effect on wood consumption from the collapse of the former Soviet Union is large. Global consumption of wood declined by over 200 million m^3. Data used to construct the upper line include the consumption of industrial roundwood by the former Soviet Union and its GDP contribution to global GDP. The lower line excludes data for the former Soviet Union. Therefore, the vertical difference between two points along the y-axis measures the former Soviet Union's consumption while the horizontal difference along the x-axis measures their contribution to world GDP. The effect of the collapse of the former Soviet Union was still evident in 1998.

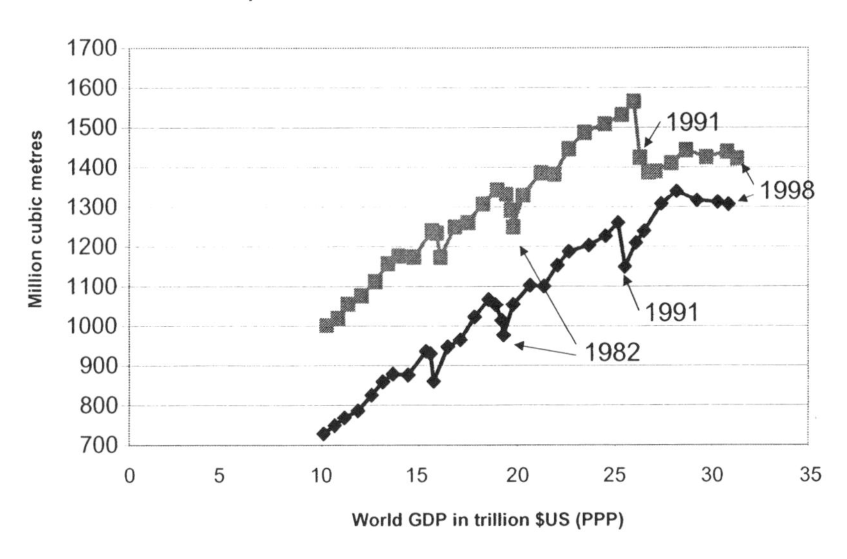

Fig. 9.2. Global consumption of industrial roundwood with and without the former Soviet Union (FSU) (sources: World Bank (2000) *World Development Indicators*; FAO (2000) *Yearbook on Forest Production*).

Second, global income elasticity measure has changed due in part to supply restraints observed in the early 1990s. Income elasticity measures the percentage change in consumption associated with a 1% change in income. This change in income elasticity can be seen as a flattening of the curve from 1995 to 1998. While the last four data points do not define a definitive change in the trend line, they do suggest a different projection than previous demand-induced reduction in consumption. The implication of the decline in income elasticity is for a smaller percentage change in the consumption of wood products as income rises. The shift in wood consumption is related to substitution from both wood and non-wood products as wood became scarcer in the early 1990s. New wood products, such as engineered wood products, have increased market share in traditional wood products markets. Further changes in the wood consumption and income growth relationship have implications for the future of global wood consumption and its relationship to nature reserves. In particular, a smaller area of forests to provide wood fibre will be needed with lower consumption trends.

9.9 The changing nature of demand for nature reserves

The study also examines the potential demand growth in nature reserves. However, an analysis of this process is more complicated. Fig. 9.3 charts a region's protected forest area (as a percentage of total forest area) with the GDP per capita. Total protected forests include all legally protected forests as defined in Sharma (1992). Countries are described by a three-letter symbol, listed in Table 9.2. The chart illustrates several interesting points. First, it suggests that as GDP per capita increases so does the percentage of protected forest area. This was expected, and is often cited in deforestation studies (e.g. Kaimonwitz and Angelsen, 1998).

% Protected Forests = 0.0399 + 0.0138 GDP Thousand per capita

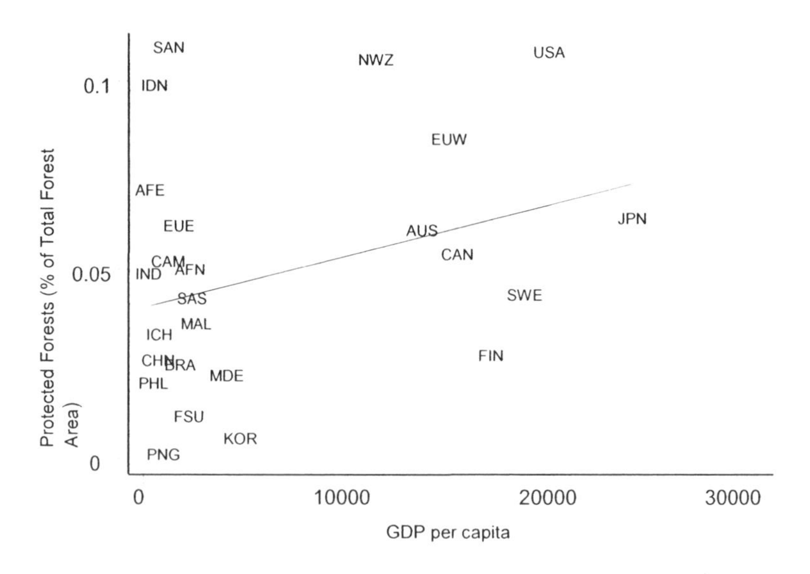

Fig. 9.3. Area of protected forest relationship with income. (sources: World Bank (2000) *World Development Indicators* for GDP per capita; Sharma (1992) for protected forest areas).

Fig. 9.3 contains two distinct groups of countries. High-income countries, such as the USA, Finland (FIN), Sweden (SWE), Canada (CAN), Western Europe (EUW), Australia (AUS) and New Zealand (NWZ), are to the right. The mean percentage of protected forest area of these countries is higher than that of lower-income countries, which include China (CHN), the former Soviet Union (FSU), Eastern Europe (EUE) and southern South America (SAS), among others. A regression line was drawn to illustrate the mean response of protected forests with GDP per capita (significant at the 95% confidence level).

Table 9.2. Regional and country definitions

Region	Code
Northern South America	SAN
Indonesia	IDN
East Africa	AFE
Eastern Europe	EUE
Central America and Mexico	CAM
India	IND
Northern Africa	AFN
Southern South America	SAS
Malaysia	MAL
Indochina	ICH
China	CHN
Brazil	BRA
Philippines	PHL
Middle East	MDE
Former Soviet Union	FSU
South Korea	KOR
Papua New Guinea	PNG
New Zealand	NWZ
United States	USA
Western Europe	EUW
Australia	AUS
Canada	CAN
Sweden	SWE
Finland	FIN
Japan	JPN

A second point has to do with how regions within the two groupings fare. Finland and Sweden are below the USA and New Zealand, for example. These two regions were also mentioned above as areas where frontier forests are at high risk. The former Soviet Union and China are also below the average for low-income countries, and listed as regions with large opportunities to preserve frontier forests, in the case of the former Soviet Union, and at risk of continuing their loss of frontier forests, in the case of China.

Third, the chart illustrates conditions during the 1980s. Here, one can speculate as to where we currently are with respect to more protected areas. It is not clear whether a chart that updates these conditions would show the positive relationship between the percentage of protected forest area and GDP per capita. For one thing, as economies expand (each region moves to the right), the percentage of protected forests may not increase substantially unless significant policy changes are employed in the regions. The reason is that forest protected areas are politically driven rather than market driven. In the USA (and most other countries), for example, changes in the management of established public lands are politically determined. While the economy expanded at an average 3% per year for the last two decades, areas in reserves have not changed. The changes in the USA relate to the management goals of existing parks and forests, which have favoured preservation over timber output. Hence, unless the changes in management goals are accounted for in the chart, the USA is likely to show a horizontal shift to the right, without any incremental increase in protected areas.

Other dynamic elements of the chart include the potential impact of trade and international mechanisms that would cause a shift in the areas protected, without any significant change in GDP. For example, GDP growth in the USA enables Joint Implementation projects and the Clean Development Mechanism to take place in other countries. Under such programmes, protected areas in low-income countries are likely to increase and represent measures that would impact protected areas without necessarily impacting their GDP. As time goes on, we may see the relationship between income levels and protected areas disappear (i.e. the line becoming flatter). Other programmes that have been implemented and result in a similar effect are debt-reducing projects. The project reduces a low-income country's debt in international markets in exchange for greater levels of conservation effort. In fact, policy goals of such international agreements are to eliminate the disparity in the value of nature reserves between low- and high-income countries. Hence, over time, the regression line would become flatter as these policies successfully eliminate value disparities among regions.

9.10 Further discussion points

The study holds that there exists excess timber in the world. As a consequence, the establishment of nature reserves may not necessarily impose a cost on the forest products sector, should the conditions of excess supply continue into the future. Hence, the global risk of impact from the establishment of large nature reserves should be small. However, excess timber in the world is not a sufficient condition to conclude that no global risk occurs. Regional timber shortages exist where demand for wood products exceed a region's production capacity. These regional shortages produce trade activity in forest products and give rise to potential conflicts in land uses (due to a timber price rise in the region and no clear change in demand for reserves). Even though there may be excess timber supply conditions at the global level, unexpected regional timber shortages can lead to large global, as well as local, effects. Hence, addressing the question of which region risks large impacts from the establishment of large nature reserves becomes important.

Second, impacts from reduced timber harvests occur to the capital employed in the forest products sector, including timber management activities as well as harvesting and processing facilities. Currently, many investments take place in the forest sector along with environmental planning that considers potential changes in forest rules and regulations. The investment decision examines the threat of a change in forest land use that affects potential timber supply. Abrupt changes in environmental policies that rapidly depreciate the value of capital investments cause the largest impact. Hence, environmental planning reduces the threat of a large impact when these plans are accurately reflected in the timber-based decision-making process.

Another assumption held by the study is that land-use values are heterogeneous across countries. That is, demands for nature reserves are characterized by curves with different demand elasticity. Globalization has linked values held in one region with actions affecting land-use decisions in another region, without considering the different demand elasticity. This has occurred since no mechanisms have been developed to capture effectively the value disparities. International agreements are the policy instrument of choice to resolve these conflicts, but it has been very difficult to implement these agreements due to the large disparity in the values held by different countries. Clean Development Mechanism (CDM) and Joint Implementation (JI), for example, are two mechanisms that are considered for carbon emission reductions. Many other on-going processes also exist. CDM and JI meet the carbon reduction targets for ANNEX I countries by providing investments in other countries, sometimes preserving forest land as a store of carbon, while US businesses pay the costs. The demand for these reserves is in one country whereas the supply comes from within the country as well as abroad. Such a framework is an attempt to institutionalize international trade as a factor to equalize disparities in values among different countries.

9.11 Conclusion

This study assesses how new nature reserves may lead to local and global effects on the forest products sector. It does so by postulating the hypothesis that the cost ranking of a region is an important factor in determining the severity of the cost impacts to the global wood fibre-using sector. A projection of the supply and demand for wood fibre was conducted to determine where wood fibre markets may be in the near future. The potential demands for new nature reserves, which distinguish regions where conflicts in fibre production and nature preservation may occur, were assessed to identify in which timber supply area and market higher costs were likely to be observed. This study makes no attempt at quantifying the benefits associated with the establishment of new nature reserves. It examines how costs to the forest products sector would change with a change in land use (timber versus reserves). Forest land-use conflicts generally arise when demand for nature reserves must be met with land previously designated for timber production. Low-cost regions with surplus production and extensive trade activity are likely to produce large local and global effects if nature reserves are established within the region at the expense of timber production. Based on an assessment of these conditions, Canada and Scandinavia were identified as regions where the establishment of new large preserves would adversely affect the global forest sector, more so than other regions, because of the low costs associated with wood fibre production. Additionally, due to their trading position, any resulting impact on their forest sector would be transmitted globally to other markets.

References

Anon. (1999) Natural forest conservation: alternative options and approaches—Australian experience with regional forest agreements. Unpublished Ms. World Bank Report.

Bryant, D., Nielsen, D. and Tangley, L. (1997) *The Last Frontier Forests: Ecosystems and Economies on the Edge.* World Resources Institute, Washington, DC.

FAO (2000) *FAO Yearbook on Forest Production.* UN Food and Agriculture Organization, Rome.

Haynes, R., McGinnis, W. and Horne, A. (1998) Where are the jobs? Lessons learned about the economic impacts of ecosystem management. In: Calhoun, J. (ed.) *Forest Policy*

Ready for Renaissance. Institute of Forest Resource, University of Washington, Seattle.

Kaimonwitz, D. and Angelsen, A. (1998) *Economic Models of Tropical Deforestation: A Review.* Centre for International Forestry Research, Bogor, Indonesia.

Nilson, S., Colberg, R., Hagler, R. and Woodbridge, P. (1999) *How Sustainable are North American Wood Supplies?* Interim report IR-99-003, International Institute for Applied Systems Analysis, Laxenburg, Austria.

Perez-Garcia, J.M. (1995) Global forest land use consequences of North American timber land withdrawals. *Journal of Forestry* 93(7), 35–38.

Perez-Garcia, J.M., and Lippke, B.R. (1997) Who wins and who loses: Trade barriers in the PNW forest sector. *Contemporary Economic Policy* XV(1), 87–103.

Perez-Garcia, J.M., Wang, Y. and Xu, W. (1999) An economic and environmental assessment of the Asia forest sector. In: Yoshimoto, A. and Yukutake, K. (eds) *Global Concerns for Forest Resource Utilization.* Kluwer Publishing Co., Dordrecht

Sedjo, R.A., Wiseman, A.C., Brooks, D. and Lyon, K.S. (1994) *Global Forest Products Trade: The Consequences of Domestic Forest Land-Use Policy.* Resources for the Future Discussion Paper 94–13, Resources for the Future, Washington, DC.

Sharma, N. (1992) *Managing the World's Forests: Looking for Balance Between Conservation and Development.* The International Bank for Reconstruction and Development, Kendall/Hunt Publishing Company, Dubuque, Iowa.

Solberg, B., Brooks, D., Pajuoja, H., Peck, T.J. and Wardle, P.A. (1996) *Long-term Trends and Prospects in World Supply and Demand for Wood and Implications for Sustainable Forest Management.* Research Report 6, European Forest Institute, Joensuu, and Norwegian Forest Research Institute, Ås.

World Bank (2000) *World Development Indicators.* World Bank, Washington, DC.

10 The Establishment of Large-scale Reserves in Temperate and Boreal Forests: Some Experiences of the USA and Prospects for Finland

R.A. Sedjo

Resources for the Future, Washington, DC, USA

Both the USA and Finland are considering making a social decision to establish a large-scale forest reserve, the dominant purpose of which would be biodiversity preservation. This chapter briefly compares and contrasts the situations and experience in the USA and Finland. The USA is experiencing a gradual transformation of its National Forest System, publicly owned land, from that of a timber reserve to a multiple-use and de facto large-scale biodiversity reserve. Despite a very contentious political climate, the de facto transition has occurred within a period of 10–15 years. In Finland, preliminary proposals have called for a fairly large area of privately owned forest to be set aside as a reserve in the south of Finland. These areas could constitute up to 10% of the total forest area in Finland. In both cases the economic implications associated with substantial reductions in timber harvests are of concern.

This chapter describes the US experience as one in which the economic disruptions appear to have been modest because substitute sources of timber were available. Additionally, a buoyant economy in the late 1990s provided generally plentiful employment opportunities for dislocated workers. The situation for Finland, while still speculative, could be similar if low-cost alternative wood sources are found, particularly from the Russian Federation.

10.1 Introduction

In relatively well-developed countries with an extensive forest sector, changes in societal values for the broad array of forest goods and services are changing

© CAB International 2005.
Forestry and Environmental Change: Socioeconomic and Political Dimensions
(eds J.L. Innes, G.M. Hickey and H.F. Hoen)

public goals for land management, regardless of land tenure. Ideally, the allocation of set-aside areas for biodiversity protection should be made at the global level. The purpose of this chapter, however, is more limited than a global assessment. It is to examine briefly the experience of the USA in its process of the gradual transformation of the National Forest System from a timber reserve to multiple-use and now to a de facto large-scale biodiversity reserve. The early transition of the uses of the National Forest System (NFS) was gradual; however, the recent transition, to a de facto biological reserve, has been quite rapid. It has occurred in a period of 10–15 years and in the face of stiff political opposition. The situation and experience in the USA is contrasted with proposals, under preliminary consideration in Finland, calling for the setting aside of a fairly large forest reserve in the south of Finland. The set-aside areas could be up to 10% of the total forest area of Finland. Since this land is almost entirely in private ownership, the establishment of a cost-effective selection of set-aside areas may be more difficult with the Finnish ownership structure compared with the NFS.

In both cases the countries are moving to the establishment of a large-scale forest reserve, the dominant purpose of which will be biodiversity preservation. In both cases the economic implications are of concern. However, in many respects the situations are quite different (Table 10.1). First, the ownership is different. In the USA, most of the land that is moving towards extensive biological reserve status, namely the NFS, is already in public ownership as a forest reserve, and has been for over 100 years.[1] The debate in the USA revolves around the values for which federal forests should be managed. In Finland, the land under consideration for the forest reserve would almost entirely be land currently in private ownership. A question relates to payments to private parties for the provision of what are being viewed as public goods.

Second, the forest condition is different. In the USA, the initial forest reserve, which was largely a primary forest, was set aside and managed largely as a 'timber reserve' to meet current and future timber needs. However, large areas, particularly areas of public ownership, have never been harvested and much of the forest still maintains its primary forest characteristics, with many others being second-growth, naturally regenerated forest. In Finland, the forests in the south that would be given forest reserve status have characteristics that tend toward those of regulated even-aged forests. Third, recent management is generally quite different in the two countries. In general, recent management in the US federal forests has been towards multiple use, with timber management being only one of many management types. Some of the forest areas have never been harvested and thus have been largely under custodial care. By contrast, the forests of the Finnish south have tended to be managed primarily for timber for a relatively long period.[2]

[1] Nevertheless, changes are also occurring on private lands in the USA reflecting changing preferences and legal constraints, e.g., the Endangered Species Act.

[2] Although forest certification is occurring in many forests in both the USA and Finland, there is some discussion in the USA as to whether certification should be applied to the federal forests since it implies future timber harvesting, which may never occur on some areas of US federal forests.

Table 10.1. The USA and Finland forest considered for large-scale reserves

	Country	
Feature	**US National Forests**	**Finland**
Ownership	Public	Private
Forest Condition	Mixed: regenerated and primary	Approximately regulated
Recent Management	Multipleuse	Primarily timber management

However, the right of common access, which is practised in Finland, Sweden and Norway, regulates free public access to all forest land. This gives the public legitimacy in the public debate and process of policy formation for forest management. Overall, the US National Forest System tends to have characteristics that are more similar to wild pre-managed forests than does the Finnish forest.

10.2 The US National Forest System

The example under examination is that of the US National Forest System. As discussed, the NFS was created as a permanent timber and water protection reserve. Over the past century this role has been modified. Today, the system has moved substantially away from a role as a timber reserve and increasingly towards a role as a biodiversity and quasi-wilderness reserve (see Sedjo, 2000).

The National Forest System, the national forest lands managed by the Forest Service (FS), covers an area of some 78 million ha of a total forest area in the USA of about 300 million ha, about 26% of the total US forest. Although forest lands within the NFS are distributed widely throughout the USA, the preponderance of its lands are found in the west, between the Rocky Mountains and the Pacific Coast. Administratively, the system is broken down into ten regions and some 150 individual national forests. The Forest Service has an annual budget in excess of $3 billion. The NFS operates within the context of a country with a large and dynamic forest industry. The USA is the world's largest industrial wood producer, accounting for over one-quarter of the world's total output.

The US forest industry lands, located predominantly in the south and the Pacific Northwest (PNW), constitute about 20% of the forest land area of the USA. The forest industry, using its own forest lands together with forest lands that are owned by a host of non-industrial owners, produces over 90% of the industrial wood in the USA. The remaining 10% comes from the National Forest System (<5%), and other public owners including federal, state and local owners.

10.3 Background of the US National Forest System

The US forest system was established in the final years of the 19th century with the 1891 withdrawal from the public domain of certain federal lands for designation as permanent 'forest reserves'. The 1891 act simply allowed the forest lands to be set aside, with no protective authority or management charge. A few years later in 1897, the Congress passed the Organic Act, which provided for protection and management authority for the forest reserves. This act mandated the permanent 'forest reserve' to undertake two tasks:

- Secure favourable conditions of water flows.
- Furnish a continuous supply of timber for the use and necessities of the people of the United States of America.

Of course, the creation of a permanent reserve implied sustainability. In 1905 the FS was created to manage these reserves, which were then called the 'National Forests'. What was to be produced was straightforward – water and timber. And the beneficiaries were well defined – the people of the USA. Within this framework the basic asset, the forests, was to be preserved and protected.

A major rationale for the creation of the NFS was the belief that the private sector could not be trusted to properly manage for the long-term availability of timber and that a timber famine was inevitable. The perceived limitations and short-sightedness of the private-sector forest industry made it an unreliable provider of timber, especially into the future when the existing timber stands were expected to be depleted. The private sector, and the forest industry of that time, was viewed as myopic. It simply exploited resources for short-term gain. The NFS was conceived of as an insurance policy to protect the American people against the short-sightedness of the private timber interests.

The period during the first half of the 20th century was one of limited direct use. Timber harvests were modest, as most of the country's timber was drawn from the vast old-growth inventories on private lands. In the economic boom of the post–Second World War period, the role of the NFS in producing timber increased. However, as this occurred, additional demands were placed on these forests, usually in the form of new legislation. These included a multiple-use sustainable yield requirement, a wilderness requirement, planning requirement and public participation requirements. Gradually, the other outputs were demanded from the NFS. The NFS was explicitly required to manage for 'multiple uses'. The NFS now became explicitly responsible for timber production, water, wildlife, recreation, forage and wilderness in a sustainable manner. This approach has been characterized as an experiment involving a 'sharing of the forest resource' among uses and users. Since the lands were publicly held, the sharing of the resource also involved strong, essentially political, battles among the various user interests. These users included timber producers, environmentalists, those involved in recreation, and others.

10.4 Recent changes

Prior to the late 1980s, timber harvests on the NFS remained high, despite the various obstacles. The situation has changed dramatically since the late 1980s. A series of legislative, judicial, regulatory and administrative changes have dramatically reduced the timber harvest levels from the NFS from about 60 million m^3 in 1988 to perhaps 12 million m^3 in 1999, a decline to about one-fifth of the former harvest in about one decade.

A key component in the transition was the Endangered Species Act (ESA) and its impact on the spotted owl. Initially the ESA required that endangered species not be destroyed. Subsequently the Act was interpreted as also requiring that the endangered species habitat be preserved. The listing of the owl as a threatened species, i.e. as a species moving towards endangerment, precipitated the battle over the harvesting of the large areas of primary old-growth forest in the NFS. Given this listing, it became increasingly difficult to legally justify harvesting old growth in the range of the owl and the various legal cases moved towards limiting harvests. Subsequently, the implications of alternative means of dealing with the owl and timber harvests were assessed by a team of scientists (Forest Ecosystem Management Assessment Team, 1993). Of course, the various alternative harvest levels had implications for employment and economic activity as well as for the owl.

Additionally, provisions in the FS's own regulations added an additional basis for challenging the harvest policy, based upon the impact of logging upon various species within the forest. For example, the viability provision that was designed to ensure that native populations were not seriously disturbed within their traditional ranges, became the basis for successful challenges in the courts against certain of the Forest Service's own practices. Furthermore, other environmental legislation required that the cumulative effects of the various actions be properly assessed. All of these requirements added to the FS's burden when trying to justify a timber harvest.

Ultimately, the result the owl and other legislation and impediments had for harvesting was a de facto decision to set aside essentially all of the remaining old growth in the National Forests of the Pacific Northwest. Also, after this decision came significant reductions in timber harvests elsewhere in the NFS. Over the period of the late 1980s and early 1990s the role of the US NFS appears to have been markedly changed away from timber harvests and indeed other multiple-use activities, such as recreation, and towards a focus on biodiversity (Thomas, 1996). Not only did timber harvests fall to one-quarter of their earlier levels, but there were continuing efforts to eliminate portions of the forest from actual or potential harvests. Perhaps most notable was the effort in the last days of the Clinton Administration to designate some 17 million ha of National Forest as roadless areas, i.e. roads were not allowed to be built into these areas. Such a designation would have severely limited the ability to have timber harvests on these lands, as well as limiting recreation and other human uses. The area would thus have become, de facto, a wilderness area. Additionally, due to concerns that logging might occur without road development, there are substantial pressures not only to designate the land as roadless but also to prohibit any timber harvests.

In summary, in the USA the creation of the NFS was initially focused on developing a timber reserve. This role has broadened over the years, but in the past decade this broadening has narrowed and the process is moving in the direction of focusing on the establishment of a rather large forest reserve dedicated primarily as a biodiversity reserve.

10.5 The role of economic analysis

Analysis and economics have played a role throughout the essentially political process regarding the role of the NFS. In most cases, however, the role has not been compelling. An assessment of the implications of different harvest levels was undertaken periodically both before the owl (Beuter *et al.*, 1976) and subsequently (Anderson and Olson, 1991). The lands involved have always been in the public domain. Also, the lands have had a formal role in watershed protection, so their role as not been limited solely to timber management. So the questions facing the NFS have always been more than simply commodity production or maximizing financial present value.

Nevertheless, substantial debates ensued over the economic implications of discontinuing the harvest of old-growth timber in the NFS of the PNW. One estimate put the opportunity costs of the old-growth timber no longer available for harvest at some $25 billion (Mead, 1990). Other studies examined the impacts, both before and after the harvesting reductions, on employment and overall economic activity (Freudenburg *et al.*, 1998; Carroll *et al.*, 1999), as well as on the well-being of local communities (LeMaster and Beuter, 1989). Still other studies used contingent valuation approaches to estimate the willingness to pay for the preservation of the spotted owl and undertook cost/benefit calculations using a contingent value approach (Rubin *et al.*, 1991).

However, being public property, the management and the role of the NFS have always been highly political in nature with a high degree of involvement by the various interest groups trying to influence the use of the land to produce the outputs and services in which the various groups had interests. The various groups tend to accept the analysis that best supports their particular positions. The outcome to date, however, has been that the reduction of harvests from the NFS experienced in the last decade has been substantial.

10.6 Actual economic impact

Despite the various concerns, the overall effect on the economy has been modest. However, it is more severe in certain regions and segments of the society, such as loggers in the PNW. Although total harvesting in the PNW has declined over the 10-year period, total US timber harvests have actually increased as other suppliers, both domestic and international, have filled the gap. Additionally, PNW lumber production has remained at levels comparable with the 1980s (Warren, 1999). Much of the maintenance of mill production can be attributable to the large redirection of export logs to domestic mills. This was facilitated by the Japanese market, which weakened appreciably during the 1990s (Sedjo *et al.*, 1994). Also the ability of the local mills to find substitute logs through more intensive harvests by private forest owners and by drawing from more distant log suppliers, e.g. the intermountain region to the east, contributed to the restructuring of log supplies. Finally, domestic consumption has looked increasingly to imports of some products (softwood lumber) from Canada (Shull and Zager, 1994).

Additionally, the overall economic health of the PNW region was buoyed by the influx of new economic activities. The region, which began to experience healthy economic growth and employment during the 1980s, continued its economic expansion due in part to the incursion of a growing software industry and new industry. Of course, the share of the forest products industry in the economy has declined as harvests fell and the wood processing industry adapted through restructuring.

The designation of the NFS away from its legally mandated multiple-use function has never been expressed in a change in the multiple-use law. Rather, within the context of that law, other environmental laws, court cases and the discretion given the FS, the de facto management has changed from multiple use to species preservation (Thomas, 1996).

10.7 The Finnish situation

Finland is an important forestry producer, and forestry is important to the economy of Finland. In 1996 the forest sector's share of Finland's gross domestic product (GDP) was 7.7% and forest employment was 4.7% of the Finnish labour force (Palo and Uusivuori, 1998). In that same year Finland enacted two pieces of legislation important to forestry. These were the new Forest Law and the new Environmental Law. The Forest Law applied to all public and private forests in Finland and put biodiversity objectives on a par with industrial wood production in the context of sustainable forestry. In addition to harmonizing Finland's environmental regulations with those of the European Union, the Environmental Law required small private landowners to inventory their lands to identify the plant and animal species present. In concept, if the habitat is critical to some species, the government could set it aside, although the landowner would be compensated (Sedjo *et al.*, 1998, p. 74).

Subsequently, a group assembled by the Ministry of Environment has released a recent paper entitled 'Need for forest protection in the south of Finland'.[3] Proposals are being considered that would establish forest setasides in southern Finland primarily for biodiversity purposes. The proposals consider setting aside a forest area up to 10% of the total area of Finland's forest.[4] Such land would be drawn from private lands. The implications of such a plan for the Finnish forest industry are being examined to determine the potential impact.[5]

[3] The discussion of the specifics of the forest set aside proposal reflect presentations and conversation during a brief visit to Finland, 15–19 October, 2000. Unfortunately, no English-language materials were available.

[4] One might argue that an economically and environmentally better approach would be to have Finland make an arrangement with Russia whereby Russia set aside an area of primary forest comparable in forest type and size to that proposed in Finland. In return the Finns could compensate Russia for maintaining this forest that would already be in quasi-primary forest condition. This would have the advantage of being achievable at lower cost to the Finns while financially benefiting Russia. Such a situation would maintain an existing natural pristine forest equal in size to the one proposed in Finland.

[5] Presentations by Lauri Hetemaki and Taneli Kolstrom to a meeting on the Socioeconomics of Forest Protection Programmes at the Finnish Forest Research Institute, 17 October 2000.

The Finns have a number of important issues similar to those faced by the PNW in the USA. As with the US experience, one of the questions is the effect of the establishment of the biological set-aside on the domestic wood industry, and on overall income, output and employment (Linden and Uusivuori, 2000; Leppänen *et al.*, 2001). If harvests are reduced, e.g. by 10%, will this decline in raw wood be made up from other sources, or will mill processing also fall? An important issue is the extent to which wood availability reductions to processing facilities due to harvest declines in Finland could be made up by increased imports from the Russian Federation, Estonia and other nearby regions. The Russian Federation is currently providing about 10% of the industrial wood being processed in Finland. To the extent that substitute wood can be utilized efficiently, the production and income losses from processes can be avoided, and the losses confined to the logging activity.

In the case of the US PNW, some of the reduction in harvests due to the policy to cease logging on old-growth forestlands has been offset through reduced wood exports and drawing wood from further locations. This allowed for a softening of the impact on the region since mill process activity remained fairly high despite the decline in regional harvesting.

A second question the Finns face is quite different to that faced in the USA. Since the land under consideration for the forest preservation area in Finland is private, some mode of compensation would probably need to be achieved. This was not a problem in the USA, as the land was already in public ownership.

10.8 Conclusion

Both the USA and Finland are considering making a social decision to establish a large-scale forest reserve, the dominant purpose of which will be biodiversity preservation. In both cases the economic implications are of concern. The experience of the USA has been that the overall economic costs of adjustment appear to have been modest because substitute sources of timber were found which allowed for the continued operation of existing lumber mills. Additionally, an exceptionally buoyant economy over that period has provided generally plentiful employment opportunities, although some employment groups, e.g. loggers, have been negatively impacted. In fact, some of the harvesting decline probably would have happened anyway, due to weak market conditions in Japan. The situation for Finland could be somewhat similar if alternative low-cost wood sources are found, as with increased wood imports from the Russian Federation.

In some respects, however, the situations are quite different. First, the ownership is different. In the USA most of the land that is moving towards biological reserve status, the National Forest System (NFS), is already in public ownership as a forest reserve, and has been for over 100 years. In Finland, the land under consideration for the forest reserve would almost entirely be land currently in private ownership. Thus the delicate problem of compensation arises.

Second, the forest condition is different. Presumably Finland would try to manage its forest back towards a primary forest status. The situation in the USA is easier in that relatively large areas of the NFS have never been harvested, with many others being naturally regenerated second-growth forest. Thus, the USA appears to be closer to the desired biodiversity reserve conditions than does Finland. However, for both countries a new social goal appears to be that of maintaining a certain portion of its forest in a natural state consistent with biodiversity values.

References

Anderson, H.M. and Olson, J.T. (1991) *Federal Forests and the Economic Base of the Pacific Northwest.* The Wilderness Society, San Francisco, CA.

Beuter, J.H., Johnson, H.N. and Scheurman, H.L. (1976) *Timber for Oregon's Tomorrow: An Analysis of Reasonably Possible Occurrences.* Research Bulletin 19, Oregon State University, Forest Resources Laboratory, Corvallis.

Carroll, M.S., McKetta, C.W., Blatner, K.A. and Schallau, C.H. (1999) A response to 'Forty years of spotted owls? A longitudinal analysis of logging industry job losses'. *Sociological Perspectives* 42(2): 325–333.

Forest Ecosystem Management Assessment Team USA (1993) *Forest Ecosystem Management: An Ecological, Economic, and Social Assessment Report of the Forest Ecosystem Management Assessment Team.* United States Department of Agriculture, United States Forest Service, Washington, DC.

Freudenburg, W.R., Wilson, L.J. and O'Leary, D.J. (1998) A response to 'forty years of spotted owls? A longitudinal analysis of logging industry job losses'. *Sociological Perspectives* 41(1), 1–26.

LeMaster, D.C. and Beuter, J.H. (eds) (1989) *Community Stability in Forest Based Economics.* Timber Press, Portland, OR.

Leppänen, J., Linden, M., Uusivuori, J., Toropainen, M. and Pajuoja, H. (2001) The economic costs of increasing forest conservation in Finland. *Global Initiatives and Public Policies: First International Conference on Private Forestry in the 21st Century.* IUFRO, Division 6 (kv.) and Auburn University, Forest Policy Center (US) and Society of American Foresters (US), Auburn, Alabama.

Linden, M. and Uusivuori, J. (2000) *Econometric analysis of forest conservation: the Finnish experience.* University of Helsinki, Department of Forest Economics, Discussion Papers 494, Helsinki, Finland.

Mead, W.J. (1990) *The Economic Consequences of Preserving Old-Growth Timber for Spotted Owls in Oregon and Washington.* Community and Organization Research Institute, University of California, Santa Barbara.

Palo, M. and Uusivuori, J. (1998) Forest based development in Finland – a unique success? In: Palo, M. and Uusivuori, J. (eds) *World Forests, Society and Environment*, Vol. I, Kluwer Publishers, Dordrecht.

Rubin, J., Helfand, G. and Loomis, J. (1991) A benefit-cost analysis of the Northern Spotted Owl. *Journal of Forestry* 89(12), 25–30.

Sedjo, R.S. (2000) *A Vision for the U.S. Forest Service, Goals for its Next Century.* Resources for the Future, Washington, DC.

Sedjo, R.A., Wiseman, C., Brooks, D. and Lyon, K. (1994) *Changing Timber Supply and the Japanese Market.* RFF Discussion Paper ENR94-25, Resources for the Future, Washington, DC.

Sedjo, R.A., Goetzl, A. and Moffat, S.O. (1998) *Sustainability in Temperate Forests.* Resources for the Future, Washington, DC.

Shull M., and Zager, L. (1994) Factors affecting the international softwood market, 1987–1993. *Monthly Labor Review* 94(2), 21–29.

Thomas, J.W. (1996) Achieving sustainable forests: is NFMA up to the task. In: Johnson, K.N. and Shannon, M.A. (eds) *National Forest Management Act in a Changing Society, 1976–1996.* Natural Resources Law Center, University of Colorado, Boulder, pp. 158–183.

Warren, D.D. (1999) *Production, Prices, Employment, and Trade in the Northwest Forest Industries, Second Quarter 1999.* PNW Resource Bulletin 199, USDA Forest Service, Portland, Oregon.

11 The New Planning Approach and Criteria Used for Sustainable Forest Management in Turkey

Ü. Asan

Department of Forest Management Planning, Faculty of Forestry, University of Istanbul, Turkey

This chapter outlines a new approach to planning for multifunctional forest management objectives in Turkey. It describes the term 'functional planning', and the criteria that should be used to determine sustainable forest management in Turkey. The functional planning approach was developed to better balance demand for timber production as well as the protective functions of the forests such as erosion control, water yield, visual effect, recreational use, wildlife habitats and biodiversity.

The concepts of 'continuous forest' and 'sustainability' are central to the functional planning approach used in Turkey. The methods used to achieve the management goals for timber production and other non-timber forest values are outlined in this chapter, followed by the advantages that this new approach to planning offers forest management in Turkey. These advantages include flexibility and the ability to systematically realize the utilization of forest resources for multiple uses. The deficiencies that will need to be addressed in implementing this new approach are also explained based on the author's experience over the last ten years.

11.1 Introduction

There are a number of forest functions that provide a range of benefits in Turkey. These include wood production, non-wood forest products, avalanche control, water yield, erosion control, soil conservation, nature and wildlife habitat protection, community health, recreation, aesthetics and the regulation of climate. A lack of clean, drinkable water and breathable air threatens the very existence of human life on Earth. Recent studies on global warming and climate change have shown that the effect of forest loss on the greenhouse effect is significant because

© CAB International 2005.
Forestry and Environmental Change: Socioeconomic and Political Dimensions (eds J.L. Innes, G.M. Hickey and H.F. Hoen)

forest areas act as a carbon sink (Brown, 1997; Raev *et al.*, 1997; Asan, 1999; Luukkanen, 1999; Brand, 2000). The benefits that forests offer become management goals when societal pressures demand any one of them. As living organisms, forest ecosystems have their own production capacity for goods and services. Therefore, forest planning must seek to strike a balance between the forest resource capacity and the yield of goods and services over time.

According to the forestry principles accepted by the United Nations Conference on Environment and Development (UNCED) in 1992 in Rio de Janeiro, Brazil, forest resources should be sustainably managed to meet the social, economic, ecological, cultural and spiritual needs of present and future generations. We must try to mitigate human actions that can lead directly or indirectly to an irreversible degradation of forest ecosystems, and the services they provide (Anon., 1997; Anon., 2002a). Consequently, forest management should be based on stable, long-term land-use policies and regulations in order to ensure that the continuity of forest goods and services realized.

The basic aims of national forest management planning in Turkey are defined as 'Sustained yield of products and services: the achievement and maintenance in perpetuity of a high-level annual or regular periodic output of the various renewable resources of the National Forest System without impairment of the productivity of the land.' In this case, both the 'sustainable forest management' and 'continuous forest' concepts are suitable for realizing these objectives (Asan and Ercan, 2002). In resolution H1 of the Helsinki Conference (1993), the concept of sustainable forest management was defined as:

> the managing and use of forests and woodlands in such a way and to such a degree that they maintain their biological diversity, their productivity, their regenerative capacity, their vitality, and their capacity, now and in the future to fulfil their relevant ecological, economic and social functions, at local, national, and international levels; and in such a way that they do not damage other ecosystems. (Leal, 1996; Eeronheime *et al.*, 1997)

From the standpoint of forest management planning, the concept of continuous forest (i.e. *Dauerwald*) should be based on maintaining a specific amount of growing-stock or basal area in the forest ecosystem (Asan and Ercan, 2002). In Turkey, this concept has grown in importance over the last 10 years (Odabasi and Özalp, 1998). It has, so far, proven to be a useful instrument for maintaining various forest functions and benefits such as water yield regulation, soil conservation, amenity, recreation and wildlife habitat protection through uneven-aged, multi-storeyed forest cover and mixed site-specific tree species (Schabel and Palmer, 1999).

In Turkey, the effects of human activities through overconsumption of forest products, and the adverse effects on forest ecosystems caused by air pollution, forest fire, agricultural land clearing, mining, or any other kinds of forest destruction have been reducing the forest area. In addition to this, the natural effects of fungal and insect disease, windfall and long-term droughts on forest area have been significant. The shortage of forest lands in Turkey has forced foresters to manage the same forest areas for both production and protective functions, corresponding to the multiple use concept employed in many other countries. The multiple-use of forest resources involves the simultaneous use of a particular forest area for two or more purposes.

Over time there has been increasing demand placed on Turkey's forest resources. This has forced forest planners to seek other approaches and tools for planning the multiple-use management concept, especially around the larger cities. Although first described in the 1960s, planning for the multiple use of forest resources could not be realized in Turkey until recently. Traditionally, some forest areas have been allocated for recreational use, watershed protection or as national parks. However, this kind of allocation does not represent multi-functional forest management in practice. In Turkey, 25% of the land area is covered with the forest. However, due to the unsuitable distribution of forest over the land, Turkey is disadvantaged in relation to many other countries. As a result, the multiple-use forestry concept has been gaining importance as the demand for water yield, erosion control and recreational use associated with forest resources has increased (Asan, 1990, 1992). Despite the need for forest protection, two-thirds of the forest estate is managed for timber production, while one-third exists on steep slopes and upper forest zones. The forests in the Turkish national park systems have been allocated protection functions. No silvicultural treatments are applied in these areas (Asan *et al.*, 1996). National parks are managed according to the individual plans prepared by special planning groups. The conventional planning systems used in Turkey manage timber production only. The functional planning approach described in this chapter is based on detailed forest function maps that are arranged to cover large geographical regions and incorporate the long-term supply and demand projections for forestry goods and services. This chapter describes the Turkish forestry situation and presents the functional planning approach as a suitable tool to allow the rational utilization of forest resources in Turkey. The procedures and principles associated with the functional planning approach are also presented.

11.2 Forest resources and forestry problems in Turkey

Turkey is a country rich in tree species. The main coniferous species are pine (*Pinus brutia, P. sylvestris, P. nigra, P. pinea, P. halepensis*), fir (*Abies nordmanniana, A. bornmülleriana, A. cilicica, A. equi-trojani*), cedar (*Cedrus libani*), spruce (*Picea orientalis*), cypress (*Cupressus sempervirens*) and juniper (*Juniperus* spp.). The main deciduous species include species of oak (*Quercus* spp.), beech (*Fagus orientalis*), poplar (*Populus* spp.), ash (*Fraxinus* spp.), chestnut (*Castanea sativa*), elm (*Ulmus* spp.), lime (*Tilia* spp.), maple (*Acer* spp.), willow (*Salix* spp.) and hornbeam (*Carpinus betulus*) (Asan *et al.*, 1996). Turkey can be divided into eight ecological zones based on the major tree species, the topographic and geomorphological features of the land and the climatic conditions. There is a strong correlation between these ecological forest zones and the productivity of the forest. The combinations of tree species vary according to the humidity and temperature in each zone. Calabrian pine (*Pinus brutia)* is the major tree species (covering approximately 3 million ha) and occurs in the warm and dry regions of the Aegean and Mediterranean Seas. The central and eastern Anatolian plains experience severe winters and irregular precipitation and it is here that vast areas of degraded oak coppices are found. These trees are predominantly exploited for fuel wood and animal fodder

Table 11.1. Forest resources in Turkey (source: Turkish Forest Management Planning Department, Ankara, 1997).

	High Forests (1000 ha)				Coppice	Forested Area
Property	Coniferous (ha)	Deciduous (ha)	Mixed (ha)	Total (ha)	(1000 ha)	(1000 ha)
Productive	5956	1414	633	8003	2545	10,548
Unproductive	3937	1178	721	5836	4319	10,155
Total	9893	2592	1354	13,839	6864	20,703
%	47,79	12,52	6,54	66,85	33,15	100,00

Table 11.2. Distribution of forest lands by management approach (source: Turkish Forest Management Planning Department, Ankara, 1997).

	Standard Coppice Forests	National Parks	Protection Forests	Protective Forests	Selection Forests	Nature Parks	Nature Protection Area	Even-aged Forests	Total
Area (1000 ha)	3844.7	298.9	3104.5	210.0	290.2	15.2	22.5	12,917	20,703
Proportion %	18.57	1.44	14.99	1.01	1.40	0.09	0.11	62.39	100.00

According to figures presented by the Turkish Forestry Service in 1997, Turkey has a forest area of 20,743,122 hectares. Of this forest area, 67% is classified as 'high' forest, although of differing density. In areas of degraded high forests, coppices cover 49% of the area, while the productive forest area covers 51%. In other words, half of the forest area in Turkey is in need of rehabilitation. The area of 'high' forest consists mainly of even-aged, one- or two-storeyed, pure or mixed, and natural stands. The uneven-aged, multi-storeyed, and vertical closure stands make up about 4.5%. The distribution of forest areas by forest form and tree species is shown in Table 11.1. The area of forest being managed under different regimes is given in Table 11.2. More than 99% of the total forest lands in Turkey are state owned. Only 20,681 hectares are classified as having communal or private ownership.

In Turkey, 8.3 million people live in rural areas. This population is divided into 17,400 villages that exist in or nearby forest areas. With an annual income of approximately US$200 per capita (2002), this population plays a vital role in all forestry-related activities. This population also makes up the largest source of labour for forestry operations. The rural community in Turkey has some extra legislative rights documented in the forest laws and regulations; however, in some parts of the country, these people have a negative effect on forests. Therefore, a significant issue that will need to be addressed to achieve sustainable forest management in Turkey is upgrading the living standard of the villagers in the rural areas. This might be addressed through various social and economic expedients such as financial support for their pastoral and agricultural projects or marketing rights on some wood and non-wood forest products (Özdönmez *et al.*, 1996; Anon., 2001).

The main tasks of the Ministry of Forestry in Turkey include the protection and improvement of the existing forest resource; enlarging the existing forest area through rehabilitation and afforestation; supplying the domestic demand for wood, non-wood forest products and other forest functions; and improving the standard of living in rural communities (Muthoo, 1997).

11.3 Forest functions and expected benefits in Turkey

Projections for timber production and consumption in Turkey show that demand for raw lumber has been increasing (Anon., 1973, 2001; Birler, 1995; Asan, 1998). The main markets for raw lumber include woodworking, construction and mining. With the recent development of the medical and chemical industries in Turkey, the demand for timber resins and other non-wood secondary forest products has also been increasing. As a result of this demand, a large area of forest has been allocated for production. The monumental trees and forests occurring mainly in the upper forest zones of Turkey are considered very important for protection. Issues such as conservation, wildlife protection, avalanche control and stream flow regulation exist in this area, as well as the forest's inherent value to scientific research. Areas of forest have also been allocated for tourism in Turkey. This process has included the designation of buffer zones along main roads to increase the aesthetic value of the landscape.

Around the larger cities of Turkey (in particular Ankara, Istanbul, Izmir and Bursa), the demand for forest recreation activities (e.g. jogging, hiking and aesthetics) has been increasing in recent years. These kinds of social pressures have caused foresters to change their management objectives in some areas of the production forests to better consider recreational and communal health usage. On the other hand, more than half of the forests in Turkey occur in steep-sloped and high-altitude regions. In these regions, issues such as water supply, erosion control and nature and wildlife habitat protection have a greater importance.

According to the classification accepted by the Forestry Service in Turkey, the forest functions considered essential to sustainable forest management include timber supply for the wood working industry, non-wood forest products supply for medical and chemical industry, erosion control and soil conservation, hydrology control and water yield, recreational facilities, communal health and amenity functions, climatic functions, nature and wildlife habitat protection, scientific research and a national defence function (Anon., 2002b).

11.4 The application of planning systems in Turkey

Turkey's forestry planning has a practical history of 80 years. Over this time, forest management planning methods have changed many times as knowledge was gained. As timber production has been the most important forest function in Turkey historically, all of the forests had been planned for timber production until the end of the 1980s. The management plans used for the regulation of sustainable timber supply and non-timber forest benefits can be divided in two groups: model and conventional plans.

The model plans used in Turkey can be further separated into three groups depending on their basic principles and planning procedures. These groups are:

- Mediterranean region planning models;
- western Black Sea region planning models;
- functional planning models.

Conventional forest management plans are prepared and revised on a 10-year cycle. They determine the annual allowable cuts based on the Turkish sustainable wood production principles. Regeneration areas and intermediate cutting blocks are separated in the plans, but no technical guidelines are given for silvicultural treatments. These plans also fail to develop better relationships between the forest enterprise and villagers living in the planning units. Model forest management plans have been developed for different purposes and are considered a major step towards sustainable forest management in Turkey. However, these plans still do not incorporate the management of livestock and grazing resources in forests and adjacent lands, an important sustainability issue in Turkey's forests.

Table 11.3. Forest functions and working circles (source: Turkish Forest Management Planning Department, Ankara, 2000).

	Forest Functions and Working Circles								
Planning Unit	Timber Production (ha)	Erosion Control (ha)	Hydrology (ha)	Recreation and Amenity (ha)	National Defence (ha)	Scientific Research (ha)	Total Forest Area (ha)	Open Area (ha)	Grand Total Area (ha)
Bahceköy	-	2248.2	1406.2	990.0	-	651.7	5296.1	112.2	5408.3
Adalar	-	60.2	-	547.5	-	-	607.7	464.3	1072.0
Gaziosmanpasa	964.0	1989.0	2792.0	674.5	1585.0	-	8004.5	45,439.5	53,444.0
Kemerburgaz	9060.5	2514.0	2480.5	421.0	28.5	-	14,504.5	8233.0	22,737.5
Istanbul	4951.0	310.0	160.0	1370.5	263.0	-	7054.5	10,259.5	17,314.0
Research Forest	-	291.7	199.9	126.9	-	80.7	699.2	40.2	739.4
TOTAL	14,975.5	7413.1	7038.6	4130.4	1876.5	732.4	36,166.5	64,548.7	100,715.2

11.4.1 The functional planning system and its application

The functional planning concept is based on fostering multipurpose forest use and was first applied to the Turkish situation in the late 1980s. Until this time, the multiple use of forest resources was included in the main planning principles, but existed in theory only. At the time of writing this chapter, six different management plans had been prepared using the functional planning system. The Istanbul–Bahceköy management plan was the prototype, developed in 1989. Four plans were then produced in 1992. The last plan was completed in 1997. Detailed information on these six plans is presented in Table 11.3. With regard to the information provided by the Turkish Forestry Service, some of the management plans covering an area more than 300,000 ha were renewed using the functional planning approach in the forest regions of Mugla, Denizli and Zonguldak, beginning in 1999.

11.4.2 Planning criteria and their adaptation for functional planning

The basic criteria taken into account during the functional planning procedure are:

1. The multi-benefical forest use concept should be taken into consideration.
2. Management goals should be based on actual forest functions covering as long a time period as possible.
3. Biological diversity and the local genetic resources in the forest lands should be conserved.
4. Wood production should be considered as a secondary management goal in the areas managed for protective forest functions, and the annual allowable cut should take this into account.
5. The natural productivity of the forest should not be decreased by over-consumption.
6. The visual amenity and silhouette of the natural landscape should not be disturbed.

These criteria were addressed through a number of approaches. Working circles were first separated based on the forest function map. By overlapping those areas defined as having more than one forest function the first planning criterion was addressed. Forest function maps were then constructed using the actual and predicted land uses in Turkey. This satisfied the second criterion for planning.

The application of natural regeneration techniques based on group and small-scale cutting schedules over long rotations made it possible to attain the third criterion. The fourth criterion was addressed through the functional or guided stand structure approach. Using this system, there is a specific stand structure designed to meet each forest function. The establishment and maintenance of these structures is achieved through silvicultural treatments designed to achieve the management objectives in the planning units. Prescribed cutting regimes, defined as a ‘functional allowable-cut term’, are used for this. Constrained wood and service production arrangements (criterion 5) have been realized through these functional allowable cuts. The visual amenity and silhouette of the natural landscape in the planning units were protected by adopting longer rotations to better follow the natural life cycle of the forest species. This was aided by adopting silvicultural treatments designed to establish and maintain multi-storeyed and uneven-aged stand structures.

11.4.3 An algorithm for functional planning

The following procedure was used to construct the six plans.

Construction of the forest functions map

As the multiple use of forest resources was the main target of the functional planning concept in Turkey, the construction of forest function maps was the most important step in developing the system. Through this map, working circle separation, silvicultural treatment prescriptions, allowable-cut assessments and cutting plan arrangements were all based on forest functions.

A number of steps were involved in creating the maps. Water was considered first. Two large and eight small dams in the planning units supply running water to the metropolis of Istanbul. Each of the dams' watersheds was determined using topographic maps. The small springs surrounding the dams that produce high-quality drinking water for the watersheds were also identified using the natural topographic boundaries. Land areas categorized with different levels of erosion risk were then classified using slope grades and soil properties. These erosion risks were broken into four groups: absent, poor, medium and severe. Following this they were reduced to two classes in order to simplify the later planning steps. The areas of forest considered important to present and future recreation purposes were determined based on actual forest use and an estimate of future demand. Forest areas in military zones were allocated a national defence function in the planning units.

On either side of the main roads in the planning unit, 50 metre buffer zones were mapped, and the necessary areas around dams and recreational places were separated for visual effect and amenity functions. The forests allocated for scientific research, or identified as having interesting plant communities that could be used for teaching and training activities, were categorized as scientific research functional areas. Forest areas having no identified functional characteristics were allocated as timber production areas. As there was no demand for non-wood forest products such as fruits, forage and other kinds of medicinal and aromatic plants, no areas were allocated for this purpose. A small area of the coppice forests was allocated for wood production in the six planning units. As a result of the increasing social pressures exerted by non-governmental organizations in Istanbul, timber production was discounted as a major forest function around Istanbul. The wood supply function was fulfilled through intermediate cutting and thinning activities while converting the coppice stands into high forests.

Each forest function was considered separately during the construction of the forest functions maps in each planning unit. The boundaries of each forest function were plotted individually, based on the topographical structure of terrain, compartment lines and the boundaries between stand types on the land (Fig. 11.1). After defining each forest function area as a different layer, their boundaries were overlaid on the same map to obtain the final product (Fig. 11.2 and Fig. 11.3). The development of forest functions maps is ongoing and depends on the increasing role of computers in forest management planning in Turkey. For example, Digital Terrain Models (DTMs) were used to determine the spatial forest functions in the 12 city groves of Istanbul Municipality. The following procedure was applied in the construction of spatial forest function maps for each one of the groves (Asan and Özdemir, 2002; Asan et al., 2003):

1. Boundaries of the planning units taken from cadastral maps were transferred onto the topographic and orthophoto maps and the areas of planning units were identified on these two maps. Digital Elevation Models (DEMs) for each of the planning units were created using elevation curves (5 metre intervals) in the unit areas. This was achieved using the ER Mapper Gridding Wizard (see Fig. 11.4).
2. Compartments in the planning units were separated by roads, ridges and streams on the ortho-photo maps. Sub-compartments (inventory units) were created using the tree species, mixture forms and ratios, average diameter at breast height (dbh) and heights of the inventory units. Initial segmentation of inventory units was done on the ortho-photos, and defined after checking them on the ground. The spatial distributions of varying plant community unit types were constructed by overlapping these two maps.

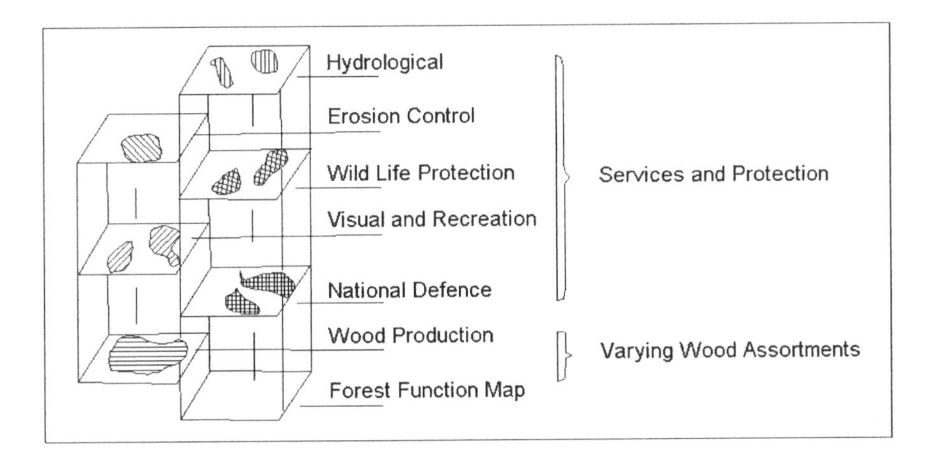

Fig. 11.1. The process of constructing a forest functions map.

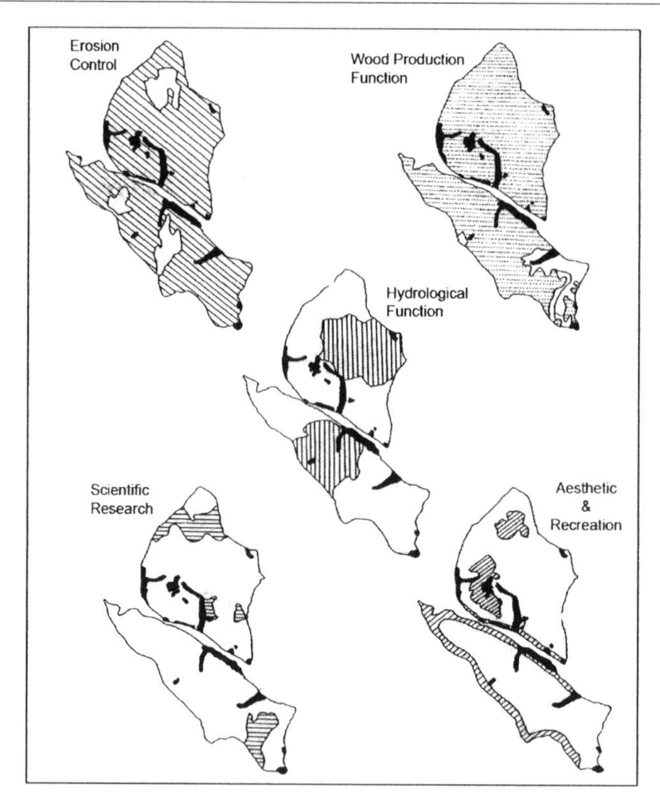

Fig. 11.2. Separation of varying forest functions in the planning unit.

3. The ortho-photos and plant community unit maps were overlaid on DEMs, and DTMs were obtained (see Fig. 11.5). Spatial forest functions expected from the groves were determined using these DTMs for each of the planning units. The following process was applied for this purpose:

   (a) *Erosion risk groups maps:* due to intensive vegetation cover, no obvious erosion was observed on the terrain. Thus, erosion risk groups were separated based on the terrain slopes and the soil properties determined on the ground.
   (b) *Hydrological function maps:* the spatial areas with important hydrological functions were determined based on the watersheds of the dams and pools in the planning units.
   (c) *Mosaic effect maps:* the visual aspects of ridges and hills, and the mixture ratios of vegetation cover, were used to determine the spatial areas having mosaic effects. Two groups were classified in the study: (1) the units consisting of one tree species or species group (coniferous or deciduous); and (2) the units covering coniferous and deciduous trees in the form of heaps or patch mixtures.

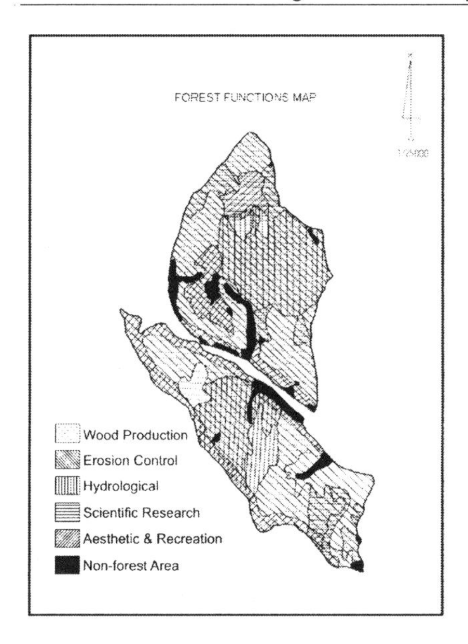

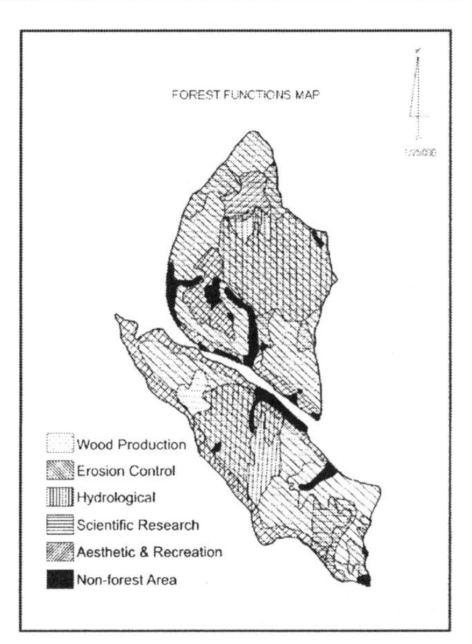

Fig. 11.3. Completing the forest functions map by overlapping individual maps

(d) *Silhouette effect maps:* the topography of the terrain (hills and ridges) was taken into account in determining the spatial areas that had a silhouette effect. The dbh and tree height within potential buffer areas were then considered for the classification and analysis.

(e) *Recreational usage maps:* the areas used for recreational purposes were separated based on the actual land-use situation, topographic feasibility and vegetation types. Present and alternative areas suitable for recreational use were all segmented on the DTMs of the city groves. Hiking and jogging paths were also defined as recreational areas.

4. Each function was considered separately at the beginning and stored on a different layer. After completing the same process for each of the forest functions, all of the layers were overlapped to obtain the final map. Each function was shown in a different colour on the forest functions map.

Separation of varying working circles

In Turkey, it is recognized that the creation of working circles during the planning process ensures the sustained flow of forest products and services. The working circle boundaries were determined functionally in the planning units. To achieve this, the forest functions were placed on the same map and any areas with two or more functions overlapped. Starting with the first compartment, the main and sub-functions were specified for each stand in the planning unit. The stands having the same forest functions or functional groups were gathered to form a working circle. The main functions were then assessed.

Determination of functional allowable-cuts

All of the data related to the planning units were included in a schedule called 'Stand Definition, Expected Functions and the Silvicultural Treatment Table'. The main cutting strategies for the stands, as well as the regeneration techniques, afforestation, thinning and conversion into high forest, were documented in this schedule. The intermediate allowable-cut amounts were prescribed in this planning system. In other words, the volume of timber to be felled within each stand was determined using the present structure and the dominant function. Due to the differences between the present and target structures of various stand types, varying allowable-cut amounts were considered for the same stand types depending on management objectives and forest functions. The appropriate silvicultural treatment, based on the form and grade of the forest, were prescribed by comparing the target and the actual structure of the stands. The forest functions were also considered.

Regulation of yield

Age class management techniques were used to regulate yield in the planning units. Long rotations, extending 300 years, and longer regeneration intervals, extended by 30 years based on the shelterwood group regeneration system, were adopted in the high forests of the planning units. A simple coppice management system (area-controlled) based on clear-cutting on a 20-year cycle was adopted in the coppice forests. However, management actions aimed at converting coppice to high forests were also adopted because of the increasing demand for recreation and amenity forest functions. Small-scale cuts were limited to 10–15 hectares, with the distribution of forest cover considered for the whole coppice forests. Because of the relatively poor site productivity in the coppice forests, and the existence of shade-intolerant tree species such as *Pinus brutia, Pinus pinea, Pinus maritima,* and *Quercus* spp. in the planning units, even-aged management was chosen as the main regulation method. In beech (*Fagus orientalis*), hornbeam (*Carpinus betulus*) and chestnut (*Castanea sativa*) stands, uneven-aged management systems were considered to be better for continuous forest objectives, yield regulation and site productivity of the planning units. The aim here was an uneven-aged, multi-storeyed forest with vertical closure and an irregular stand structure. As a result, any of the management methods based on volume or basal area control at the stand level could be used to regulate yield.

Construction of the coupe map

The final stage of the functional planning system involved constructing coupe maps to show the individual regeneration and intermediate cutting blocks. The forest stands and compartments chosen in the yield cutting plans were identified to show regeneration areas on the cutting maps. Allowable-cut calculations, made using the Cotta formula (Osmaston, 1968), were presented for only two planning units (Bahceköy and Adalar). Since there were no stands in the other planning units that had reached the specified cutting age, only intermediate cutting plans were prepared. After calculating the total intermediate cuts of the planning units, the annual amounts were obtained by dividing the total with the accepted cutting cycle. By using this method, cutting blocks were determined by compiling the stands that gave approximately the same annual yield. The boundaries of cutting blocks were also shown on the cutting maps.

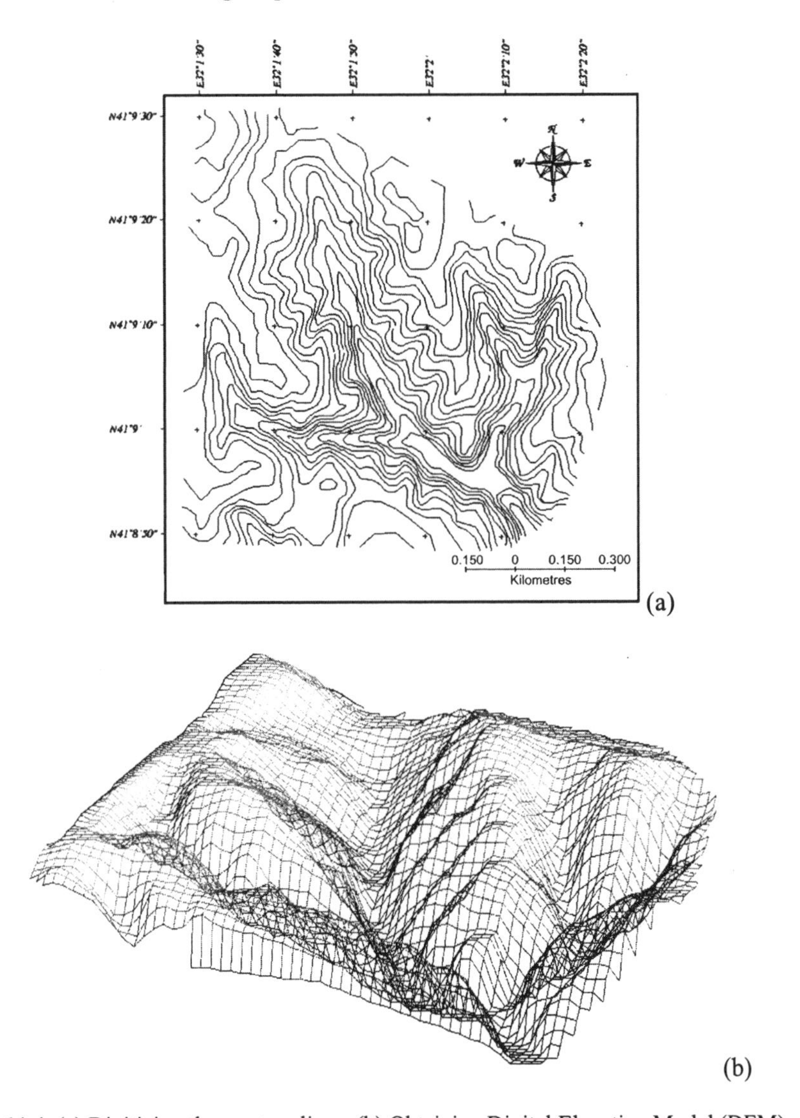

Fig. 11.4. (a) Digitizing the contour lines; (b) Obtaining Digital Elevation Model (DEM).

11.4.4 Evaluation of the functional planning system

The planning system introduced here under the name of 'functional forest management planning' is a flexible and suitable tool for multipurpose forest management in Turkey. In 2002, the Turkish Forestry Service published a bulletin that clarified the criteria to be used when defining various functional areas, and described the silvicultural treatments that will be adopted in these areas (Anon., 2002b). Although use of the functional forest management planning system has been increasing, it has only been adopted in a very limited area. As a result, we can expect that a range of potential problems may appear in future studies. The following points need to be considered before applying this planning approach to the whole country. Potential forest allocation needs to be considered before planning for the regulation of yield. This includes the following division:

1. The forests that could be allocated for timber harvesting are identified as production forests.
2. The forests that should be protected against hazards are identified as conservation (non-production) forests.

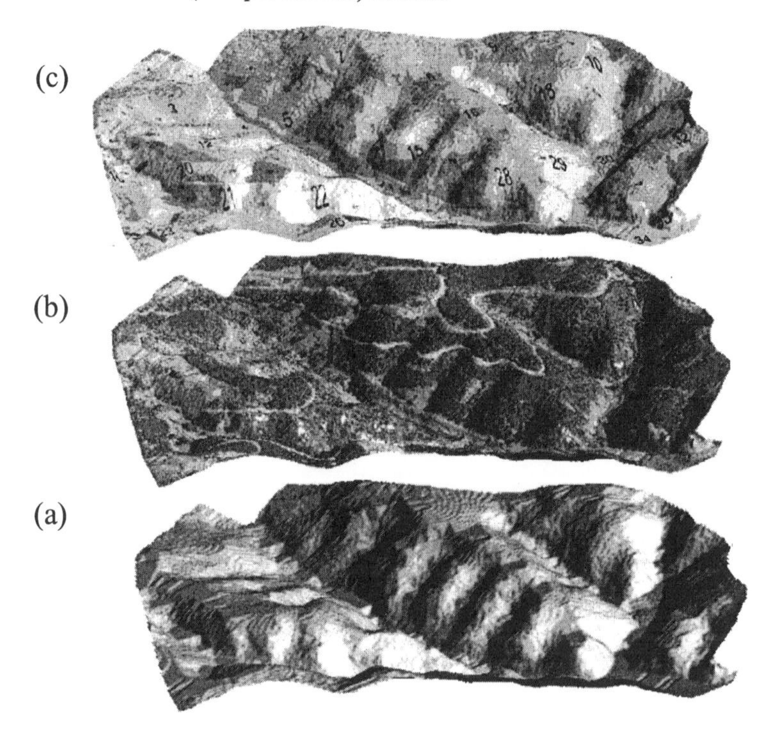

Fig. 11.5. (a) Digital terrain model; (b) draped with ortho-photo; and (c) spatial plant community map.

3. The forest areas to be managed for their protective functions are identified as protection (constrained-production) forests.

In the planning units, the upper forest zones with relic and endemic tree species were included in the non-production working circles. Forest areas having non-profitable operational conditions, due to their presence in deep valleys or on steep-sloped rocky areas, were also put into this working circle. Very limited interventions, designed to maintain the forest, should be undertaken in these circles. Depending on their condition, these areas can also be separated for 'nature protection' or 'wildlife conservation'. The remaining forests in the planning units were then divided into the varying working circles related to forest forms and functions. Stands containing light-demanding tree species, with either a one- or two-storeyed stand structure, were included in the even-aged management system. Stands consisting of shade-tolerant tree species, such as multi-storeyed fir, spruce and beech forests, were considered in the uneven-aged management system, with Möller's *Dauerwald* (continuous forest) approach being adopted. In the Turkish examples, the continuous cover approach was adopted for forests managed for erosion control, watershed regulation, wildlife protection, landscape amenity and recreation functions (Asan and Ercan, 2002). Although the *Dauerwald* concept is being accepted as the antithesis of qualification and regulation (Guldin, 1996), a suitable growing-stock or basal area interval should be maintained to assure the sustainability of the many forest functions in these planning units. The sustainability of the timber yield and the other forest functions can be realized either on a stand basis or on a working circle basis in the functional planning system.

A variety of problems were seen during the development of the functional planning system in Turkey. First of all, it was understood that the forest functions map should be constructed for as large an area as possible. Since the water dams have vast watersheds, the assessment of the small areas around the dams was inadequate for determining the hydrological functions. In the case of large watersheds consisting of a few planning units, the construction of a small-scale forest functions map was inappropriate, and proved to be inadequate for sound and reliable planning. Another issue was that erosion hazard could not be easily understood in the planning areas with dense forest cover. In the case of intense silvicultural operations such as clear-cutting or short-term regeneration, soil erosion can occur immediately. Consequently, the assessment of the actual erosion situation, based on soil properties and slope degree, is insufficient for a correct evaluation unless the precipitation form (amongst other things) is taken into account. The visual effect and common health functions associated with harvesting the forest cannot be directly measured using quantitative methods.

There was no demand for non-wood forest products during the planning activities. As a result, no yield regulation plans were arranged for these in the six plans. Multipurpose production plans, based on various wood assortments such as saw-log, pole, fuel and industrial wood, and various non-wood forest products such as resin, styrax, leaf, mushroom, forage and medicinal and aromatic plants, could be added into the functional planning system as required. These plans could be completed independently or combined with the timber production plan. Operational research techniques could also be applied to the planning units to optimize the total benefits. Preliminary studies by Gül (1998a) and Misir (2001) have shown that linear programming techniques can produce very good models that are suitable for functional planning. On the other hand, because of the absence of a suitable area for industrial plantations in the study area, there is no information on this type of forestry in the six plans analysed here. In fact, forests classified as timber production have been considered in two new categories as 'normal production forests', and 'industrial plantation forests'. Planning strategies and intensities are different in these two approaches to forestry. Plantation forestry is an inevitable production activity for countries experiencing shortages of timber (Birler, 1995; Anon., 1998; Asan, 1998). Therefore it will ultimately need to be taken into account during site evaluations for functional planning in Turkey.

There is no single document that outlines the optimal stand structure required to maximize the various forest functions. Decision-making related to the target structure of forest stands is therefore a problem that needs to be solved. We used general silvicultural principles such as tree species and mixture ratios, crown closure, stem quality, thinning grades, etc. to determine the functional allowable cuts in our analysis. The absence of sound and reliable information related to the financial resources and production potential of forests was identified as another handicap that restricted the effectiveness of an economic analysis of the planning units. The determination of the main functions and the order of the sub-functions highlighted problems due to the lack of economic analysis. All of the deficiencies described here made it impossible to use operational research techniques during the planning activities. Although there are many factors that make it difficult to apply the functional planning system to the whole country, the functional planning approach can be applied at a wider scale year by year (Güzenge *et al.*, 2002). With a number of improvements to the system, and with appropriate training of the relevant Turkish Forestry Service personnel, the whole forest resource in Turkey could eventually be managed using this scheme.

11.4.5 Improvement of functional planning

There are a number of problems that need to be addressed to ensure that the functional management planning system is effective in Turkey. Firstly, the functions expected from the forests need to be presented based on the social, economic, ecological and physical factors in the region. This requires reliable terrain data, which can be obtained from a geographic information system (GIS). The most common demands on these functions then need to be estimated over a period of time (the longer the better) using techniques such as trend analysis. The supply of forest resources and the demands for different forest functions must then be balanced for each forest region over the projection. The management objectives that need to be achieved in each planning unit must then be determined based on the balance of outputs for each case. This process allows either single or combined forest functions to be presented clearly and confidently for any given stand. A detailed site map provides a good guide for the various silvicultural treatments required to achieve the targeted structures in single stands.

Reliable operational reporting is essential for good planning. Regular reporting circumstances need to be created and enhanced in the forestry enterprises. In the proposed functional plans, the reporting of natural sustained yield is currently the only guaranteed operational information. However, sustained yield data should also be incorporated. Either the natural or economic continuity of the allowable cuts can be guaranteed by including alternative cutting policies in the functional plans. The use of economic analysis procedures can facilitate the creation of different management options to be incorporated into the planning activities. All of the thematic maps given in the functional plans were constructed by hand at the beginning of our case studies. However, it is also possible to use programs such as ARC-INFO or INTERGRAPH for this purpose. ARC-INFO was applied to the construction of forest functions and the other thematic maps in the management plans for Kerpe Research Forest around Izmit (Asan and Ercan, 2002), and the ER Mapper program was used for the construction of 12 management plans for the city groves belonging to Istanbul Municipality (Asan and Özdemir, 2002). Very good results were obtained in these projects.

Another aspect of functional planning in Turkey that can be improved is the use of quantitative assessment techniques for the prediction of various forest functions such as recreation values and scenic quality. Since these six projects were the preliminary prototypes of the functional planning approach, no quantitative assessment techniques were used to combine the various forest functions. Operational research techniques such as goal programming, linear programming, integer programming and simulation models would provide very suitable decision-making information. Gül (1998a, b), Gül and Kurdoglu (2002), and Misir (2001) have developed varying models for this purpose.

11.5 Conclusion

The Pan-European Criteria and Indicators for Sustainable Forest Management, adopted at the Helsinki Conference held on 24 June 1994, are outlined below (Anon., 1997):

1. Maintenance and appropriate enhancement of forest resources and their contribution to global carbon cycle.
2. Maintenance of forest ecosystem health and vitality.
3. Maintenance and encouragement of productive functions of forests as wood and non-wood forest products.
4. Maintenance, conservation and appropriate enhancement of biological diversity in forest ecosystems.
5. Maintenance and appropriate enhancement of protective functions in forest management (notably soil and water).
6. Maintenance, conservation and appropriate enhancement of the other socioeconomic functions.

I propose that the adaptation of each of these criteria in Turkey is possible through spontaneous functional planning. A functional planning system is a simple and practical tool for planning the allocation of resources and compares favourably to the other planning models that are available, offering a number of advantages. First, the multiple values of a forest area, in practice, can be effectively realized using a forest functions map. The conservation of biological diversity, the protection of the natural environment and aesthetics and preserving the continuity of the diverse array of forest functions in Turkey possible, particularly if functional allowable cuts and planning strategies for regeneration are used. Planning costs using this method are not expensive compared to other approaches.

The planning procedure, described here under the term 'functional planning', is obviously not a perfect system. It still needs improvement and development in order to improve its efficiency in the Turkish environment. The functional planning approach requires the support of a GIS, and must be based on operational research methods such as goal programming and simulation models. This will permit quantitative decision-making and facilitate economic feasibility reports to be compiled regarding different cutting strategy options. In the Turkish situation it is currently impossible to determine if the procedure offered here is going to deliver sustainable forest management from the standpoint of the modern planning concept, although the studies made by Gül (1998a, b) and Misir (2001) appear to offer some hope that this will be possible.

References

Anon. (1973) *Our Forestry in the 50th Anniversary of Republic of Turkey*. 187/145, Ministry of Forestry, Ankara (in Turkish).

Anon. (1997) *Sustainable Forest Management in France*. Ministry of Agriculture and Fisheries, Countryside and Forest Department, Paris, France.

Anon. (1998) *Final Report of the Workshop on Evaluation of the Works of Afforestation with Fast-growing Tree Species*. Proceedings, 083, Ministry of Forestry, Ankara, pp. 357–361 (in Turkish).

Anon. (2001) *VIIIth Five-Year Development Plan*. Private Commission Expert Report on Forestry, Ankara (in Turkish).

Anon. (2002a) *The State of Canada's Forests 2001–2002*. Natural Resources Canada, Ottawa, Canada.

Anon. (2002b) *Forest Functions, the Criteria Used in Definition of Functional Areas, and the Silvicultural Principles which Should be Adopted in these Areas*. Bulletin 6273/1961, Ministry of Forestry, Ankara (in Turkish).

Asan, Ü. (1990) *Multiple use of Forest Resources and Functional Planning*. Review of the Faculty of Forestry, University of Istanbul, Series B, 3, Instanbul, pp. 67–85 (in Turkish).

Asan, Ü. (1992) *Functional approach in dividing of working circles. Review of Forestry Engineering* 5, 30–31 (in Turkish).

Asan, Ü., Yesil, A. and Destan, S. (1996) A new approach to the rational utilization of the forest resources in Turkey. In: *Proceedings of the Second Balkan Scientific Conference on the Study, Conservation and Utilisation of Forest Resources, Sofia, Bulgaria, 3-5 June 1996*. Forest Research Institute, Bulgarian Academy of Sciences, Sofia, Bulgaria; Vol. I., pp. 110–115.

Asan, Ü. (1998) Industrial plantations and their application in Turkey. *Proceedings of Evaluation on the Works of Afforestation with Fast-Growing Tree Species*. Report 083, Ministry of Forestry, Ankara, pp. 25–37 (in Turkish).

Asan, Ü. (1999) Climate change, carbon sinks and the forests of Turkey. In: *Proceedings of the International Conference on Tropical Forests and Climate Change: Status, Is-*

sues and Challenges, Sponsored by University of the Philippines Los Baños, Integrated Environmental Management for Sustainable Development – Department of Environment and Natural Resources, United Nations Development Programme, 25 November 1997, Makati, Philippines, pp. 157–170.

Asan, Ü. and Ercan, M. (2002) A new concept and its application in forest management planning (with an example from Kerpe). In: *Proceedings of the Symposium on the New Concepts and Targets in Forest Management Planning, Faculty of Forestry of Istanbul University, 18-19 April 2002.* Published privately, Istanbul, pp. 8–22 (in Turkish).

Asan, Ü. and Özdemir, I. (2002) Determination and mapping of the spatial forest functions in the groves of Istanbul. In: *Proceedings of the Symposium on the New Concepts and Targets in Forest Management Planning, Faculty of Forestry of Istanbul University, 18-19 April 2002.* Published privately, Istanbul, pp. 67–76 (in Turkish).

Asan, Ü., Yesil, A., Özdemir, I. and Özkan, U.Y. (2003) Importance of the participatory approach, and the place of Digital Terrain Models in determining spatial forest functions. In: *Proceedings of the Second National Forestry Congress: The Management of Turkey's Forests and Participation, 19-20 March, 2003.* The Association of Turkish Foresters, Ankara, Ankara, pp. 162–173 (in Turkish).

Birler, A.S. (1995) *Importance of Industrial Plantations Established with Fast-Growing Tree Species in the Protection of Natural Forests and the National Economy.* Poplar and Fast-Growing Tree Species Research Institute, Ýzmit, Turkey (in Turkish).

Brand, D. (2000) The combined challenge of forestry and terrestrial carbon management. In: *Proceedings of the XXI IUFRO World Congress,* Vol. 1, pp. 558–564. Forest research Institute of Malaysia, Kuala Lumpur.

Brown, S. (1997) Forests and climate change: roles of the forest lands as carbon sinks. In: *Proceedings of the XI World Forestry Congress*, pp. 117–129. FAO, Rome.

Eeronheimo, O., Athi, A. and Sahlberg, S. (1997) *Criteria and Indicators for Sustainable Forest Management in Finland.* Ministry of Agriculture and Forestry, Helsinki.

Gül, A.U. (1998a) *Realisation of Functional Planning by Using Linear Programming.* Habilitation Thesis, 62 pp. (in Turkish).

Gül, A.U. (1998b) A linear programming model for functional planning. In: *Proceedings of the Symposium on the New Concepts and Targets in Forest Management Planning, Faculty of Forestry of Istanbul University, 18-19 April 2002.* Published privately, Istanbul, pp. 59–66 (in Turkish).

Gül, A.U. and Kurdoglu, O. (2002) Prediction of scenic quality and biodiversity. In: *Proceedings of the Symposium on the New Concepts and Targets in Forest Management Planning, Faculty of Forestry of Istanbul University, 18-19 April 2002.* Published privately, Istanbul, pp. 212–219 (in Turkish).

Güzenge, E., Mizrakli, A. and Yalcin, S.A. (2002) An application of the functional planning system in KDZ. EREGLI Forest Enterprise. In: *Proceedings of the Symposium on the New Concepts and Targets in Forest Management Planning, Faculty of Forestry of Istanbul University, 18-19 April 2002.* Published privately, Istanbul, pp. 95–102 (in Turkish).

Guldin, J.M. (1996) The role of uneven-aged silviculture in the context of ecosystem management. *Western Journal of Applied Forestry* 11(1), 4–12.

Leal, L. (1996) Pan-European process on criteria and indicators for sustainable forest management. In: *Summary Report of the Intergovernmental Seminar on Criteria and Indicators for Sustainable Forest Management.* Helsinki, Finland, pp. 53–57.

Luukkanen, O. (1999) Forest plantations or managed natural forests for carbon sequestration in tropical countries. In: *Proceedings of the International Conference on Tropical Forests and Climate Change: Status, Issues and Challenges.* Sponsored by University of the Philippines Los Baños, Integrated Environmental Management for Sustainable Development – Department of Environment and Natural Resources, United Nations Development Programme, November 25, 1997, Makati, Philippines, pp. 73–82.

Misir, M. (2001) A model for multiple-use forest management planning by means of geographical information system (GIS) and goal programming (with the example ormanüstü planning unit). PhD Thesis, University of Istanbul (in Turkish).

Muthoo, M.K. (1997) *Forests and Forestry in Turkey*. Food and Agriculture Organization, Ankara, Turkey.

Odabasi, T. and Özalp, G. (1998) *Silvicultural and Managerial Systems and Sustainable Forestry Considerations*. Ministry of Forestry, Department of Silviculture, Ankara (in Turkish).

Osmaston, F.C. (1968) *The Management of Forests*. George Allen and Unwin Ltd., London.

Özdönmez, M., Istanbullu, T., Akesen, A. and Ekizoglu, A. (1996) *Forest Policy*. Report 3968/435, Istanbul University, Turkey (in Turkish).

Raev, I., Asan, Ü. and Grozev, O. (1997) Accumulation of CO_2 in the above-ground biomass of the forests in Turkey and Bulgaria in recent decades. In: *Proceedings of the XI World Forestry Congress,* Vol. 1, pp. 131–138. FAO, Rome.

Schabel, G.H., and Palmer, S.L. (1999) The Dauerwald: its role in the restoration of natural forests. *Journal of Forestry* 97(11), 20–25.

12

Application of Criteria and Indicators for Sustainable Forest Management to a GIS-Based Multiple-criteria Decision-making Approach for Forest Conservation Planning in the Kinabalu Region, Sabah, Malaysia

M.H. Phua and M. Minowa

Department of Forest Science, The University of Tokyo, Japan

Conservation planning using a species approach involves the prioritization and selection of conservation areas based on biological criteria. Therefore, there is a need to adopt appropriate criteria for forest conservation planning. We propose that forest conservation planning refers to globally accepted criteria and indicators (i.e. C&I) for sustainable forest management. Criteria related to 'biodiversity conservation' and 'soil and water conservation' are two functions adopted and applied to forest conservation planning using a geogrpahic information system (GIS)-based multiple-criteria decision-making approach (MCDM). The biodiversity conservation criterion comprises indicators of ecosystem diversity and species diversity, whereas the soil and water conservation criterion consists of indicators related to landslide, flood and drought prevention functions. In addition, a criterion containing indicators for potential timber species distribution, proximity to settlements and roads is used as a measure of potential forest threats. The application of C&I to the GIS-based MCDM approach has enabled forest conservation priorities to be considered in terms of the important forest functions. It also integrated two decision-making processes (i.e. 'biodiversity conservation' and 'soil and water conservation') into a single process.

© CAB International 2005.
Forestry and Environmental Change: Socioeconomic and Political Dimensions
(eds J.L. Innes, G.M. Hickey and H.F. Hoen)

12.1 Introduction

To date, conservation efforts have targeted particular species and have often resulted in the establishment of protected areas for the 'charismatic megavertebrates' that attract public attention, have symbolic value and play a key role in ecotourism (Primack, 1993). This approach has been criticized as being a very expensive and inefficient conservation approach because an organism per se cannot be conserved (Barnes *et al.*, 1998). Although there has been a gradual shift from a single-species approach to a multi-species approach, ecosystems should be the target for conservation efforts (Franklin, 1993). An ecosystem approach can conserve a much larger number of species in a self-maintaining unit. This is especially true for tropical forests, where ecosystems are too diverse for the species-based approach.

In general, conservation planning involves the evaluation and selection of conservation areas using a set of criteria, a process termed 'criteria-based evaluation' (Bibby, 1998). Forest conservation planning can be defined as activities designed to evaluate the forest using appropriate criteria and techniques at the regional or landscape scale, and to prioritize the forest so that important conservation areas are selected. This implies that a lower priority will be given to some forests. However, forest conservation planning is not meant to deny the conservation value of any forest; rather, it will identify those forests that, in relation to the agreed goals and actions, do not have such an urgent need for conservation (Williams, 1998). While the ecosystem-based or forest-based conservation approach is becoming increasingly recognized, there are questions concerning the most appropriate criteria and evaluation techniques.

12.1.1 Criteria for forest conservation planning

Conservation planning using the species approach adopts criteria that are usually related to the species, such as its distribution. The species approach asks questions such as 'what is the minimum set of areas within Madagascar to represent all species of lemurs?' to achieve at least *n* representations within the area sets. The species approach has been criticized in terms of the 'meaning of complete representation' and 'viability' (Williams, 1998). Hotspots of species richness (i.e. area with highest number of species) and rarity are among the most popular biological criteria for conservation planning. Most of the criteria used in conservation planning are of biological origin. Almost one-third of 131 studies that were examined by Smith and Theberge (1986) employed biological criteria for evaluating conservation areas.

When planning forest conservation, the criteria used for the species approach may be inappropriate. There is no definition for the 'appropriateness' of criteria, although they should be explicit and quantifiable (Bibby, 1998). Forests, as an entity, contain elements such as biodiversity, soil and water resources.

After the 'Earth Summit' in 1992, C&I were pioneered by the International Tropical Timber Organization (ITTO) as guidelines and criteria for the sustainable management of natural tropical forests at national and management unit levels. The ITTO efforts were followed by other initiatives such as the Helsinki process for European forests and the Montreal process for temperate and boreal forests outside Europe. Because these initiatives have all evolved independently, and serve geographically distinct areas, there are disparities between them. Although some countries are still not participating in the development of C&I, the existing initiatives have led to the formulation of a globally accepted set of C&I that contain the most important tropical forest functions.

12.1.2 GIS-based multiple-criteria decision-making (MCDM) and its applications

MCDM approaches emerged among urban and rural planners in the 1970s in response to the limitations associated with traditional neoclassical environmental economics when dealing with decision-making in the planning process (Carver, 1991). The MCDM analysis involves a process of assigning values to alternatives that are evaluated along multiple criteria. However, it was only in the early 1990s that research on the integration of MCDM approaches with GIS began for solving spatial problems. In England, a nuclear waste site-search study using the combined GIS-MCDM approach showed the compatibility of integrating GIS with the MCDM approach (Carver, 1991). The GIS-based MCDM has been used for solving planning problems that involve conflicting objectives such as the land-use allocation problem (Janssen and Rietveld, 1990; Eastman *et al.*, 1995; Yeh and Li, 1998).

In conservation planning, GIS techniques are widely used for identifying the distribution of species during the selection of conservation areas. Jones *et al.* (1997) applied GIS techniques for predicting the distribution of wild bean relatives by analysing the preferred climate of beans. In Kenya, GIS techniques have contributed to the prioritization and selection of potential conservation areas by using relationships between environmental factors and the distribution of birds (Muriuki *et al.*, 1997). Smallwood *et al.* (1998) conducted an indicators assessment approach using GIS techniques for multiple-species habitat conservation planning in California, USA. In the USA, the development of so-called 'gap analysis' emphasizes ecosystems in conservation planning, applied at a regional or landscape scale (Scott *et al.*, 1993). However, the use of GIS-based MCDM in forest conservation planning is rare. Forest conservation planning deals with the evaluation of forest conservation priorities and involves many criteria considered by multiple decision-makers. This is most often done by assuming rational decision-making processes where the selection of conservation areas is based on scientific results.

12.1.4 Objective

This chapter proposes the use of C&I in forest conservation planning at a regional scale. In this case, the C&I are applied to GIS-based MCDM for forest conservation planning in the Kinabalu region, Sabah, Malaysia.

12.1.5 Study area

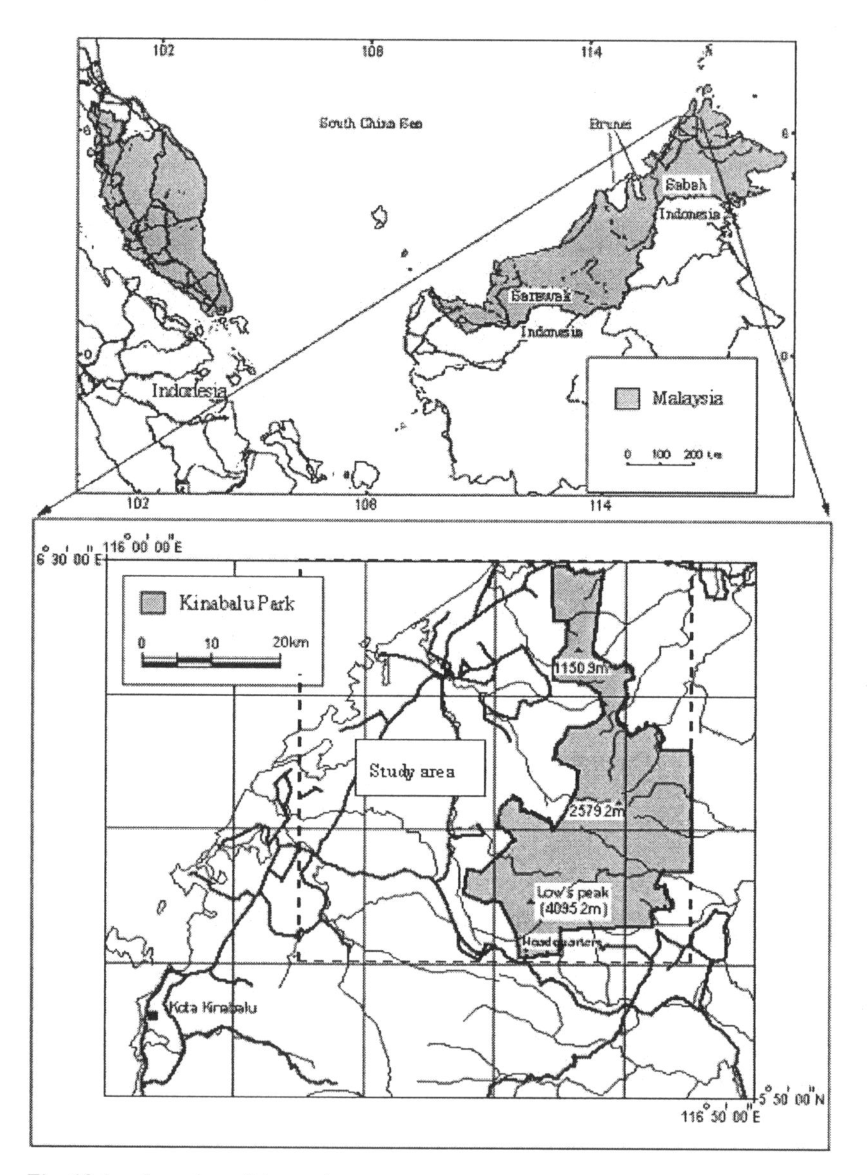

Fig. 12.1. Location of the study area.

Conservationists often find that they have contradictory views on which forest to conserve. While some stress the importance of forests in remote areas, because of their pristine ecological integrity, others emphasize forests adjacent to human settlements because of the increased threats to the forest. If the forest conservation priority is interpreted as the relative urgency of taking conservation actions, then forests adjacent to human settlements should be emphasized. The Kinabalu region is located in the northeastern part of Kota Kinabalu, the capital of the state of Sabah, Malaysia (Fig. 12.1). This region ranges from low-lying flood plains in the coastal areas to mountains reaching an altitude of 4095 m. The highest peak between the Himalayas and Mount Whelm in Irian Jaya is located within the Kinabalu Park (753.7 km^2), gazetted in 1964. Forest ecosystems in this region are very diverse, ranging from mangroves in coastal areas and lowland rainforests to montane rainforest and subalpine forest. Ultrabasic forest, originating from a particular geological substrate, is also present. The mean air temperature ranges from over 30°C in the lowland areas to 10°C at about 3700 m.

12.2 Materials and methods

12.2.1 GIS-based MCDM approach to forest conservation planning

As an MCDM problem, forest conservation planning can be structured using the Analytical Hierarchical Process (Saaty, 1980). This consists of a goal at the top level, and criteria and indicators at the subsequent levels. A GIS-based MCDM approach to forest conservation planning can then be formulated as:

$$C^k = \sum_{i=1}^{I} w_i x_i^k$$

where C^k designates the forest conservation priority attained by the kth alternative, w_i is a weight and x^k is the kth alternative, a pixel containing i indicators which are GIS layers.

The goal of the Analytical Hierarchical Process is to evaluate forest conservation priorities in the region. Given the hierarchy, for each level, a pairwise comparison matrix is constructed to rate the relative importance of each element (according to their impacts on the level above) and weights (w_i) computed. Weights are assigned by a representative of the conservation community, a key decision maker in the management of the Sabah Parks and the organization that administers all of the parks in the state. The hierarchy and assigned weights are shown in Fig. 12.2.

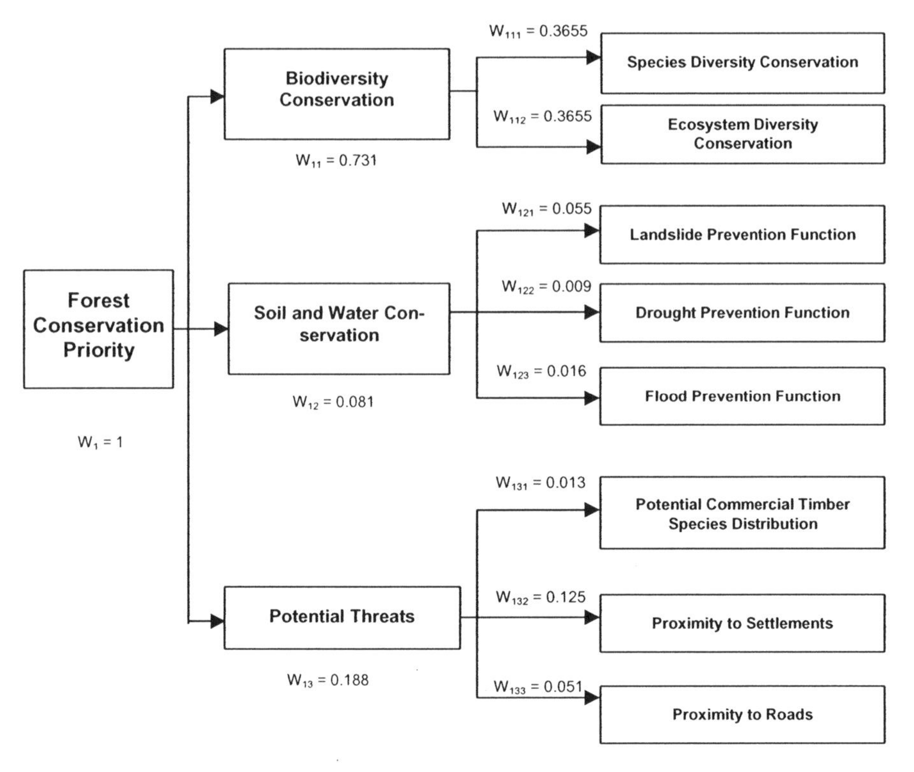

Fig. 12.2. Hierarchy of forest conservation planning in Kinabalu Region, Sabah, Malaysia, and weighting by a conservationist.

Table 12.1. Comparison of the C&I for sustainable forest management (adopted from Hiroshima, 1997).

Criterion related to:	Montreal Process	Helsinki Process	ITTO
Biodiversity conservation	Criterion 1	Criterion 4	Criterion 5
Maintenance of forest productivity	Criterion 2	Criterion 3	Criterion 4
Forest health and vitality	Criterion 3	Criterion 2	Criterion 3
Soil and water resources	Criterion 4	Criterion 5	Criterion 6
Global carbon cycle	Criterion 5	Criterion 1	-
Cultural and socioeconomic benefits	Criterion 6	Criterion 6	Criterion 7
Legal, economic and institutional frameworks	Criterion 7	Under consideration	Criterion 1

This is the case using one decision maker. In the case of multiple decision makers:

$$C_m^k = \left(C_1^k, C_2^k, \ldots, C_M^k\right)$$

where:

C_m^k: kth alternative of forest conservation priority of decision maker m, m = 1,..,M

12.2.2 Adopting and deriving the C&I for the GIS-based MCDM approach

First we examine the C&I associated with three initiatives; the ITTO, Montreal and Helsinki Processes. Table 12.1 compiles the criteria in common for the three initiatives. Only one criterion related to global carbon cycling is not shared by the three initiatives. All of the criteria, except those related to cultural and socioeconomic benefits and legislation, may be evaluated using GIS and remote sensing techniques. The criteria related to biodiversity conservation and soil and water conservation functions, quantifiable with GIS and remote sensing techniques, are the two common criteria adopted. However, the sustainable forest management (SFM) indicators are mainly based on statistical descriptions, designed for application at the national and forest management unit levels. For the GIS-based MCDM approach to forest conservation planning, spatial variations are of importance. Therefore, spatially quantifiable indicators need to be developed.

12.2.3 Biodiversity conservation function

In the Mount Kinabalu area, it has been found that species richness (Kitayama, 1992) and the Fisher's alpha index (Kitayama, 1996) decrease as altitude increases. The influence of geological substrates on the altitudinal pattern of biodiversity was reported by Aiba and Kitayama (1999). In this study, the Fisher's alpha diversity index, calculated from field data using the formula below, is used as the indicator of biodiversity conservation function (Kitayama, 1992, 1996; Aiba and Kitayama, 1999):

$$S = \alpha \ln\left(1 + \frac{N}{\alpha}\right)$$

where S is the total number of species per stand, α is the Fisher's alpha diversity index and N is the total number of individuals per stand.

Altitude governs the generally decreasing pattern of biodiversity with increasing altitude in both geological substrates (adjusted $R^2 > 0.94$). The coefficients of the linear regression models are significant at 1% and 5% for the non-ultrabasic and ultrabasic forests, respectively. Altitude was derived as a digital elevation model (DEM) by digitizing contour maps (1:50,000) at 60 m intervals. Ultrabasic and non-ultrabasic substrates were digitized based on a geological map (1:100,000) and a vegetation map (1:50,000). Polygons of the two substrates were used to extract the DEM into ultrabasic and non-ultrabasic DEMs where the species diversity conservation functions on non-ultrabasic and ultrabasic substrates were derived using the empirical relationships. For more details, refer to Phua and Minowa (2000).

The empirical functions do not apply to the mangrove forest as there is no altitudinal distribution pattern. Field survey data for mangrove forests on Palawan Island, the Philippines, were acquired (source: Japan Overseas Forestry Consultants Association, 1999) for calculating the Fisher's alpha index. As Palawan Island is immediately to the north of Sabah, the species composition, as well as the natural structure of the mangrove forests, was assumed to be the same as in the western part of Sabah. The mangroves of Sabah and other parts of Borneo, together with the Philippines, Papua New Guinea, the Malay Peninsula, Singapore, Java, Sumatra, Sulawesi, Halmahera, Bali, Lombok and Timor, belong to the '*Rhizophora mucronata* area' of the Southeast Asia and Pacific regions (Nakamura, 1992). Both Sabah and Palawan Island are located in the same longitudinal zone (114° to 120° E), where, in total, only 20 species of mangrove are found (Tomlinson, 1986).

Ecosystem diversity has a multi-scale dimension. The main difficulty in mapping ecosystem diversity is how to rank the 'importance' of existing ecosystems. Ecosystems in the Kinabalu region can be regarded in terms of forest types at a regional scale. Vegetation maps (Kitayama, 1991) cover only Kinabalu Park so forest types of other areas have to be mapped. For that, a geology map was used to extract the ultrabasic substrate areas, known to contain many rare species. The lowland rainforest and mangrove forests were mapped using a Landsat-TM image acquired on 8 April 1996. Geometric and radiometric corrections were undertaken prior to supervised classification using the maximum likelihood algorithm. Overall classification accuracy was 87.4%, which is considered to be very good.

Ecosystem diversity can be viewed in terms of species diversity among the landscape ecosystems using the conventional alpha or beta diversity indices (Barnes *et al.*, 1998). Besides the diversity of richness at higher organization levels (Fisher's alpha diversity index of a forest type or FI), a shape index (SI) and an area index (AI) were included to represent ecosystem diversity as an ecosystem diversity index (EDI):

$$EDI_j = FI_j * SI_j * AI_j$$

where:

FI is the Fisher's alpha diversity index

SI is the Shape index, $SI = \lambda - (\lambda - 1) * \left(4\sqrt{a_j}\right) / g_j$

λ=2 is used to limit the SI values in a range between 1 and 2,

g_j is the perimeter of forest type j,

a_j is the area of forest type j,

AI is the ratio of the area of forest type j to the total forest area

j is the forest type: 1, 2, …., J

Means of the Fisher's alpha index were computed for the lowland rainforest, montane rainforest, subalpine forest and ultrabasic forest. The alpha index computed using the data from Palawan Island was used to represent both the species and ecosystem diversity of the mangrove forest. The SI indicates the complexity of an ecosystem as the degree of deviation from a square (the 'simplest' shape of a forest ecosystem). It is identical to the shape index used by Tanaka (2000), except a square was used as the simplest shape, instead of a circle. The AI suggests that bigger areas constitute a higher level of ecosystem processes than smaller areas.

12.2.4 Soil and water conservation functions

For this study, the soil and water conservation functions refer to landslide prevention, flood prevention and drought prevention functions. The landslide prevention function of a forest can be considered in terms of landslide hazard. The more hazardous an area, the more important it is to keep the forest in place. Forests are known to have an important role in preventing landslides by binding the soil with its roots. Generally, the drought and flood prevention functions of forest refer to the water retention capability of the forest. Forests play an important role in recharging groundwater by increasing the rate at which rainwater infiltrates into the soil and releases the water.

An evaluation matrix was constructed to evaluate the soil and water conservation functions. Environmental factors were based on an extensive review of the related literature. After accounting for data availability, the environmental factors of slope, rainfall, soil depth, geology and topography were chosen.

We then developed a weighted linear combination of GIS layers, widely used in GIS-based studies such as in Kato *et al.* (1997). The weights and scores used in the evaluation matrix for the GIS evaluation of the soil and water conservation functions were placed in a weighted linear combination form as follows:

Landslide prevention function = (0.3*Slope) + (0.2*Annual Rainfall) + (0.2*Soil Depth) + (0.15*Geology) + (0.15*Topography)

Flood prevention function = (0.2*Slope) + (0.2*Annual Rainfall) + (0.25*Soil Depth) + (0.15*Geology) + (0.2*Topography)

Drought prevention function = (0.1*Slope) + (0.25*Annual Rainfall) + (0.3*Soil Depth) + (0.15*Geology) + (0.2*Topography)

Slope layer was derived from the DEM. The surficial geology layer was constructed using geological maps (1:100,000). A soil depth map was generated by classifying the DEM into ‘thick’, ‘medium’ and ‘thin’ soil classes. Annual rainfall data from 13 climate stations were acquired and imported into the GIS as a point layer where Voronoi polygons were generated. The average, minimum and maximum rainfall in the study area are 3274 mm, 2451 mm and 5486 mm respectively. Topography as categories of divergence, convergence and slope were derived from the DEM. Topography was derived using a GIS procedure on the DEM with the widely used D8 algorithm (see Phua and Minowa, 2000).

Table 12.2. Feasible forest area for conservation

Feasibility class	Not protected area (ha)	(%)	Not protected area selected with a smallest 10% of the d_p (ha)	(%)
Not feasible	103,761	91.53	12,418	94.71
Feasible	4349	3.84	440	3.36
Highly feasible	5254	4.64	253	1.93
Total	113,364	100	13,111	100

12.2.5 Potential threats

The suburban forest in this region is exposed to various threats induced by human presence. These threats can be measured in terms of proximity to settlements and roads that provide access to deforestation activities. In forestry, the available timber resource often attracts logging activities. Therefore, the distribution of commercial species can indicate the threat of logging activity. The distribution of commercial species (Sabah Forestry Department, 1989) was empirically derived based on altitude. At least 40 species were found in the region, up to an altitude of about 2000 m.

12.3 Results

12.3.1 Feasibility of establishing a new protected area

Assuming that a new 'totally protected area' is to be established, the main constraint would be land ownership. Three feasibility classes can be defined in terms of land ownership types: highly feasible, feasible and infeasible for government, semi-government and privately owned forests, respectively. For government-owned forest, it is feasible to establish a new protected area because the forest can be easily transferred from one government agency to another. Although there may be constraints, forests owned by a semi-governmental body can also be transferred to a new management body. For the forests under private management, establishing a new protected area is often an unrealistic expectation, except under extraordinary circumstances.

We excluded Kinabalu Park from subsequent analyses because it is strictly protected. We therefore attempted to identify a new protected area for 113,364 ha of forest where privately owned forest constitutes 91.5% (see Table 12.2). Forests in the highly feasible class had relatively low C^k scores, irrespective of the weights assigned. For forests in the feasible and infeasible classes, the C^k scores ranged from 0.69 to 0.86 (see Fig. 12.3).

A compromise programming technique was used to integrate the forest conservation priority of the three decision-makers. This technique identifies the alternatives which are closest to the ideal point by using some measure of distance. The ideal point is usually an alternative with an unattainable level. We defined the ideal point as 1 for each forest conservation priority map:

$$C^* = \left(C_1^*, C_2^*, \ldots, C_M^*\right)$$

Using the distance metrics (described by Zeleny, 1982), d_p (separation distance measure) can be written as:

$$d_p = \left[\sum_{m=1}^{M} \left(C_m^* - C_m^k\right)^p\right]^{1/p}$$

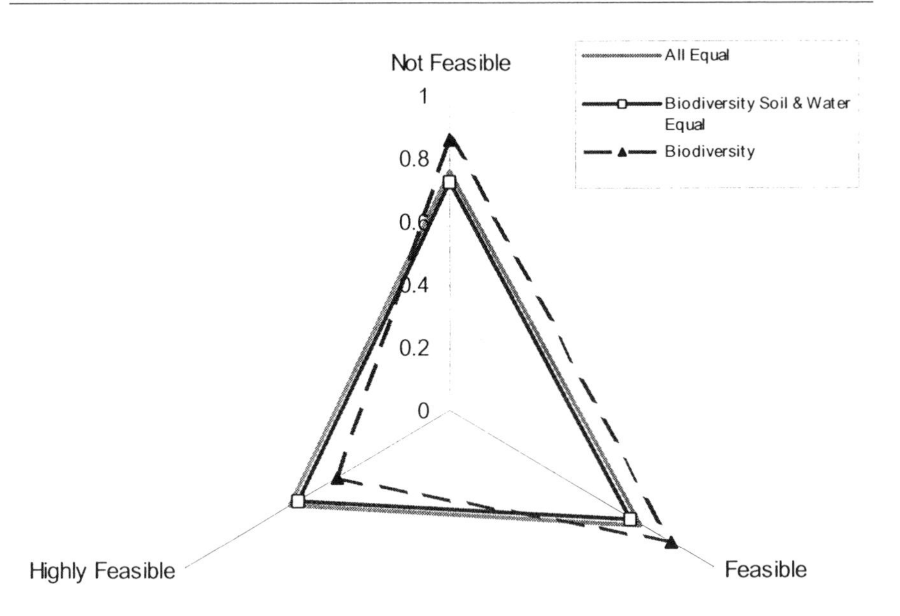

Fig. 12.3. Forest conservation priority of multiple decision makers by feasibility class in Kinabalu Region.

In this study, Euclidean distance ($p = 2$) was used. Therefore, the smaller the separation distance measure, the closer the alternative is to the ideal point. This alternative is called the 'compromise solution'.

The mean distance of the highly feasible class was almost 0.5 times higher than that of the infeasible class (0.29). The mean distance of the feasible class (0.32) was between the two classes. A threshold was then determined to select the forest to be conserved. We simulated the selection using the smallest 10% of the distance where 11.6% of the area was selected. This process selected forests near streams, some forest bordering the Kinabalu Park, with the forest stretching from the western coastal lowlands towards the Kinabalu Park in the east. Once again, we examined the feasibility of forest conservation within the selected forests. A similar pattern was revealed with the semi-government-owned forest reflected in the forest area ratio in the area under analysis. While the infeasible class dominated at 94.7%, the highly feasible class constituted only 1.9% of the selected forests, much lower than the forest area ratio (Table 12.2). This indicated that forests in the highly feasible class (government owned) were not very important to forest conservation. This is clearly shown in the low C^k scores, the proportion of government-owned forests in the top 10% of the d_p and also their insignificant area coverage. These findings indicate that extra political will is needed if establishment of the new protected area is to be attempted.

12.3.2 Consensus of decision makers in forest conservation planning

Forest conservation planning involves assessing the different preferences of decision-makers on different criteria. The weights assigned by conservationists (Sabah Parks) that actively engage in forest conservation activities represent the conservation sector. The weights assigned to the biological criterion (73%) outran the weights assigned to the soil and water conservation criterion (10%). This may be the typical preferential pattern of conservationists in forest conservation planning. In addition, we simulated a representative from the soil and water conservation sector with a high level of concern for biodiversity, soil and water conservation functions (i.e. equal weights are assigned to the indicators) while the weights for the criterion of potential threats were kept invariable. We also simulated a 'rational' decision maker by assigning equal weights to all criteria and indicators.

We found that the forest conservation priority of each decision-maker deviates from other decision-makers due to different preferences on criteria and indicators. The deviation can then be measured by modifying the separation distance measure into a distance measure between every two decision-makers (i.e. measuring how close the forest conservation priorities are between the decision-makers). While mean, minimum and maximum values of the consensus between the decision makers are 0.1686, 0.0017 and 0.3857, respectively, the distribution reveals that a high degree of consensus was found mainly in hilly and mountainous areas, especially within Kinabalu Park.

12.4 Discussion

We strongly recommend using C&I for sustainable forest management for the planning of protected forest areas at a regional scale. However, it is important to note that these C&I have not been specifically developed for conservation planning objectives. It is therefore important to include other criteria where relevant. The biological criterion in this study is a hotspot of richness but employs a more sophisticated biodiversity index. This is a straightforward measure of the number of species and their arrangements in a spatial unit, irrespective of the particular species. Some may argue that comparing biodiversity across many forest types is not meaningful. This argument is particularly true for non-tropical areas, where forest types and species richness may be less diverse. In the case of tropical forest areas, it is practically impossible to derive a species diversity map if a distinctive environmental gradient does not exist. In the study area, the species richness gradually decreases with increasing altitude gradient, and its pattern depends on geological substrates (Kitayama, 1992, 1996; Aiba and Kitayama, 1999). It is, therefore, reasonable to express species richness as an index for species diversity.

In comparison, rarity, a frequently used criterion in species conservation planning, is not practical for forest conservation planning in the tropics. The rarity concept embraces both a spatial and a numerical dimension. It may be an evolution property such as habitat specificity and small natural range or density (Bibby, 1998). Its use in conservation planning requires reliable data on the range and/or the numbers of individual species, information that is rarely available for tropical environments. In a narrower sense, endemism due to habitat specificity may be included at an ecosystem level. In the study area, only the 'ultrabasic' forest was differentiated from other forest types. This information can be included as a binary mask at the final stage to show forest ecosystem types with high levels of endemism.

There are many ways to evaluate forest functions that depend on factors such as purpose, data availability and scale. A regional scale was attempted in this study due to limitations associated with the scale and availability of data (particularly maps). The regional scale is appropriate for forest conservation planning because it is often the spatial distribution of forest conservation priorities that is important to planners.

To determine the biodiversity value of a particular area, a great deal of time, money and expertise are needed – an acute problem in the tropics. Point data from field surveys are an invaluable source of data for GIS mapping at a regional scale, provided that an environmental gradient exists.

Forest types may also have played a role in both soil and water conservation functions. However, the relationships between various forest types and functions have not yet been well established. Forest types can only be used as a factor layer if the relationships between forest properties and the indicators of soil and water conservation are well established. In this study, the role of the forest was treated as a prerequisite for all of the functions to be realized. Therefore, biodiversity conservation function in non-forested areas was not considered.

Conservation planning using the species approach assumes a rational decision-making process, where there is either one decision-maker or many decision-makers that share the same preferences. This is not, generally, a realistic expectation. In reality, many decision makers are involved in planning. Forest conservation planning should therefore be implemented using a forest ecosystem approach, as it enables integration of the results into land-use planning. Inclusion of the soil and water conservation function has broadened biodiversity conservation to the forest, as an entity itself. This has integrated the need for the conservation of soil and water resources into a single decision-making process.

12.5 Conclusion

As the approach to conservation shifts from species-based to ecosystem-based, the criteria used must also change from solely biological criteria to better consider important forest functions. In this regard, globally accepted C&I should be used. We have demonstrated an application of C&I in forest conservation planning by adopting the criteria of biodiversity conservation and soil and water conservation. It is conceivable that high conservation priority areas represent important forest functions that maintain ecological vitality and balance, and ensure the safety of society. This implies integration of biodiversity and soil and water conservation sectors into a single decision-making process. The GIS-based MCDM approach enabled a realistic decision-making situation to be considered in forest conservation planning. As land is a finite resource, setting it aside from other competing uses without considering the role of decision makers is futile. Therefore, forest conservation planning using the GIS-based MCDM approach has gained legitimacy through the application of criteria and indicators.

Acknowledgements

Portions of this chapter have previously been published by the same authors in the article 'A GIS-based multi-criteria decision making approach to forest conservation planning at a landscape scale: a case study in the Kinabalu Area, Sabah, Malaysia' in the journal *Landscape and Urban Planning* (2005), volume 71, pp 207-222. Reproduced with permission of Elsevier.

References

Aiba, S. and Kitayama, K. (1999) Structure, composition and species diversity in an altitude–substrate matrix of rain forest tree communities on Mount Kinabalu, Borneo. *Plant Ecology* 140, 139–157.

Barnes, B.V., Zak, D.R., Denton, S.R. and Spurr, S.H. (1998) *Forest Ecology*, 4th edn. John Wiley and Sons, Inc., New York.

Bibby, C.J. (1998) Selecting areas for conservation. In: Sutherland, W.J. (ed.) *Conservation and Action.* Blackwell Science, Oxford, pp. 176–201.

Carver, S.J. (1991) Integrating multi-criteria evaluation with Geographic Information Systems. *International Journal of Geographic Information Systems* 5(3), 321–339.

Eastman, J.R., Jin, W., Kyem, P.A.K. and Toledano, J. (1995) Raster procedures for multi-criteria/multi-objective decisions. *Photogrammetric Engineering and Remote Sensing* 61(5), 539–547.

Franklin, J.F. (1993) Preserving biodiversity: species, ecosystems, or landscapes? *Ecological Applications* 3, 202–205.

Hiroshima, T. (1997) A study on evaluation methods of sustainability in forest planning: evaluation of long term forecasting of government based on criteria and indicators. Masters Thesis, The University of Tokyo (in Japanese).

Janssen. R. and Rietveld, P. (1990) Multicriteria analysis and Geographic Information Systems: an application to agricultural land use in the Netherlands. In: Scholten, H.J. and Stillwell, J.C.H. (eds) *Geographic Information Systems for Urban and Regional Planning.* Kluwer Academic Publishers, Dordrecht.

Japan Overseas Forestry Consultants Association (1999) *Final Report of the Study on Mapping and Land Cover Assessment of Mangrove Areas in The Republic of Philippines.* Japan Overseas Forestry Consultants Association, Tokyo.

Jones, P.G., Beebe, S.E., Tohme, J. and Galwey, N.W. (1997) The use of geographic information systems in biodiversity exploration and conservation. *Biodiversity and Conservation* 6, 947–958.

Kato, Y., Yokohari, M. and Brown, R. (1997) Integration and visualization of the ecological value of rural landscape in maintaining the physical environment of Japan. *Landscape and Urban Planning* 39, 69–82.

Kitayama, K. (1991) *Vegetation of Mount Kinabalu Park, Sabah, Malaysia, Map of Physiognomically Classified Vegetation.* East-West Center Press, Honolulu, Hawaii.

Kitayama, K. (1992) An altitudinal transect study of the vegetation on Mount Kinabalu, Borneo. *Vegetatio* 102, 149–171.

Kitayama, K. (1996) Patterns of species diversity on an oceanic versus a continental island mountain: a hypothesis on species diversification. *Journal of Vegetation Science* 7, 879–888.

Muriuki, J.N., De Klerk, H.M., Williams, P.H., Bennun, L.A., Crowe, T.M. and Berge, E.V. (1997) Using patterns of distribution and diversity of Kenyan birds to select and prioritize areas for conservation. *Biodiversity and Conservation* 6, 191–210.

Nakamura, T. (1992) Ecological characters of mangroves in Pacific Areas. In: Nakamura, T. (ed.) *Integrated Research on Mangrove Ecosystems in Pacific Islands Region.* Japan International Association for Mangroves, Tokyo, 413 p.

Phua, M.H. and Minowa, M. (2000) Evaluation of environmental functions of tropical forest in Kinabalu Park, Sabah, Malaysia using GIS and remote sensing techniques: implications to conservation planning. *Journal of Forestry Research* 5, 123–131.

Primack, R.B. (1993) *Essentials of Conservation Biology.* Sinauer Associates Inc., Sunderland, Massachusetts, USA.

Saaty, T. (1980) *The Analytical Hierarchy Process.* McGraw-Hill, New York, USA.

Sabah Forestry Department (1989) *Forestry in Sabah.* Sabah Forestry Department, Kota Kinabalu, Sabah, Malaysia.

Scott, J.M., Davis, F., Csuti, B., Noss, R., Butterfield, B., Groves, C., Anderson, H., Caicco, S., D'Erchia, F., Edwards, T.C., Ulliman, J. and Wright, R.G. (1993) Gap analysis: a geographic approach to protection of biological diversity. *Wildlife Monographs* 123, 1–41.

Smallwood, K.S., Wilcox, B., Leidy, R. and Yarris, K. (1998) Indicators assessment for habitat conservation plan of Yolo County, California, USA. *Environmental Management* 22(6), 947–958.

Smith, V.K. and Theberge, J.B. (1986) A review of criteria for evaluating natural areas. *Environmental Management* 10, 715–734.

Tanaka, K. (2000) Land use diversity index by Moment method. In: *Proceedings of the 111th Conference of the Japanese Forestry Society.* The Japanese Forestry Society, Tokyo.

Tomlinson, P.B. (1986) *The Botany of Mangroves.* Cambridge University Press, Cambridge.

Williams, P.H. (1998) Key sites for conservation: area-selection methods for biodiversity. In: Mace, G.M., Balmford, A. and Ginsberg, J.R. (eds) *Conservation in a Changing World.* Cambridge University Press, Cambridge.

Yeh, A.G. and Li, X. (1998) Sustainable land development model for rapid growth areas using GIS. *International Journal of Geographic Information Systems* 12(2), 169–189.

Zeleny, M. (1982) *Multiple Criteria Decision Making.* McGraw-Hill, New York, USA.

Social Revolutions in Forest Management

13

J. Romm

University of California, Berkeley, USA

The classical science and management of forests developed on the assumption that forest land ownership – public, private or collective – defined the legitimate scope of forest control. Choices of actions and outcomes were confined within the boundaries and interests of owners; public policies were directed primarily towards these variables. As land ownership was a territorial imprint of the social distribution of power and wealth, forestry regimes typically reflected the motives and capacities of powerful groups. Today, the valued contributions of forests have come to reach regional – and even global – scales, far beyond the control of any owner, and the diversification, intensification and broadening of public influences are challenging the merits of forest privilege throughout the world. In these circumstances, forestry regimes are under great pressure to change if their public acceptability and the forests they regulate are to be sustained. Sole lines of ownership will not bind future regimes. Rather, primarily transboundary relations that entitle, enable and engage the influence of all they affect will unify them. Their scientific, managerial and policy requirements will differ substantially from regimes of the past.

13.1 Classical forestry

Forestry has a history of connection with the territorial sovereignty of a state. It has been used as a purposeful instrument of the frontier for those who rule land and people. Almost 2500 years ago, the Hindu statesman, Kautilya, prescribed the duties of the Forest Superintendent as the King's agent, the forms of claims that were legitimately undertaken or transferred for use of forest resources, and the penalties and punishments for breached boundaries and rules. Crucial features obviously have changed in the intervening millennia. The science of forests has improved the predictability of the consequences of forest actions. Professionalization has created a self-regulating body of expertise upon which

© CAB International 2005.
Forestry and Environmental Change: Socioeconomic and Political Dimensions
(eds J.L. Innes, G.M. Hickey and H.F. Hoen)

states and societies rely. Massive population growth, innovation in transport, communication, information and processing technologies, market development and urbanization have dramatically changed the context, scale, possibilities, expectations, content and influences of forest management. In the past century alone, nation-states have replaced mercantile empires of industrial, metropolis, nations and resource-supplying colonies; after a half-century of primacy, they now share influence with an emerging global system and its newly empowered rural communities and urban middle classes. Nevertheless, Kautilya's prescriptions continue to resonate for contemporary observers. This is because, in addition to its science, art, technology and finance, forestry is a system of human and territorial control that has been remarkably resistant to change.

The character of forest control varies greatly around the world. Where frontiers continue to exist, as in vast tropical and boreal regions of the world, governments may use private or quasi-public concessions to organize lands and people residing beyond the established reaches of state power. In areas of established hinterland authority, public, corporate, collective, community and tribal systems of ownership may divide the forest map, often in conflicting overlapped claims. In more densely settled areas, forests assume the patchwork-and-ribbon forms of farms, watershed headwaters, riparian, canal and roadside strips, parks, preserves and plantations. Patterns of political authority may be unitary, federated, localized, feudal, egalitarian or communal, and these differences modify the tangible imprints of different modes of control on the landscape. Moreover, the dynamics of relations among levels of governance turn, bend and flex these imprints over time.

To such infinite variations, forestry and foresters bring science, art, technique and financial calculus to the choices among actions and outcomes. They have also sustained the boundaries and rules of those with the power and wealth to control territory or to assume the right to do so. In other words, they bring authority. The tension between dynamic functions and authoritative boundaries of forests is the essence of the forestry problem.

13.2 Issues

Several tensions between forest boundaries and functions drive the dynamics that are forming future trends in forest management. The most notable sources of tension include the declining strength of conventional forest boundaries, the increasingly metropolitan formation of forest functions and values, and the globalization of forest flows, influences and interests.

13.2.1 Boundaries

In addition to their designation of public ownership, government and quasi-public forest boundaries typically serve as interior borders of the state. While the external borders of a nation regulate the international movements of goods, people and influence, forest borders typically serve the same function with regard to domestic movements. The effect is to contain pressures for settlement and use within a smaller realm of territorial opportunity than would exist in the absence of interior borders. If forest borders favour some groups relative to others, they contribute to the structure of social power as well. The capacity to sustain forest borders is weakening through loss of their economic value and changing distributions of social power.

The regulation of timber flows, land clearance and livelihood, like the maintenance of buffer stocks of food and minerals, or of reserves of unemployed people, governs market quantities and prices of wood, land and labour, and the consequent patterns of national society, economy and settlement. Strong borders have been used, for example, to increase the prices for timber and land, and to reduce wages, outside the jurisdictional forest; weak borders have opposite effects. The sectoral, class and urban–rural conflicts over forest borders are not simply about the values of forests within given boundaries, a presumption of conventional forest management, but also about the leverage that forest controls exert on economic, social and environmental possibilities throughout a nation. These conflicts are intensifying in part because the globalization of market influences weakens the market leverage that states can actually wield, while popular expectations of state power have not diminished accordingly. Ironically, the national leverage of forest borders is gaining recognition at precisely the time in which the capacities of nations to exercise it are diminishing relative to global and local forces.

Forest borders have direct social functions as well. They block access to some classes, races and ethnic groups in ways that expand relative opportunities for others. They may assert state authority for lands to which minority groups or local communities perceive themselves to possess prior claim. In some countries, forest borders exclude certain groups; in others, they harden territorial sovereignties over land that minorities, dissidents, separatists or adjacent nations may claim. In other words, forest borders distribute social opportunity and, more often than not, are instruments for sustaining social order. The pervasive violence and political passion around forest borders reflect their significance in political, social and economic relations. These conflicts are intensifying as global democratization, shared information and the emerging power of non-governmental organizations strengthen the claims among groups and classes who have been denied entitlements, benefits and influence in the past.

These boundary issues, and the conflicts associated with them, are not merely annoyances or aberrations, as forestry tends to treat them. They are fundamental parts of the structure of forestry itself.

13.3 Systems

Forest functions increasingly are not confined to hinterland territories, for example, as production sites for timber or as nature preserves. Forests form and protect the environmental infrastructure of cities, the air, water, food, fibre, climate and energy sources of urban and industrial life, the natural and aesthetic templates of metropolitan systems. Throughout the world, growing urban middle classes tend to buy forest and settle on lands that farmers vacate to work in cities. They also build environmental resistance to uses and claims on forests, not infrequently at the expense of landless, resource-dependent and hinterland minority communities. The conflicts between environmental and equity interests in forests are just beginning to join the central concerns of forestry.

Metropolitan forest influences spread outward over conventional resource-defined jurisdictions, covering apparent differences between city, country and forest while increasing inequalities between urban and rural layers of the people who occupy a given place. The relationships between city, country and forest have become more fundamental than their differences; the distinction between 'urban' and 'rural' is increasingly unhelpful. Yet boundaries continue to sever these relations, fragmenting and distorting forest systems and actions in ways that address neither their consequences nor the social instabilities associated with them.

In this context, there is increasing tension, and related social disorder, between the structures of forests that are defined for ownership and control (e.g. timberlands and reserves), and the structures of systems (e.g. river basin, climate or metropolis) that would internalize the functional values of forests in contemporary contexts. Foresters' responsibilities both for territory and for function are increasingly in conflict in social regimes that privilege both territorial boundaries and functional values, each of which has very different managerial implications.

13.3.1 Scales

The global opening of nations' external borders lets forces pour through into metropolitan systems and forests over which nations exert relatively little control. As national barriers to trade, migration and environmental responsibility weaken, the burdens on national forestry systems have increased in ways that are as yet extremely difficult to accommodate. In the Pacific Basin, for example, where at least half of the world's wood trade occurs, an archipelago of world cities is forming around the rim of the Basin – its nodes tightening together, like the closing mouth of a net – relative to their hinterlands. Within nations, the proliferating trials of alternative systems of protective boundaries – community forestry, river basin councils and metropolitan authorities, airshed and habitat regimes and indigenous forest management agencies – reflect efforts to compensate for weakening external and internal borders with new modes of forest organization and social order. Other kinds of boundary innovations, such as the distinction between certified and uncertified forests, and among increasing variations of public right in private forests, also respond to desires for protection in the open winds of a globalizing system. Global forestry arrangements and localized forestry systems squeeze and weaken but do not as yet replace state capacities to regulate changes in forest conditions and consequences.

In sum, there are growing disparities between: (1) dynamic social structure and rigid forest boundaries; (2) metropolitan forest functions and conventional forest territories; and (3) global and local objectives that are relative to a state's capacity to regulate forest conditions. The tensions are shaping wholly new kinds of forests. Their issues form the most forceful challenges to future forest management. Maps of forests still show the sharp parcels and unambiguous forest jurisdictions of the past. A map by the criteria of today's forest debates would show instead the systems of tree aggregations that stretch from the core of great cities along streams and roads through farmland homesteads, orchards and plantations, and through natural forest to mountainous ridges, each piece functionally interdependent with the next. They would show the global flows of wood, wildlife, carbon and pests and diseases. They would show the diversifying assertiveness of localized forms of forest control that typically conflict with established managerial regimes and capacities. The science, management capacity and modes of governance for such systems do not yet exist.

13.4 Trends

Social forces shape forests as surely as climate, topography and soil. These forces are changing at an unprecedented rate, breaking, bending or covering the established structures of forestry. They appear in the kinds of emerging relationships that seem to confuse and clutter conventional forestry, the 'social static' that gets in the way of foresters and forestry agencies doing 'what we know is best'. They appear in the novel forms of forests that are proliferating around the world in response to newly empowered implicit social criteria. They appear in the agrification of timber production and the increased attention to forests as sources of food and medicines. The turning social kaleidoscope is shaping a landscape that has little relation to the bold lines of conventional forest authority.

13.4.1 Novel management units

As examples of the scale of social innovation, let us consider a few of the now commonplace categories of forests that were viewed as esoteric only a few decades ago: agroforests, fodder forests, homestead forests, community forests, social forests, energy forests, carbon forests, habitat forests, science forests, riparian forests, medicinal forests, food forests, roadside forests, biodiversity forests, sacred forests, urban forests, farm forests, school forests and tribal forests. Moreover, any one of these categories contains almost infinite socioecological possibilities. Agroforestry, for example, includes the intensive rich homegardens of Sri Lanka, Java, Bangladesh and southern China, the commercial orchard-based cropping systems of Thailand and California, the grains-plus-timbers and fuel-fodder-forage-livestock systems of India and, throughout the world, the multiple-use farm mosaics, rotations and cycles that contain tree crops as primary elements. Each of the other categories takes on similarly infinite arrays of biological and social possibilities.

This kind of diversification indicates a social revolution in the concepts of forests, the structures of forestry and the primary units of forest management. It is a pragmatic response on a grand scale to the break-up of historic blocks of land control and to the associated vacuum of authority, security and management into which new social capacities then move. Social innovations are filling the void, with not a little discomfort in the forestry profession. They could lead to destructive fragmentation of the forest unless new forms of cohesive governance and management emerge.

13.4.2 The metropolitan forest

Another aspect of the revolution is the emerging dominance of the metropolitan forest as the organizing principle of forest systems. In contrast to the jurisdictionally defined hinterland forest, the metropolitan forest is the landscape of trees that serves and is shaped by the actions, influences and needs emanating outward from urban centres. It includes the spreading parks, homegardens and tree-lined streams and roads, the urban-owned plantations and orchards on lands farmers cultivated before moving to city jobs, the patchwork of farm forests and tree-bordered fields, the forests that protect water supplies and control floods, the distant preserves that provide recreational services and the satisfaction of environmental values for a growing urban middle class, and the timberlands that supply the materials needed to build, fuel, furnish, feed and inform growing urban areas and populations. Although the conventional concept of 'forest' is still a tree-covered territory beyond the edge of settlement, today's functional reality is a patterned aggregation of trees that moves outward from cities through the forces of urbanization and industrial growth that it sustains in fundamental ways. Rather than being characterized by distance, isolation, altitude or ownership, the metropolitan forest transects the social landscape from city and sea through plantations and fields to the ridge tops. Its needs for science, management, and governance differ dramatically from those of relatively homogeneous territorial ownerships and jurisdictions.

13.4.3 Global forest flows and institutions

The third aspect of the revolution is the movement of forests into the mainstream of international relations. This is so recent that it continues to be surprising. Forests have emerged as the metaphor for sustainable development, the worldwide pursuit of economic growth that is socially equitable and environmentally sound, and the logic of forestry has become a potentially central paradigm for this pursuit. In matters of climate change, forests are approached as the counterbalance to the impacts of industrialization on atmospheric carbon. Biodiversity, typically a euphemism for natural forests, has become a basis for global preservation strategy, as well as debt-for-nature exchanges and other opportunities to monetize existence values of nature. Timber extraction has been at the forefront of evolving multinational economic strategies, by nations as well as corporations, and global movements of forest products are perhaps the most visible expressions of the world's growing economic, social and environmental interdependence.

Today, logs from forest plantations in Chile and New Zealand cross the Pacific to mills in the forest regions of Korea and California, transforming the landscapes at one end while reducing forest pressures on the other. Electricity generators in North America and Europe pay for carbon-offset plantations in the southern hemisphere. Tibetan villagers in northern Yunnan, tracking Tokyo market prices by telephone, pick forest mushrooms one day that are in Tokyo markets the next. Himalayan villagers talk by email with forest action groups throughout the world. Networks of non-governmental organizations monitor timber harvests everywhere while others monitor the treatment of indigenous forest peoples. These developments are transforming the agendas of science, the negotiating texts of international affairs and the opportunities for localized and minority populations now in instant touch with one another. These developments have just begun.

13.5 Managing forests as functions rather than as territories

Modern trends are revolutionizing the social context in which forests function and, thus, are redefining the nature of what forests are thought to be. Forestry is adapting in some ways, reacting in others. The climate change discussion, for example, has mobilized scientific talent to an extent that was unimaginable even a decade ago. The social diversification of localized forest management units has had an equally profound effect. The pool of forest-oriented scientists is a magnitude greater today than just a short while ago. While some have sought to protect the borders around classical forest science, most have come to embrace the new opportunities to lead advances in scientific explanation and in modes of managerial integration.

The social context of forestry is forming new alignments of power in the landscapes that are more difficult to accept. The growing disparity between system and boundaries, function and structure and science and authority is one of the greater threats to forestry as a sustaining process of adaptive mediation between nature and society. The territorial habit of classical forestry no longer suffices to serve social interests, nor do the assumptions of authority in its management regimes. Management has become much more than a problem of optimization within fixed boundaries, objectives and constraints. The processes with which foresters work increasingly involve localized systems (e.g. agroforests), community forests and biodiversity reserves that value and manipulate different cause–consequence relations than those of the classic timber forest. They call for diversified forms of governance and management as well as new pathways in science.

The concept of the metropolitan forest itself encourages the diversification of forest governance from the classical territorial departments to regional flows that unify city, country and frontier. Production forest management, still oriented towards timberland ownerships and jurisdictions, would have remarkably different requirements, for example, if oriented instead towards the flows between timberlands, timber-dependent communities, workers and mills, urban housing, publications, furniture and packaging industries and the people these should serve.

Water forest management, thus far constrained to soil and water conserving strategies of upland vegetative manipulation, would focus instead on relations between watersheds, reservoirs, streamside woods, fishers, farmers and irrigation systems, cities, deltas and wetlands. This connective forest includes different and more diverse constituencies, specializations, market influences and needs for coordination that participate in territorial forest management. An analogous argument can be made for preservation or stabilization forestry, which would simultaneously protect the habitat systems that link varied landscapes and regulate the distribution of opportunities among people, places and activities. Such developments would divide, diversify and specialize currently unitary territorial departments that no longer provide the governance, science, finance, education and management which emerging functional systems of forests require.

The globalization of forestry creates the need for a third type of managerial form. International pressures shape the forest as surely as the logger and cultivator. Effective national forestry increasingly is the knowledgeable mediation of exchanges between international and local interests with regard to, for example, carbon banking, biodiversity investments, timber and medicinal trade and social justice. Together with the social diversification of localized forests and the metropolitan formation of functional forest systems, national mediation will call forth regimes of forest management that depart dramatically from the territorial conventions of the past.

13.6 Future developments

This sweeping perspective neglects the immense differences that exist among nations and regions. It may appear to devalue the site and social specificity to which effective on-the-ground forestry must respond. It is intended instead to suggest that the primary future developments of forestry will create capacities to operate at larger scales and with different systems and instruments than have characterized the previous history of forest management. These developments will not displace the need for continuing advances in the management of specific properties for specific interests or in national responses to unique circumstances. Rather, the extraordinary diversification of desired qualities and services, and of the publics that want them, requires capacities to create conditions in which the aggregate impacts of diverse choices are likely to satisfy the scales and complexities of public interests that forest functions are expected to satisfy. The kinds of explanation, the sources of leverage and the modes of implementation that these aggregate scales and systems involve demand capacities which the historic patterns of forestry development alone cannot create.

Current developments indicate future directions as well as the chasms of need and opportunity ahead. Plantation forestry is reducing the world's dependence on wild lands as sources of wood fibre. By the end of the century, biotechnological advances will have freed the world of this dependence and dramatically confined the area and reconfigured the distributions and compositions of production forests. Urban forestry is inching gradually outward from its aesthetic projects towards an understanding of the climatic, hydrological and structural functions of cities that forests can satisfy. Moreover, even so-called agrarian or 'Third World' countries have achieved, or are approaching, the urbanization of more than half of their populations. By the end of the century, the central concept of forestry will focus on the sustainable well-being of cities and the roles of forests in regulating and supporting urbanization processes. Global arrangements are forming today to acknowledge and guide the impacts of forest conditions on climate and biodiversity. By the end of the century, global systems of finance and governance will influence such conditions as a matter of course, and will deal as well with hydrology, human health and nutrition and social equity. The explanatory and implementational jumps that such developments require will not supplant forestry as we know it but must greatly expand and diversify its commitments, scales and capacities.

In addition to scientific advances – these alone present an immense challenge – two convergent paths of institutional development become apparent. The first involves systems of exchange that connect those who need and those who affect forest conditions. The growth of forest certification, debt-for-nature swaps, carbon banks, forest food and medicinal markets and payments for watershed services exemplifies the emergence of institutions of financial exchange that, operating over and above site and social diversity, permit people to support the forest functions they need by influencing the choices of those who affect them. This path of development is in its early stages. By the end of the century, relatively predictable regimes of exchange – regimes that deliver their desired outcomes with acceptable regularity – will comprehend all of the values of forests currently viewed as qualitative.

The second path is in systems of governance. These include regimes that adequately represent and act on behalf of all people who affect and are affected by forest conditions – that is, those that exclude no one. Initial examples are emergent watershed organizations, community-based systems of resource management, metropolitan environmental management, green national development strategies and international rule-making regimes. These reflect early stages of innovation in relatively binding relations among diverse interests, classes and levels of governance. These relations produce neither harmony nor narrow efficiency, but collective capacities to resolve conflicts and adapt well to error, stress and catastrophe. By the end of the century, timber and housing policies, preservation and social policies and forest, climate, energy and water policies will be bound together in modes of governance that fully represent the diverse interests these interacting policies affect.

The agenda is large. The agenda of a century ago appeared no less daunting. It was revolutionary then. It is revolutionary now. The accomplishments of the past set the foundation for the next stages, not as a source merely of extensions, but of confident capacity to recognize that a vastly new world calls for fundamental advances in the concepts and instruments of forestry.

14 Recent Developments in Decision Analysis for Forest Management

G.A. Mendoza

Department of Natural Resources and Environmental Sciences, University of Illinois, USA

Decision analysis in forestry has undergone a series of changes over the last decade. This chapter provides an overview of the changes and developments in forestry, in terms of both management philosophies and methodologies. In presenting these developments, two types of decision-making techniques are reviewed. One is based on 'hard systems' or scientific method and the other is based on a problem structuring technique. The history and evolution of management philosophies are also reviewed, leading to the current management paradigm, which can best be described as participatory. Current techniques in participatory decision-making are then presented, particularly those that follow the value exploration or problem structuring approach.

14.1 Introduction

Forest management, in its purest sense, involves the use of forests to meet the goals and objectives of its owner, whether public or private. In this context, decision analysis in forest management implies the use of decision methods that identify the best management strategy to achieve the goal in the most efficient manner. Decision analysis in forestry was conceived, and subsequently evolved, while adhering to this basic definition of forest management.

To more meaningfully describe the recent developments of decision analysis in forestry, it is necessary to briefly review its history and development. This will provide perspective, and may also reveal significant developments, both in theory and practice, that have led to the current state of forest management today.

Perhaps one of the best ways to examine the history of decision analysis in forestry is to look at the scope and contents of leading forest management books due to their unique and significant role in shaping forest management as a profession. In the third edition of their *Forest Management* book, Davis and Johnson (1987) stated that they revised the previous editions of Kenneth Davis's book, *Forest Management: Regulation and Valuation* because of three fundamental shifts. These were:

© CAB International 2005.
Forestry and Environmental Change: Socioeconomic and Political Dimensions
(eds J.L. Innes, G.M. Hickey and H.F. Hoen)

> 'the new dominance of economic and social determinants of forest management choices, necessity to consider all timber management decisions in the context of a larger, socially defined multiple use management problem and the need to justify, based on quantitative analysis, that management recommendations, decisions, and plans satisfy owners constraints and are the best alternative choices.'

Clearly, these fundamental shifts were socially driven, and methodologically guided by the principles of scientific management, the dominant paradigm at the time the book was written. The subtitle of the fourth edition (Davis *et al.*, 2000): *Forest Management: To Sustain Ecological, Economic, and Social Values*, reflects yet another set of fundamental shifts in forest management philosophies. As the subtitle indicates, these changes were triggered and fuelled primarily by increasing ecological and social concerns.

The evolution of decision analysis in forestry is also reflected and documented by other forest management books (see Leuschner, 1984; Buongiorno and Gilles, 2003), technical publications and conference proceedings. Such evolution may be characterized and categorized historically into five management eras. These are: (1) the classical forest management era (1900–1960), where forest management was essentially timber-dominated and consequently management techniques were timber-driven; (2) the multiple-use–sustained yield era (1960–1980) characterized primarily by two dominant management tenets, namely, multiple use and sustained yield; (3) the scientific forest management era (1980–1990), characterized by the adoption or use of systematic and quantitative management tools; (4) the ecosystem management era (1990–2000), best described as holistic management that recognized the forest as an ecosystem; and (5) the sustainable forest management era (2000 to date) reflecting the view that forests should be managed so that health and productivity are sustained through time.

These eras provide a snapshot of the historical development of forest management and the duality of its theory and practice. On one hand, the underlying theory informs its practice. On the other hand, societal views and perspectives tend to guide and dictate its practice. For example, in the USA, the use of quantitative decision analysis tools in forestry was driven, primarily, by the need for scientific forest management (mandated by the National Forest Management Act of 1976). As a result, changes in societal perspectives, motivated primarily by increasing ecological and environmental concerns, led to the development of forest ecosystem management techniques. These tend to be more robust, open-ended and often more qualitative compared to the quantitative management techniques developed in the 1980s.

This chapter describes recent developments in forest management decision analysis, specifically related to methodological developments and management philosophies and paradigms. In presenting the methodological developments, emphasis is given to more recent developments, particularly as they relate to the role of analytical models in decision-making analysis. Readers are referred to the relevant literature for more details. When discussing management philosophies and paradigms, arguments are made for a new and evolving decision-making paradigm, namely, participatory decision-making. Finally, some preliminary methods consistent with the participatory decision-making paradigm are presented.

14.2 Decision analysis: recent methodological developments

In its most basic form, decision-making has been generally stated as the problem of choosing the 'optimal' solution from a set of alternative options to attain an objective. From a traditional management science perspective, the product of the decision-making process is typically the most technically correct and rational option based on the decision environment. This is typically captured by a formal, often highly structured, model. The decision process is justified and guided by three inherent, and often overlooked, assumptions: (1) *rationality* – optimal decision or solution is determined by the option that yields the 'highest' achievement of the objective; (2) *objectivity* – the analytical model used in searching the solution is neutral and unbiased; and (3) *reductionism* – the decision problem can be decomposed into its essential elements.

Following these assumptions, decision analysis in forest management was initially formalized using quantitative methods like mathematical programming. In its most basic form, the decision making model is formulated as:

$$\text{Max } Z = \Sigma\, c_{ij} X_j \qquad (1)$$

Subject to:

$$\Sigma\, a_{ij}\, X_j < B_i; \qquad i = 1, 2, \ldots m \qquad (2)$$

The formulation above is the traditional linear programming (LP) model. The single objective is represented by Z; the management alternatives, or options, are represented by X_j, and the model (2) represents the operational, resource and management constraints the decision-maker operates on. The basic LP model has been adopted extensively in many harvest scheduling applications used in the timber industry. It has also been applied in land management planning processes (e.g. in the national forests of USA, where a large-scale model called FORPLAN (Kent, 1989) was used). From the basic LP model, a number of refinements and improvements were made to enhance the model's capability to address a wider range of problems and issues.

14.2.1 Multi-objective programming

The model presented in (1) and (2) is best suited for decision problems that involve one decision maker with one objective. This is clearly a major limitation in forest planning because, typically, most forest management problems involve a number of stakeholders or decision-makers with their own goals or objectives. Therefore, an important refinement and subsequent development of the simple LP model is the incorporation of multiple objectives. Mendoza and Prabhu (2000a, b) describe the simplest form of a multi-objective programming (MOP) model as follows:

$$\text{Optimize } Z_1 = f_1(x_1, x_2, x_3, \ldots, x_n);$$
$$\text{Optimize } Z_2 = f_2(x_1, x_2, x_3, \ldots, x_n) \qquad (3)$$
$$\text{Optimize } Z_k = f_k(x_1, x_2, x_3, \ldots, x_n)$$

where, $Z_1, Z_2, \ldots, Z_k$ are the objectives

A number of overview papers on multi-objective programming and its application to forest planning are described by Pukkala (2002), Howard (1991), and De Steiguer *et al.* (2003). MOP models have enabled the incorporation of other objectives (e.g. aesthetics, ecological, economic, biophysical and environmental concerns) in the identification of best management strategies. This simultaneous consideration of multiple objectives offered a significant improvement from the single-objective, mainly economic-based type of decision-making.

14.2.2 Multi-criteria analysis, multi-attribute and multi-stakeholder decision models

The MOP models described in (3) assume that the alternative management regimes, including the objectives, can be represented using decision variables that are continuous. However, it is quite common that decision alternatives are a discrete set of options and must be addressed as such (e.g. prioritizing a set of alternatives, or measuring decision-makers' preferences). A number of books and review articles have been published describing the theory and application of multi-criteria decision-making (see Schmoldt *et al.*, 2001; De Steiguer *et al.*, 2003). There is also an extensive literature describing the application of multi-criteria decision-making (MCDM) techniques in forest management planning (see Prato, 2000; Mendoza and Prabhu, 2003a, b).

Of the many MCDM techniques described in the literature, perhaps the most commonly used is the Analytical Hierarchical Process, originally developed by Saaty (1980). Its basic framework is to decompose a decision problem by following a hierarchical structure and to pursue the evaluation process using pairwise comparison of alternatives. Examples of AHP applications in forestry have been reported by Kangas (1992), Pukkala and Kangas (1996) and Schmoldt *et al.* (2001). MCDM has also been used as a framework for pursuing participatory decision-making (Schmoldt *et al.*, 1995; Schmoldt and Peterson, 2000).

14.2.3 Fuzzy models, expert systems and decision support systems

Uncertainty is pervasive in forest decision-making due to the inherent complexity of the forest ecosystem. To accommodate this uncertainty, two general methodologies have been introduced, namely fuzzy set theory and expert systems. Fuzzy set theory is an extension of classical set theory, based on two-valued logic: i.e. in or out. Fuzzy sets, on the other hand, were formulated by Zadeh (1965) based on the simple idea of introducing a degree of membership. This simple theoretical construct has interpretative relevance to the treatment of risk and uncertainty in forest management. For example, because it is highly unlikely that precise estimates related to forest conditions or options can be made, it is more meaningful to characterize assessments about forests in terms of degrees of soundness or 'health', instead of the dichotomous evaluation as good or bad, or in or out. Recent applications of fuzzy set theory in forestry include: forest sustainability assessments (Varma *et al.*, 2000; Mendoza and Prabhu, 2004), timber harvest scheduling (Bare and Mendoza, 1992) and forest planning and decision-making (Mendoza and Sprouse, 1989).

Decision support systems are computer-assisted models offering a wide range of robust methodologies that have also been developed and applied in forest management. Many of these methodologies were developed to address complex issues, particularly with respect to ecosystem management and adaptive management. Excellent reviews and discussions of these types of models are reported by Reynolds *et al.* (2000), Mowrer *(*1997), and Rauscher *(*1999).

14.2.4 Heuristic algorithms

Many of the solution algorithms, particularly optimization models such as those previously described, are techniques with well-defined procedures designed to search an optimal solution. Typically, the models are 'closed-form'; hence, algorithms can be developed to seek an optimal solution. However, there are often decision problems that possess unique or peculiar functional forms where 'true' optimal solutions cannot be determined using traditional optimization models. Similarly, the computational requirement of traditional methods can be exceedingly large. A number of heuristic algorithms have been developed and applied in forestry. Murray and Church (1995) and Weintraub *et al.* *(*1995) offer excellent reviews of these heuristics. Examples of these heuristics are: tabu search (Bettinger *et al.*, 1997, 1999), simulated annealing (Lockwood and Moore, 1992; Bos, 1993) and random search (Nelson and Brodie, 1990). Other heuristic methods applied in forestry are: 2-opt decision choice (Bettinger *et al.*, 1999), SNAP II (Bettinger *et al.*, 1996) and combinatorial algorithms (Weintraub *et al.*, 2000).

14.2.5 Spatial analysis and optimization

The models previously described are capable of identifying optimal solutions to a properly formulated mathematical programming problem. In general, the solutions generated are in terms of optimal values of the decision variables. However, forest management strategies are often spatially explicit and geographically sensitive. For example, adjacency concerns, fragmentation, off-site effects and edge effects are spatially relevant concerns that must be addressed when generating solutions. Therefore, the optimal solutions may prove to be less desirable when spatial considerations, including aesthetics, are factored into the analysis. Moreover, spatially explicit management strategies (e.g. those pertaining to wildlife population concerns and timber age-class distribution that satisfies multi-species wildlife population objectives) should be accommodated. Hof and Bevers (2000) describe different spatial optimization models that are sensitive to spatial dynamics, particularly for wildlife and timber. Other models sensitive to 'locational' issues include Church *et al.* (1998) and Snyder and Revelle (1997). In addition to spatial optimization, other models linked to geographic information systems (GIS) have also been developed and used in forestry. Examples of these applications include: Kyem (2002), Jankowski (1995) and Jankowski *et al.* (1997).

14.2.6 Multi-agent systems

Multi-agent systems (MAS) are an emerging discipline that has evolved from the more general fields of decision support systems, game theory and artificial intelligence. As its name implies, MAS is a general approach that takes into account the presence of multiple agents (actors or stakeholders), each with their unique views, perspectives and behaviour. Each agent acts or reacts (makes decisions) as he/she pursues his/her objectives rationally, or according to his/her own rules and behavioural patterns.

MAS is a robust approach for analysing and simulating complex systems involving multiple agents with mechanisms for coordination of independent agents' behaviours. The most significant component of MAS is an agent (or actor). The way an agent acts is called its 'behaviour' (Stone and Veloso, 1997). MAS focuses on systems in which many intelligent agents interact with each other. The agents are considered to be autonomous entities whose interactions can be either cooperative or selfish. That is, the agents can share a common goal, or they can pursue their own interests (Sycara, 2000).

Huston and DeAngelis (1988) were perhaps the first to put forward the notion that individuals affecting and affected by a resource are uniquely situated with their own set of beliefs, behaviours and patterns of localized action, reaction and interaction. Following this notion, Hogeweg and Hesper (1990) proposed the use of individual-oriented modelling as an approach to understanding ecological systems. Building from these seminal works, MAS has been applied to the modelling of natural resource management. One of the first applications was on a common property management regime that is pervasive among developing nations (particularly with agriculture and forest-related resources). In this context, much of the initial development and application of MAS was done by Bousquet *et al.* (1998). Several authors have since applied MAS to a number of cases studies, including irrigation systems (Barreteau and Bousquet, 2000); resource sharing regimes (Thebaud and Locatelli, 2001); natural resource management (Rouchier *et al.*, 2001); environmental management (Bousquet *et al.*, 2002); and forest management (Purnomo *et al.*, 2005).

14.3 Decision analysis: emerging management and decision-making paradigms

The methodologies described in Section 2 are generally prescriptive, algorithmic and referred to by Checkland (1981) as the 'hard systems approach' to decision making. Such an approach is deeply rooted in the principles of quantitative 'systems analysis'. Its modus operandi is that once the goals are defined, a systematic evaluation of alternatives follow; assisted by a number of powerful analytical techniques such as different forms of mathematical programming. This general procedure follows four fundamental assumptions: (1) there is a desired state of the 'system' S_0 which is known a priori; (2) there is a present state S_1 which is also known; (3) there are alternative ways of getting from S_0 to S_1; and (4) the process of decision-making is to find the best means to get from S_0 to S_1.

The hard systems approach is consistent with the scientific forest management philosophy, which was popular in the 1980s. This management paradigm is rooted in the positivist and reductionist view of management science that presupposes the following: the management situation can be decomposed or reduced into elements or management units, each unit can be managed following a set of options and, finally, the decision problem is to identify the best option for each management unit. This management philosophy was challenged and criticized in the 1990s. Most of the questions and criticisms that have been raised stemmed from the need to accommodate ecological concerns and other environmental objectives. These concerns were considered too broad, wide-ranging, and too encompassing to be addressed adequately by a reductionist approach. This perception resulted in a general shift in management paradigm, from a reductionist view to something more holistic or system-based.

Along with the shift in management philosophy in forestry, the scale and emphasis of management also changed in the 1990s: from being production oriented and management unit-based, to emphasizing forest health and ecological integrity. Decision analysis shifted from one that is optimization-oriented to something less structurally demanding. The role of analysis in decision-making also changed, from identifying optimal alternatives to the development of socially acceptable and ecologically sound strategies. Moreover, the emphasis of decision-making also changed, from problem solving to something more related to problem structuring. The goal of analysis has therefore moved away from optimality and has been replaced by something more pragmatic such as producing 'reliable' and 'satisfactory' strategies.

This evolution has continued beyond the ecosystem management era in the 1990s. More recently, the ultimate goal of sustainable forest management (SFM) has caught the attention of many forest managers and now occupies the centre stage of the forest management debate worldwide. SFM, as a management philosophy, not only views the forest as an ecosystem, it also draws attention to the primacy of socioeconomic needs and the importance of incorporating the human dimensions of forest management. The pivotal role of the forest as a source of, and instrument for, economic development among forest communities has been underscored particularly in developing nations. Consequently, more socially oriented management strategies, such as community-based forest management, joint forest management and social forestry are now the favoured management philosophy among public forests in many developing nations.

These changes have also shifted the debate surrounding planning and decision-making in forestry: from the allocation of scarce resources to sustainable development. From a decision-making perspective, one of the most significant developments brought forth by SFM is the recognition that society or forest communities should be given more direct and active roles in making decisions about the management of the forests. Consequently, issues concerning power, fairness and competency have become highly relevant to community-based management and even more important than concerns about rationality, objectivity or technical soundness. Moreover, the process of decision analysis is considered more significant than the product of the analysis. Subsequently, management philosophies that deal with empowerment, or management initiatives that enable local communities, have become relevant to forest management particularly among the developing nations.

14.4 Recent developments in participatory decision analysis

The use of participatory decision-making approaches has become the primary guiding principle in the management of forest resources worldwide. Over the last decade, a significant amount of effort has been devoted to developing methodologies that support the practice of participatory management. Many of these approaches have taken different forms and theoretical constructs. Some of the well-known concepts and principles include: participatory rural appraisal (PRA) (Chambers and Guijt, 1995), community-based resource management, participatory action research (PAR) (Selener, 1997), co-management, joint management, etc. While these methods are somewhat different in their overall approach, they have fundamental similarities in terms of their general procedure and the nature of the issues they are designed to address. These generally include: multiple stakeholders and their multiple interests, plurality of perspectives and the empowerment of local communities. Common to these approaches is the prerequisite for direct and active involvement of stakeholders in the decision-making process.

14.4.1 Problem structuring approaches

The concept of problem structuring is different from the traditional problem-solving approach. The primary motivation for its development stems from the recognition that traditional approaches are often too restrictive and therefore unable to capture the breadth and complexity of problems that are dominated by human concerns. As stated in Section 3, the classical management science approach has its origins in the scientific management paradigm that attempts to create an objective representation of the perceived 'reality' of a real world phenomenon. This analysis is often a one-off decision based on the presumed objective model. Rarely, if ever, can an objective model be designed from a problem beset with uncertainty and complexity such as those dominated by humans, whose actions, behaviours, desires and decisions could not be predicted, let alone fully understood. Consequently, alternative approaches have been proposed with a range of methodologies whose basic aim is not to identify an 'answer' or develop an objective model of reality, but to facilitate an enriched process that is transparent and participatory. Rosenhead (1989) calls this alternative paradigm the 'problem structuring' approach. Schon (1987) made an explicit distinction between the two contrasting paradigms when he stated:

> 'In the swampy lowland, messy, confusing problems, defy technical solutions. The irony of this situation is that the problems of the high ground tend to be relatively unimportant to individuals or society at large, however great their technical interest may be, while in the swamp lie the problems of great human concern.'

There are a number of issues arising from participatory decision-making that suit a problem structuring approach. The first is the plurality of perspectives from stakeholders and interest groups. An ideal participatory process must be able to provide an environment where adequate consideration of all perspectives is achieved. Exploring values and options in an open environment is one of the strengths of the problem structuring methods. This feature of structuring methods has its origins in the concept of 'value focused' thinking (Keeney, 1992). Independently, other authors have advocated and subsequently developed methodologies that conform to the original value-focused idea. Some of these methodologies include: soft systems methodology (Checkland, 1988); Strategic Options Development and Analysis (Eden, 1988, 1989), JOURNEY (Eden and Ackerman, 1998), Decision Explorer (Banxia, 2000), Visual Interactive Sensitivity Analysis (VISA) (Belton *et al.*, 1997; Belton and Stewart, 2001) and MACBETH (Bana e Costa and Vansnick, 1997; Bana e Costa *et al.*, 1999). Most of these approaches have been developed as decision support systems (DSS) under a computer-assisted, user-friendly environment.

All of the methods previously described recognize the importance of value exploration as a crucial initial step of the process. Initially, each method recommends an open-ended process to generate a rich picture of the problem at hand. This contrasts with the traditional approaches, which often begin with alternative development right away. Keeney (1992) distinguished between: (1) value-focused thinking, which focuses on eliciting values prior to alternatives; and (2) alternative-focused thinking, which identifies alternatives early in the process, focused on distinguishing and choosing alternatives. Keeney (1992) also argued that a decision process should be driven by value-focused thinking because alternative-focused thinking tends to anchor the thought process prematurely and thereby stifle creativity and innovation (Belton and Stewart, 2001). Consistent with this open articulation of perspectives or values, Eden (1988) proposed the concept of cognitive mapping as an effective way to explore the values, issues, concerns, perspectives, goals, objectives or 'worldviews' (Checkland, 1981) of stakeholders.

In addition to the rich generation of ideas, concepts, perspectives, issues and concerns made possible through open exploration, another feature of problem structuring methods that makes them well suited for participatory decision-making is the pre-eminence given to the elements of a facilitated process, namely the facilitator, analyst, intended user and the decision maker/s. Facilitation is a key concept in problem structuring. It is pivotal throughout the initial stages of both value exploration and decision exploration. This is important for two reasons: (1) stakeholders' perspectives and worldviews must be adequately considered; and (2) each stakeholder must be comfortable and given a chance to voice opinions or ideas at any time during the process. Hence, support provided by problem structuring methods should be such that they stimulate thinking and enhance learning and understanding. Mendoza and Prabhu (2003a) describe some of these methodologies in the context of community-managed forests. The following sections describe three general methodologies that are consistent with the problem structuring approach, namely cognitive mapping, qualitative system dynamics and fuzzy cognitive mapping. For the sake of brevity, these methods are described briefly only for the purpose of illustrating their general characteristics. The relevant literature is cited for more detailed descriptions of the methods.

14.4.2 Cognitive mapping

Cognitive maps (CMs) were first introduced by Axelrod (1976) as a way to represent complex decision problems, composed of dynamic entities that are interrelated in complex ways and usually include feedback links. These complex entities are represented as nodes and the causal links are represented by edges or arrows with the direction of the arrow representing the direction of influence. CMs are essentially structured ideas laid out purposely for understanding the basic relationships and dynamics of a system. The process begins with the generation of concepts with the active participation of all stakeholders. This process is very similar to participatory rural appraisal techniques (Chambers and Gujit, 1995). However, cognitive mapping goes beyond simply listing the essential concepts. The concepts are organized into a 'map' showing the relationships and interactions between, and among, the ideas (see Eden and Ackermann, 1998).

Mendoza and Prabhu (2003b) describe the application of cognitive mapping at a community-managed forest in Zimbabwe. In this case study, three groups representing three villages were convened to assess the sustainable management of the Mafungautsi forest, whose boundary encompasses the community forest. The modelling process started with an open-ended discussion of the issues and factors affecting the management of the forest. In addition to 'listing' these issues or factors, the villagers were also asked to indicate the connections or relationships between these factors using lines, arrows and nodes. The process was facilitated by a team of local scientists who were familiar with the forest, its history and ecology. Each idea or factor was discussed and debated. Often, original ideas were revised, restated or sharpened to make them more meaningful and relevant to the overall objective of sustainable forest management. Each connection, denoted by arrows, also went through group scrutiny.

The cognitive map generated by the villagers served as an excellent learning and communication tool. The resulting map increased the villagers' awareness and appreciation of the extent and complexity of managing their forest in a sustainable manner. They found it instructive to see the issues in a holistic or systems-wide view instead of a simple 'listing', as is often done using other participatory methods such as participatory rural appraisal (Chambers and Gujit, 1995) and participatory action research (Selener, 1997). Specifically, three significant analytical results were pursued. First is the concept of the 'domain' of a factor, reflecting the extent of influence or tactical significance of a factor. This is determined from the cognitive map by examining the number of nodes affected by, or directly connected to, a given factor. Another significant concept is the 'centrality' of an indicator. This reflects the 'strategic significance' of a factor, which can be determined by examining the scope of influence of a factor through its direct and indirect connections. Finally, the third concept is the 'criticality' of a node, which is determined by examining the number of 'critical nodes' connected to a factor.

14.4.3 Qualitative system dynamics

The CM approach is essentially a first attempt to structure the essential elements or components of a system. Clearly, the objective of developing a cognitive map is to lay out the overall relationships of the factors or elements of a system. For some applications, this may be a sufficient level of analysis, given the inherent complexity of the management problem. However, in some situations, particularly where there is more information, knowledge or experience about the different factors or elements, it may be possible to structure the cognitive map as 'influence diagrams'. In other words, the relationships are described in terms of causalities between nodes connected by an arrow. In this case, the concept of system dynamics is appropriate (Forrester, 1961, 1999).

System dynamics is a general term associated with the study of the dynamic behaviour of a variety of complex systems (Coyle, 2000). Typically, influence diagrams, using nodes and directed arrows, are used to denote this dynamic behaviour. In addition, the relationships, sometimes referred to as feedback loops or causality diagrams, are either positive or negative. The potential advantages of qualitative inference diagrams showing causal loops were described by Wolstenholme (1999) as follows:

> Causal loop qualitative model enhances linear and 'laundry list' thinking by introducing circular causality and providing a medium by which people can externalize mental models and assumptions and enrich these by sharing them. Furthermore, it facilitates inference of models of behavior by assisting mental simulations of maps.

Purnomo *et al.* (2004) used a number of influence diagrams to examine the criteria and indicators of a community-managed forest in Indonesia. In this study, selected members and representatives from two villages were asked to serve as a team that will examine the sustainability of their forests. Following the principles of participatory action research, the villagers were asked to participate in a historical examination of their forests and, in the process, generate a set of relevant indicators that could be used to evaluate and monitor the sustainability of their forests. During the group modelling process, a qualitative system dynamics model was generated following participative or collaborative modelling procedures (see Richardson and Anderson, 1995; Vennix, 1996).

14.4.4 Fuzzy Cognitive Mapping

The two soft system dynamics models previously described offer qualitative frameworks with which resource management systems can be examined. Both models are highly transparent, simple and easily understandable even for practitioners who are unfamiliar with models and model-building processes. For some applications, these broad insights and general inferences may be sufficient. For other resource management situations, deeper analysis and more detailed evaluation of results may be desirable.

General limitations associated with qualitative analysis are: (a) lack of knowledge or understanding of the dynamics of the resource management system; (b) lack of data; or (c) the elements being inherently qualitative and not amenable to quantification or highly structured description. Because of these limitations, it is not possible to pursue 'simulation' analysis that could allow for more in-depth evaluation of scenarios or 'what-if' type of analysis. Mendoza and Prabhu (2003b) have shown how 'proxy variables' or surrogate values may be used to transform a qualitative system dynamics model into a quantitative system dynamics model amenable to simulation analysis. However, this transformation unnecessarily imposes a system dynamics framework that may or may not be appropriate for certain resource management situations. In other words, the system of elements may be a set of interacting components with unclear or fuzzy cause and effect relationships that cannot be structured in a stock-and-flow framework of quantitative system dynamics. Also, the knowledge concerning the elements themselves and their interactions is often too general such that their explicit functional relationships cannot be determined or modelled formally.

An alternative to the quantitative system dynamics framework is Fuzzy Cognitive Mapping (FCM). As the name implies, FCM is an extension of traditional CM. FCM was originally developed by Kosko (1986) to describe a CM model with three distinct characteristics: (1) signed causality indicating positive or negative relationship; (2) the strengths of the causal relationships are fuzzified; and (3) the causal links are dynamic, where the effect of a change in a concept/node affects other nodes, which in turn affect other nodes in the 'path'. The first characteristic implies both the direction and the nature of the causality. In addition to signed causality, the second characteristic assigns a number or value to reflect the strength of the causality or the degree of association between concepts. Finally, the third characteristic reflects a feedback mechanism that captures the dynamic relationship of all the nodes, which may have temporal implications.

Following the conventions of Kosko (1986), the interconnections between two nodes C_i and C_j are linked by the strength or weight of the causality, e_{ij}, which can take a value in the range of -1 to 1. Values of -1 and 1 represent full negative and full positive causality, respectively, while zero denotes no causal effects. In a simple FCM model, the values of e_{ij} are discrete (–1, 0, 1), implying full negative, no relationship or full positive causalities. However, most complex systems do not exhibit simple FCM characteristics. Most often, the strength of the causalities is 'fuzzy', which can be modelled with e_{ij} between 0 and 1.

There are at least two different methods that can be used to analyse FCMs, namely rule-based and fuzzy methods. The rule-based approach relies on having some knowledge about the influence or causality between and among concepts, which can be expressed as rules. These influence relations are best described in terms of 'if ... then' causality. Cause and effect relationships between and among concepts can be developed based on the observed interactions between these indicator variables. Cause and effect can be direct (between two concepts) or indirect (effect is through a path). Reynolds *et al.* (2003) describes an application of this approach in using fuzzy or qualitative variables to assess ecosystem sustainability.

The fuzzy approach is based on the strength of causality, denoted by e_{ij}. Moreover, e_{ij} also allows simulation analysis of, or makes inferences about, the behaviour or general dynamics of the system. The values or status of the nodes should also be noted. In ordinary cognitive maps, these nodes are not defined explicitly in terms of their values or conditions. In rule-based cognitive maps, the concepts or nodes are associated with some state or condition variables. In FCM, the values of the nodes are 'fuzzified', meaning that, depending on their 'computed' values, their 'output' value is between 0 and 1. The interactive or causality relationships of an FCM model are dynamic and 'cumulative'; hence, analyses can have a temporal dimension. That is, the status or condition of the system can be monitored through time, both long term (steady state) and short term (transient state). Hence, regardless of the time frame, the cumulative effects on the 'affected' node j is dependent on the cumulative effects of the 'affecting' nodes i as follows:

$$C_j = \Sigma\, C_i e_{ij} \qquad (4)$$

The analytical constructs in equation (4) provide a framework for analysing the behaviour of the dynamics system consisting of nodes and their relationships (denoted by e_{ij}). The 'computed' values of C_j represent the cumulative effects of all affecting nodes. Kosko (1986) originally conceived a simple FCM where the nodes take only values of 0 (inactive) or 1 (active). Furthermore, different transformation functions can be used to adjust the computed value of C_i to either 0 or 1 (Kosko, 1986). Recent applications of FCM to participatory planning and decision-making in ecosystem management include Ozesmi and Ozesmi (2003) and Hobbs *et al.* (2002).

14.5 Conclusion

This chapter has described some potential methods that can be used to pursue collaborative or participatory decision-making techniques, arguably the most recent development in forestry decision analysis. Decision analysis in forestry has evolved following the changes in forest management philosophies and societal views, perspectives and objectives. Management methodologies have also evolved to conform with the changes in management paradigms.

In general, decision analysis in forestry may be categorized into two types; one that is best characterized as prescriptive, algorithmic and highly structured, and the other best described as descriptive, soft and qualitative. Both methods are continually evolving and undergoing changes and improvements so that they can better deal with the new issues and problems facing forest management.

Between these two general methods, the latter has become more popular and more widely applied, in part because of its affinity to the participatory management approach. It is now widely accepted that participatory methods are the most effective to achieve sustainable resource management, particularly among local forest communities in the developing nations. Increasingly, local communities are demanding more input and influence in the manner that public forests are managed. This requires more active and direct participation from stakeholders that are affecting, or affected by, the forest.

References

Axelrod, R. (ed.) (1976) *Structure of Decision.* Princeton University Press, Princeton, NJ.

Bana e Costa, C.A. and Vansnick, J.C. (1997) Applications of the MACBETH approach in the framework of an additive aggregation model. *Journal of Multi-Criteria Decision Analysis* 6, 107–114.

Bana e Costa, C.A., Ensslin, L., Correa, E. and Vansnick, J.C. (1999) Decision support systems in action: integrated application in a multi-criteria decision aid process. *European Journal of Operational Research* 113, 315–335.Banxia Software LTD (2000) [http://www.banxia.com]

Bare, B.B. and Mendoza, G.A. (1992) Timber harvest scheduling in a fuzzy decision environment. *Canadian Journal of Forest Research* 22, 423–428.

Barreteau, O. and Bousquet, F. (2000) SHADOC: a multi-agent model to tackle variability of irrigated systems. *Annals of Operations Research* 94, 139–162.

Belton, V. and Stewart, T. (2001) *Multiple Criteria Decision Analysis: An Integrated Approach.* Kluwer Academic Publishers, Norwell, MA.

Belton, V., Ackermann, F. and Shepherd, I. (1997) Integrated support from problem structuring through to alternative evaluation using COPE and V.I.S.A. *Journal of Multi-criteria Decision Analysis* 6, 115–130.

Bettinger, P., Johnson, N. and Sessions, J. (1996) Forest planning in an Oregon case study: defining the problem and attempting to meet goals with a spatial analysis technique. *Environmental Management* 20(4), 565–577.

Bettinger, P., Sessions, J. and Boston, K. (1997) Using tabu search to schedule timber harvests subject to spatial wildlife goals for big game. *Ecological Modeling* 94, 111–123.

Bettinger, P., Boston, K. and Sessions, J. (1999) Intensifying a heuristic forest harvest scheduling search procedure with 2-opt decision choice. *Canadian Journal of Forest Research* 29, 1784–1792.

Bos, J. (1993) Zoning in forest management: a quadratic assignment problem solved by simulated annealing. *Journal of Environmental Management* 37, 127–145.

Bousquet, F., Bakam, I., Proton, H., and Le Page, C. (1998) Cormas: common pool resources and multi-agent systems. *Lecture Notes in Artificial Intelligence* 1416, 826–838.

Bousquet, F., Barreteau, O., d'Aquino, P., Etienne, M., Boissau, S., Aubert, S., Le Page, C., Babin, D. and Castella, J.-C. (2002). Multi-agent systems and role games: collective learning processes for ecosystem management. In: Janssen, M. (ed.) *Complexity and Ecosystem Management: The Theory and Practice of Multi-agent Approaches.* Edward Elgar, Cheltenham, UK.

Buongiorno, J. and Gilles, K. (2003) *Decision Methods for Forest Resource Management.* Academic Press, Burlington, MA.

Chambers, R. and Guijt, I. (1995) PRA five years later – Where are we now? *Forest, Trees and People* 26/27, 4–14.

Checkland, P.B. (1981) *Systems Thinking Systems Practice.* Wiley, Chichester.

Checkland, P.B. (1988) Soft systems methodology: overview. *Journal of Applied Systems Analysis* 15, 27–30.

Church, R., Murray, A. and Weintraub, A. (1998) Locational issues in forest management. *Location Science* 6, 137–153.

Coyle, G. (2000) Qualitative and quantitative modeling in system dynamics: some research questions. *Systems Dynamics Review* 16(3), 225–244.

Davis, L. and Johnson, N. (1987) *Forest Management*, 3rd edition. McGraw-Hill, Columbus, Ohio.

Davis, L., Johnson, N., Bettinger, P. and Howard, T. (2000) *Forest Management. To Sustain Ecological, Economic, and Social Values*, 4th edition. McGraw-Hill, Columbus, Ohio.

De Steiguer, J.E., Liberti, L., Schuler, A. and Hansen, B. (2003) *Multi-Criteria Decision Models for Forestry and Natural Resources Management: An Annotated Bibliography.* USDA Forest Service General Technical Report NE-307, USDA Forest Service Northeastern Research Station, Newtown Square, Pennsylvania.

Eden, C. (1988) Cognitive mapping: a review. *European Journal of Operational Research* 36, 1–13.

Eden, C. (1989) Using cognitive mapping for strategic options development and analysis (SODA). In: Rosenhead, J. (ed.) *Rational Analysis of a Problematic World.* John Wiley and Sons, Chichester and New York.

Eden, C. and Ackermann, F. (1998) *Making Strategy: The Journey of Strategic Management.* Sage Publications, London.

Forrester, J.W. (1961) *Industrial Dynamics.* MIT Press, Cambridge, Massachussetts.

Forrester, J.W. (1999) *System Dynamics: The Foundation Under Systems Thinking.* Sloan School of Management, Massachusetts Institute of Technology, Cambridge, Massachussetts.

Hobbs, B.F., Ludsin, S., Knight, R.L., Ryan, P.A., Biberhofer, J. and Ciborowski, J.J. (2002) Fuzzy cognitive mapping as a tool to define management objectives for complex ecosystems. *Ecological Applications* 12, 1548–1565.

Hof, J. and Bevers, M. (2000) Direct spatial optimization in natural resource management: four linear programming examples. *Annals of Operations Research* 95, 67–81.

Hogeweg, P. and Hesper, B. (1990) Individual-oriented modelling in ecology. *Mathematical and Computer Modelling* 13, 83–90.

Howard, A. (1991) A critical look at multiple criteria decision-making techniques with reference to forestry applications. *Canadian Journal of Forest Research* 21, 1649–1659.

Huston, M. and DeAngelis, D. (1988) New computer models unify ecological theory. *Bioscience* 38, 682–691.

Jankowski, P. (1995) Integrating geographical information systems and multiple criteria decision making methods. *International Journal of Geographical Information Systems*

9, 251–273.

Jankowski, P., Nyerges, T.L., Smith, A., Moore, T.J. and Horvath, E. (1997) Spatial group choice. A SDSS tool for collaborative spatial decision making. *International Journal for Geographical Information Science* 11, 576–602.

Kangas, J. (1992) Multiple use planning of forest resources by using the analytic hierarchy process. *Scandinavian Journal of Forest Research* 7, 259–268.

Keeney, R.L. (1992) *Value-Focused Thinking: A Path to Creative Decision Making.* Harvard University Press, Cambridge, MA.

Kent, B. (1989) *Forest Service Land Management Planning: Introduction to Linear Programming.* USDA Forest Service General Technical Report RM-173, USDA Forest Service Rocky Mountain Forest and Range Experiment Station, Fort Collins, CO.

Kosko, B. (1986) Fuzzy cognitive maps. *International Journal of Man–Machine Studies* 1, 65–75.

Kyem, P.A.K. (2002) Using GIS to support multi-objective decision-making in forest management: an experience from Ghana, West Africa. In: Pukkala, T. (ed.) *Multi-objective Forest Planning.* Kluwer Academic Publishers, Dordrecht.

Leuschner, W.A. (1984) *Introduction to Forest Resource Management.* John Wiley & Sons, New York.

Lockwood, C. and Moore, T. (1992) Harvest scheduling with spatial constraints: a simulated annealing approach. *Canadian Journal of Forest Research* 23, 468–478.

Mendoza, G.A. and Prabhu, R. (2000a) Multiple criteria analysis for assessing criteria and indicators in sustainable forest management: a case study on participatory decision making in a Kalimantan forest. *Environmental Management* 26(6), 659–673.

Mendoza, G.A. and Prabhu, R. (2000b) Multiple criteria decision making approaches to assessing forest sustainability using criteria and indicators: a case study. *Forest Ecology and Management* 131, 107–126.

Mendoza, G.A. and Prabhu, R. (2003a) Qualitative multi-criteria approaches to assessing indicators of sustainable forest resource management. *Forest Ecology and Management* 174, 329–343.

Mendoza, G.A, and Prabhu, R. (2003b) *Enhancing Participatory Planning of Community-managed Forests using Problem Structuring Models and Approaches: Experiences from a Case Study.* Department of Natural Resources and Environmental Sciences. Working paper 2003–2. University of Illinois, Urbana, Illinois.

Mendoza, G.A. and Prabhu, R. (2004) Fuzzy methodologies for assessing criteria and indicators of sustainable forest management. *Ecological Indicators* 3(4), 227–236.

Mendoza, G.A. and Sprouse, W. (1989). Forest planning and decision-making under fuzzy environments: an overview and illustration. *Forest Science* 35, 481–502.

Mowrer, H.T. (1997) *Decision Support Systems for Ecosystem Management: an Evaluation of Existing Systems.* USDA Forest Service General Technical Report RM-GTR-296, USDA Forest Service Rocky Mountain Forest and Range Experiment Station, Fort Collins, CO.

Murray, A. and Church, R. (1995) Heuristic solution approaches to operational planning problems. *OR Spektrum* 17, 193–203.

Nelson, J. and Brodie, D. (1990) Comparison of random search algorithm and mixed integer programming for solving area-based forest plans. *Canadian Journal of Forest Research* 20, 934–942.

Ozesmi, U. and Ozesmi, S. (2003) A participatory approach to ecosystem conservation: fuzzy cognitive maps and stakeholder group analysis in Uluabat Lake, Turkey. *Environmental Management* 31, 518–531.

Prato, T. (2000) Multiple attribute evaluation of landscape management. *Journal of Environmental Planning and Management* 60, 325–337.

Pukkala, T. (ed.) (2002) *Multi-objective Forest Planning.* Kluwer Academic Publishers, Dordrecht.

Pukkala, T. and Kangas, J. (1996) A method for integrating risk and attitude into forest planning. *Forest Science* 42(2), 198–205.

Purnomo, H., Mendoza, G.A. and Prabhu, R. (2004) A model for collaborative planning for community managed resources based on qualitative soft systems approach. *Journal of Tropical Forest Science* 16(1), 106–131.

Purnomo, H., Mendoza, G.A., Prabhu, R. and Yasmi, Y. (2005) Developing multi-stakeholder forest management scenarios: a multi-agent system simulation approach. *Forest Policy and Economics* 7(4), 475–491.

Rauscher, H.M. (1999) Ecosystem management decision support for federal forests in the United States: a review. *Forest Ecology and Management* 114, 173–197.

Reynolds, K., Bjork, J., Hershey, R.R., Schmoldt, D., Payne, J., King, S., DeCola, L., Twery, M. and Cunningham, P. (2000) Decision support for ecosystem management. In: Johnson, N.C., Malk, A.J., Sexton, W.T. and Szaro, R.C. (eds) *Ecological Stewardship: A Common Reference for Ecosystem Management.* Elsevier Science Ltd., Amsterdam, pp. 687–721.

Reynolds, K.M., Johnson, K.N. and Gordon, S. (2003) The science/policy interface in logic-based evaluation of forest ecosystem sustainability. *Forest Policy and Economics* 5, 433–446.

Richardson, G.P. and Anderson, D.F. (1995) Teamwork in group model building. *System Dynamics Review* 11, 113–137.

Rosenhead, R. (ed.) (1989) *Rational Analysis for a Problematic World: Problem Structuring Methods for Complexity, Uncertainty and Conflict.* John Wiley and Sons, Chichester and New York.

Rouchier, J., Bousquet, F., Requier-Dejardins, M. and Antona, M. (2001) A multi-agent model for transhumance in North Cameroon. *Journal of Economic Dynamics and Control* 25, 527–559.

Saaty, T. (1980) *The Analytic Hierarchy Process.* McGraw-Hill, New York.

Schmoldt, D.L. and Peterson, D. (2000) Analytical group decision making in natural resources: methodology and application. *Forest Science* 46(1), 62–75.

Schmoldt, D.L., Peterson, D. and Silsbee, D. (1995) The analytic hierarchy process and participatory decision making. In: Strome, T.C. and Daniels, M. (eds) *Decision Support 2001, Proceedings of the 4th Biennial International Symposium for Advanced Technology in Natural Resource Management.* American Society of Photogrammetry and Remote Sensing, Bethesda, MD.

Schmoldt, D., Kangas, J., Mendoza, G. and Pesonon, M. (2001) *The Analytic Hierarchy Process in Natural Resource and Environmental Decision Making.* Kluwer Academic Publishers.

Schon, D. (1987) *Educating the Reflective Practitioner: Toward a New Design for Teaching and Learning in the Professions.* Jossey-Bass, San Francisco.

Selener, D. (1997) *Participatory Action Research and Social Change.* The Cornell Participatory Action Research Network, Cornell University, New York.

Snyder, S. and Revelle, C. (1997) Multi-objective grid packing model: an application. *Location Science* 5(3), 165–180.

Stone, P. and Veloso, M. (1997) *Multiagent Systems: A Survey from a Machine Learning Perspective.* Computer Science Department, Carnegie Mellon University, Pittsburgh, PA.
[http://www-2.cs.cmu.edu/afs/cs/usr/pstone/public/papers/97MAS-survey/revised-survey.html]

Sycara, K. (2000) Multi-agent systems (a subtopic of agents). American Association for Artificial Intelligence (AAAI), http://www.aaai.org/AITopics/html/multi.html,

Thebaud, O. and Locatelli, B. (2001) Modeling the emergence of resource sharing conventions: An agent-based approach. *Journal of Societies and Social Simulation* 4(2), http://www.soc.surrey.ac.uk/JASS/4/2/3.html.

Varma, V.K., Ferguson, I. and Wild, I. (2000). Decision support system for sustainable forest management. *Forest Ecology and Management* 128, 49–55.

Vennix, J. (1996) *Group Model Building: Facilitating Team Learning Using System Dynamics.* John Wiley and Sons, Chichester, England.

Weintraub, A., Jones, G., Meacham, M., Magendazo, A., Magendazo, A. and Malchuk, D. (1995) Heuristic procedures for solving mixed-integer harvest scheduling-transportation planning models. *Canadian Journal of Forest Research* 25, 1618–1626.

Weintraub, A., Church, R., Murray, A. and Guinard, M. (2000) Forest management models and combinatorial algorithms: analysis of state of the art. *Annals of Operations Research* 96, 271–285.

Wolstenholme, E.F. (1999) Qualitative vs. quantitative modeling: the evolving balance. *Journal of Operational Research Society* 50, 422–428.

Zadeh, L. (1965) Fuzzy sets. *Information Control* 8, 338–353.

Index

1995+ Group, 156

Abies
  bornmülleriana, 195
  cilicica, 195
  equi-trojani, 195
  nordmanniana, 195
Abiotic damage, 128
Acacia, 16
  mangium, 4
Access, 185
Acer, 195
Adaptive management, 119, 120
Aesthetic values, 21
AF&PA. *See* American Forest and Paper Association
Afforestation, 6, 135, 139, 197
AFS. *See* Australian Forestry Standard
Agenda 21, 8
Agroforestry, 233
Air pollution, 17
Albania, 131, 132, 134, 135, 136
Alberta-Pacific Forest Industries Inc, 120
Albizia
  falcataria, 115
  white. *See Albizia falcataria*
Amazona aestiva, 82
Amenity forest, 204
American Forest and Paper Association, 151
American Tree Farm System, 152, 156
Ammonia, 128
Analytical Hierarchical Process, 217, 242
Apteryx australis, 114
Aracruz Cellulose S.A., 5
Argentina, 82, 175
Artificial intelligence, 244
ASIO system (Sweden), 111
AssiDomän, 161
Association of Consulting Foresters, 153
ATFS. See American Tree Farm System
Atmospheric pollution, 2
Australia, 8, 17, 25, 147, 162, 173, 175, 178, 179
  Victoria, 104–106
Australian Forestry Standard, 153
Austria, 149
Authenticity, 17
Avalanches, 26

Bangladesh, 67, 233
Bedding, 115
Belarus, 128, 134, 136
Belgium, 138, 149
Benefit-cost analysis, 31–51
Biodiversity, 21, 54, 82, 102, 112, 114, 117, 184
  hot spots, 23, 118, 214
  protection zones, 101
  reserves, 110
Birdlife International, 23
Bolivia, 148, 154, 162
  Lomerio, 154
Bosnia-Herzegovina, 132
Brazil, 4, 5, 148, 179, 194
Brown bear. *See Ursus arctos*
Bulgaria, 128, 129, 134, 136
Burundi, 76

Cameroon, 9, 18
Canada, 8, 20, 25, 97, 114, 122, 124, 143, 144, 147, 148, 150, 152, 154, 160, 173, 175, 176, 178, 179, 181, 188
  Alberta, 114, 119, 120
  British Columbia, 7, 10, 98, 99–104, 114, 145, 158, 162
    Caribou-Chilcotin, 99
    Clayoquot Sound, 98

Kamloops, 101
Okanagan-Shuswap, 101
Stillwater, 103, 106
Vancouver Island, 103
New Brunswick, 119
Ontario, 144, 162
Canadian Council of Forest Ministers, 150
Canadian Standards Association, 150–151, 156, 158, 162
Capital asset pricing model, 65
Capital availability, 139
Carbon
banking, 236
conservation, 10
emission reductions, 181
global cycle, 219
sequestration, 3, 58, 235
tax, 56
Carpinus betulus, 195, 204
Castanea sativa, 195, 204
CCFM. *See* Canadian Council of Forest Ministers
Cedrus libani, 195
Center for International Forestry Research, 20, 23
Certification, 8, 27, 143–164
costs, 154
group, 154
market access, 156
price premiums, 155
Certified Forest Products Council, 156
Chain of custody, 150, 154, 158
Charcoal, 4
Charismatic megavertebrates, 214
Chile, 25, 154, 173, 174, 175, 235
China, 4, 172, 173, 174, 175, 178, 179, 233
Ussuri River Watershed, 108
Chiquitano forest enterprise, 154
Choristoneura fumiferana, 120
CINTRAFOR Global Trade Model, 170
Civil law, 7
Clean Development Mechanism, 179, 181
Clear-cutting, 204
Climate change, 17, 21, 66
Club Pro Forêts, 156
Cognitive mapping, 247, 248
Cognitive maps, 248, 249
Collins Pine Co., 154, 159
Co-management, 246
Commission on Resources and the Environment, 99
Common law, 7
Communal health, 197
Community forestry, 108, 248, 249
Community-based forest management, 245
Community-based resource management, 246
Confederation of European Paper Industries, 146
Congo Basin, 25
Connectivity, 21
Consejo Boliviano para la Certificación Forestal Voluntaria, 146
Conservation, 140
Conservation ethic, 6
Conservation International, 23
Conservation planning, 214
Conservation value, 21
Conservation Zone, 106
Contingent valuation, 188
Continuous-cover forestry, 9
Convention on Biological Diversity, 8, 20, 22
Convention on Long-Range Transboundary Air Pollution, 128
Cooperative Fairness, 49
Coppice management, 204
Corporations, 46
Corruption, 8
Costa Rica, 154, 163
Côte d'Ivoire, 77, 79, 87
Cotta Formula, 204
Council on Environmental Quality, 16
Countries in transition, 125–140
Criteria and indicators, 10, 20, 21, 215, 219, 249
Croatia, 134, 135, 136, 138, 148, 149

CSA. *See* Canadian Standards Association
Cuba, 76
Cupressus sempervirens, 195
Czech Republic, 128, 131, 133, 134, 135, 136, 138, 140, 149
Czechoslovakia, 127, 128, 129, 134

Dauerwald, 194, 207
Debt for Environment Swaps, 83
Debt-for-nature exchanges, 234
Decision analysis, 239–252
Decision exploration, 248
Decision Explorer, 247
Decision Support Systems, 243
Deforestation, 5, 75–93
  and tenure security, 80
  causes, 80
  tropical rainforests, 3
DEFORPA, 2
Denmark, 78, 138, 149
Deutsche Gesellschaft für Technische Zusammenarbeit, 163
Devolution, 6
Diamond Raya Timber, 155
Diminishing marginal utility, 68
Discount factor, 67
Discount premiums, 63
Discount rate, 34, 35, 42, 67
Discounting, 32, 53–72
Disease, 131
Dispersed-patch–clear-cutting, 104
Dobris Assessment, 132
Dry-Zone Africa Process, 22

Ecological integrity, 245
Economic rights, 48
Ecosystem diversity, 220
Ecosystem management, 9, 240, 245
Ecosystem-based management, 117, 119, 120
Ecotourism, 68, 140, 214
Efficiency analysis, 33
Empowerment, 246
Endangered species, 115
Endangered Species Act, 187
Endemic bird areas, 23
Engineered wood, 177
Environmental change, 54
Environmental Management System, 147, 155
Environmental services, 21
Estonia, 126, 128, 132, 134, 136, 138, 148, 149, 155, 190
Eucalyptus grandis, 115
European Commission, 2
European Union, 128, 135, 151, 163, 189
  Eco-Management and Audit Scheme, 151
Expert systems, 243

Facilitation, 248
Fagus orientalis, 195, 204
Far East Leopard. *See Panthera pardus*
Fertilization, 115
Fertilizers, 4
FFCS. *See* Finnish Forest Certification Scheme
Fibre farms, 120
Finland, 2, 22, 132, 141, 144, 145, 149, 150, 151, 156, 158, 162, 172, 175, 178, 179, 184, 185, 189–190
Finnish Association for Nature Conservation, 151
Finnish Central Union of Agricultural Producers and Forest Owners, 151
Finnish Forest Certification Council, 146
Finnish Forest Certification Scheme, 151, 162
Finnish Forest Industries Federation, 151
Fire ecology, 17
Fire Protection Zone, 106
Fisher's alpha index, 219, 220
Flooded gum. *See Eucalyptus grandis*
Forest borders, 231
Forest conversion, 79
Forest cover change, 76, 132
Forest decline, 2, 17, 126
Forest degradation, 79
Forest Enterprise (UK), 148

Forest health, 21, 245
Forest Management Areas (Australia), 106
Forest management plans, 106, 108, 153, 198
Forest Management Unit, 106
Forest quality, 6, 15–27
  definition, 17
Forest quantity, 16
Forest Resource Assessment, 17
Forest Stewardship Council, 23, 144, 145, 146, 147–149, 150, 151, 152, 153, 154, 155, 156, 157, 158, 159, 160, 161, 162, 163
Forest stewardship plan, 102, 103, 106, 108
Forest tenure, 6, 7
Forestry Continuing Studies Network, 10
Forests and water, 26
FORPLAN model, 241
FORREX, 10
Fragmentation, 21, 79
France, 22, 149, 150, 156
  DEFORPA, 2
Fraxinus, 195
Frontier forests, 24
FSC. *See* Forest Stewardship Council
Fuzzy cognitive mapping, 248, 250, 251
Fuzzy set theory, 243

Gabon, 18
Game theory, 244
Gamma discounting, 38
Gap dynamics, 120
Gender, 8
General Management Zone, 106
Genetically improved seedlings, 115
German Democratic Republic, 127
Germany, 18, 83, 129, 138, 147, 148, 149, 150, 163
  Fichtelgebirge, 2
Ghana, 91
Global 200 ecoregions, 23
Global Forest and Trade Network, 156
Global Forest Resource Assessment 2000, 4
Globalization, 180, 236
Governance, 8, 237, 246
Greece, 138
Green revolution, 87
Green Tag, 152–153
Greenpeace, 147, 149, 150, 158
Guyana, 109–111
  Forestry Commission, 110

Habitat fragmentation, 104
Habitat reserves, 114
Haiti, 76
Hallmark Cards, 156
Hancock Natural Resource Group Inc., 5
Hardboard, 4
Helsinki Process, 20, 215, 219
Herbicides, 4
Heuristic algorithms, 243
Hevea brasiliensis, 4
High Conservation Value Forests, 20, 23, 24, 161
Home Depot, 152, 156
HomeBase, 156
Honduras, 22, 160
Honey, 3
Hotelling equation, 36
Hungary, 125, 127, 128, 132, 134, 135, 136
Hunting, 140
Hyperbolic discounting, 58

IKEA, 156
Illegal logging, 8, 132
Independent Forestry Ltd., 154
India, 26, 87, 179, 233
Indigenous peoples, 8, 21, 145
Indonesia, 4, 155, 179, 233, 249
Influence diagrams, 249
Innovative Forest Practice Agreement, 104
Integrated management, 114
Intensive management, 115
Intensive production, 115
Interest rates, 41

Intergenerational discount rate, 59, 61
Intergovernmental Panel on Climate Change, 2, 25
International Forest Industry Roundtable, 158
International Forest Products, 162
International forestry dialogue, 134
International Labour Organization Convention on Indigenous and Tribal Peoples, 8
International Organization for Standardization, 145, 146–147, 154, 155
International Tropical Timber Organization, 20, 22, 153, 215, 219
Intramarginal forest resources, 81
Invasive species, 17
Investment returns, 55
Ireland, 138
ISO 14000, 23
Italy, 138, 149, 150

Japan, 25, 147, 149, 173, 176, 179, 188, 190, 213
J.D. Irving Limited, 119, 160
Joint forest management, 246
Joint Implementation, 179, 181
JOURNEY, 247
Juniperus, 195

Kaliningrad Region, 128
Kenya, 215
Kimberly-Clark Ltd., 5
Kiwi. *See Apteryx australis*
Korea, 179, 235

Labour costs, 44
Lagrangian function, 85
Laminate strand lumber, 4
Land allocation, 98
Land allocation theory, 89
Land and Resource Management Plans, 98
Landscape biodiversity, 111
Landscape matrix, 117
Land-use change, 3
Latvia, 126, 128, 130, 134, 136, 138, 148, 149
Laws of trespass, 7
Legal framework for forestry, 135
Lembaga Ekolabel Indonesia, 153
Linear programming model, 241
Lithuania, 128, 132, 134, 136, 138
Living Forests (Norway), 146
Lomé Convention, 163
Lowes, 156
Low-intensity silviculture, 111
LRF Skogsagarna, 146
Luxembourg, 138

MACBETH, 247
Macedonia, 132, 137
Malaysia, 162, 179, 220
  Sabah, 10, 106–108, 213–226
Malthusian explanations of population growth, 80
Management Institute of British Columbia, 10
Mangrove forests, 220
Marginal utility of income, 67
Market discount rate, 42
Market failure, 42
Market incompleteness, 44
Market reform, 38
Mazuma Crafts Ltd., 154
Medium-density fibreboard, 4
Meridian Institute, 152
Metropolitan forest, 234
Mexico, 163, 179
Millennium Ecosystem Assessment, 3
Ministerial Conference on the Protection of Forests in Europe, 20, 22, 133
Moldova, 126
Mondi Business Paper, 5
Montreal Process, 20, 22, 27, 150, 215, 219
Mountain caribou. *See Rangifer tarandus*
Multi-agent systems, 244
Multi-criteria decision-making, 215, 242

Multiple-use forestry, 32, 110, 184, 189, 195, 199, 240
Mutual recognition, 158
Mycorrhizal-inoculated stock, 115

Nampak Ltd., 5
NAPAP, 2
National Association of Forest Industries (Australia), 153
National Forest Management Act, 240
National forest plan, 109
National Forest System (USA), 184–187
National Timber Certification Council, 162
National Woodlands Association, 153
Natural disturbance, 111, 112, 117, 119
Nature reserves, 171
Near-to-nature forestry, 9
Negative exponential discounting, 58
Nepal, 76
Net present social value, 90
Net present value, 55
Netherlands, 138, 144, 149, 156, 160, 162, 163
New Zealand, 8, 16, 17, 18, 25, 98, 106, 115, 147, 160, 164, 173, 174, 175, 178, 179, 235
Nitrogen oxides, 128
Non-timber forest products, 3, 4, 21, 26, 108, 109, 140, 197
Non-wood products, 19
Nordic Forest Certification Programme, 148
Norway, 2, 75, 132, 146, 149, 185
Nothofagus forest, 106

Oil palms, 16
Old-growth forest, 17, 25, 103, 115, 139, 145, 161, 162, 187, 188. *See also* High Conservation Value Forest
Optimization, 245
Optimization models, 243
Orang-utan. *See Pongo pygmaeus*
Organic Act, 186
Oriented strand board, 4

Panthera
  pardus, 108
  tigris, 108
Papua New Guinea, 179, 220
Pareto Frontier, 49
Participatory action research, 246, 249
Participatory decision-making, 241, 242, 246, 247, 248, 251
Participatory rural appraisal, 246, 248, 249
Participatory techniques, 5, 249
Particleboard, 4
Paulownia tomentosa, 4
PEFC. *See* Programme for the Endorsement of Forest Certification
Peru, 4
Pesticides, 4
Pests, 128, 131
Philippines, 179
  Palawan Island, 219
Picea orientalis, 195
Pine
  Caribbean. *See Pinus caribea*
Pinus
  brutia, 195, 204
  caribea, 115
  halepensis, 195
  maritima, 204
  nigra, 195
  pinea, 195, 204
  radiata, 106
  sylvestris, 195
Plantations, 4, 15, 17, 105, 106, 107, 115, 116, 208, 235, 236
Podocarp forest, 106
Poland, 127, 128, 131, 132, 134, 135, 137, 148, 149
Political ecology, 5
Pollution, 128
Pongo pygmaeus, 108
Poplar Farm Program, 120
Populus, 120, 195
Poverty and deforestation, 80

Private property rights, 5
Privatization, 129, 138
Production silviculture, 115
Production Zone, 106
ProForest, 23
Programme for the Endorsement of Forest Certification, 23, 145, 146, 149, 150, 151, 153, 156, 158, 159
Property restitution, 131
Property rights, 7, 34
Protected areas, 27, 114
Public forests, 32
Public ownership, 230
Public sector services, 44
Pulpwood, 4

QUAD approach, 117
Qualitative system dynamics, 248, 249
Quercus, 195, 204

Radioactivity, 128
Rainforest Action Network, 149, 152, 162
Rainforest Alliance, 147
Random search, 243
Rarity, 214
Rattan, 3, 108
Rawlsian Difference Principles, 49
Rayonier, 160
Recreation, 103, 197, 204
Reduced-impact logging, 6
Regional Forest Agreements, 106
Rehabilitation, 139, 197
Resilience, 18
Restructuring of wood products industry, 189
Return on investment, 55
Rhizophora mucronata, 220
Riverside Forest Products Ltd, 103–104
Romania, 127, 128, 131, 134, 137
Rubberwood. *See Hevea brasiliensis*
Russia, 121
Russian Federation, 108, 109, 126, 128, 129, 131, 132, 133, 134, 137, 139, 148, 173, 174, 175, 190
  Siberia, 108
  Ussuri River Watershed, 108

Salix, 195
Sappi Ltd., 5
Scandinavia, 17
Scientific Certification Systems, 23, 159
Scientific forest management, 240
SFI. *See* Sustainable Forestry Initiative
Shadow price corrections, 43, 44, 45
Shadow pricing, 86
Siberian tiger. *See Panthera tigris*
Simulated annealing, 243
Singapore, 220
Site preparation, 115, 116
Slovakia, 131, 133, 134, 137, 138, 139
Slovenia, 126, 128, 131, 134, 135, 137, 138, 139
Social discount rate, 42, 43, 45
Social forestry, 246
Soft law, 9
Soft systems methodology, 247
Soil Association, 23
Soil erosion, 26
South Africa, 4, 148
Soviet Union, 6
Spacing, 115
Spain, 138, 149
Special Management Zone, 106
Special Protection Zone, 106
Spruce budworm. *See Choristoneura fumiferana*
Sri Lanka, 233
Standards Council of Canada, 150
State control, 135
Strategic Options Development and Analysis, 247
Submarginal forest resources, 81
Sulphur dioxide, 128
Super-intensive management, 116
Supply chain, 144
Sustainability, 71
Sustainable forest management, 245
Sustainable Forestry Initiative, 151–152, 156, 158, 162
Sustained yield, 240

Sweden, 111–112, 138, 144, 145, 146, 147, 148, 149, 156, 157, 159, 160, 161, 172, 175, 178, 179, 185
Swedish Society for Nature Conservation, 148
Switzerland, 7, 18, 26, 81, 149
  Solothurn, 154
Systems approach to forestry, 245

Tabu search, 243
Tajikistan, 126
Tanzania
  Tanga Region, 92
Tarapoto Process, 22
Tax evasion, 91
Tenure, 17
Thailand, 10, 233
Thinning, 115
Tilia, 195
Time preference function, 57
Tolko Ltd., 103
Tourism, 103, 108, 140
Trade in wood products, 171
Tree Farm, 152
TRIAD planning, 98, 103, 112–120
Tropenbos Foundation, 20
Turkey, 9, 193–210
Two-valued logic, 243

UK Woodland Assurance Scheme, 146
Ukraine, 127, 128, 134, 137
Ulmus, 195
Undercut, 131
Uneven-aged management, 204
United Kingdom, 18, 54, 73, 123, 138, 144, 148, 149, 154, 156, 160, 161, 162, 215
  Department for International Development, 163
United Nations
  Conference on Environment and Development, 17, 144, 194
  Convention on Long-Range Transboundary Air Pollution, 25
  Economic Commission for Europe, 17, 20, 25
  Food and Agriculture Organization, 16, 20, 24, 76
  Forest Resources Assessment, 20–22
  Universal Declaration of Human Rights, 8
Urban forests, 232, 236
Ursus arctos, 108
US National Forestry Association, 152
USA, 5, 8, 31, 78, 81, 115, 148, 152, 155, 156, 159, 160, 162, 169, 175, 176, 178, 179, 183, 184, 190, 229, 239, 240, 241
  California, 154, 215, 233, 235
    Almanor Forest, 154
    Chester, 159
  Colorado
    Aspen, 159
    Vail, 159
  Maine, 119
  Massachusetts
    Quabbin Forest, 161
  NAPAP, 2
  National Forests, 6
  New Mexico
    Santa Fe, 159
  Oregon
    Lakeview, 159
  Pacific Northwest, 167, 169, 174, 176, 185, 187, 188, 190
  Pennsylvania, 154, 159
  Texas
    Austin, 159
Utility maximization, 49

Valuation, 140
Value exploration, 248
Value pluralism, 50
Value theory, 32
Value-focused thinking, 247
Vietnam, 18
Visual Interactive Sensitivity Analysis, 247

Welfare analysis, 32
Welfare economics, 32, 33, 39
Weyerhaeuser, 103, 106, 115

Wilderness, 187
Wildlife corridors, 115
Woodland degradation, 75
Woodmark certification, 163
Working circles, 203
World Bank, 153, 162
World Conservation Union, 18, 27
World Resources Institute, 23, 24, 174
World Wide Fund for Nature, 17, 20, 23, 145, 147, 148, 149, 151, 156, 158, 162
WWF. See World Wide Fund for Nature
WWF International, 17

Yield regulation, 204
Yugoslavia, 128, 129, 132, 133, 134

Zambia, 154
Zimbabwe
  Mafungautsi Forest, 248
  Save Valley, 82
Zoning, 97–121

cabi-publishing.org/bookshop

Browse Read and Buy

ANIMAL & VETERINARY SCIENCES
BIODIVERSITY CROP PROTECTION
HUMAN HEALTH NATURAL RESOURCES
ENVIRONMENT PLANT SCIENCES
SOCIAL SCIENCES

Reading Room

Bargains

New Titles

Forthcoming

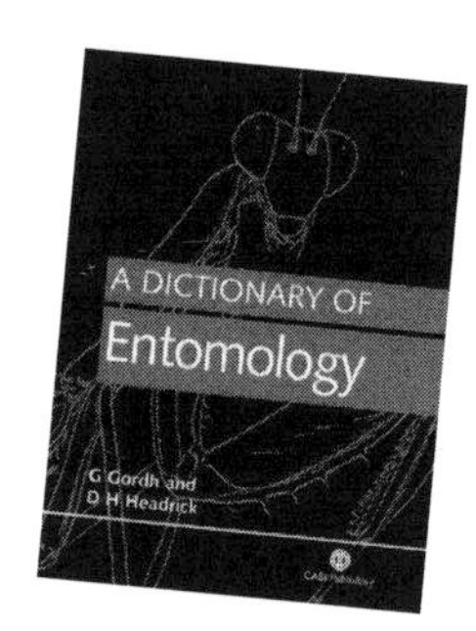

FULL DESCRIPTION

BUY THIS BOOK

Tel: +44 (0)1491 832111 Fax: +44 (0)1491 829292